Practical Well Control

By Ron Baker

published by

THE UNIVERSITY OF TEXAS

CONTINUING EDUCATION

PETROLEUM EXTENSION SERVICE

1998

Library of Congress Cataloging-in Publication Data

Baker, Ron, 1940-
    Practical well control/by Ron Baker.—4th. ed.
        p.   cm.
    ISBN 0-88698-183-2
    1. Oil wells—blowouts.   I. Baker, Ron.   II. Title.
TN871.2.F547 1989 622'.3382—dc20   89-39589
                                                              CIP

First Edition published 1980. Fourth Edition 1998
Fifth Impression 2005
Printed in the United States of America

Catalog No. 2.80040
ISBN 0-88698-183-2

Contents

Figures

Tables

Foreword

Control of formation pressures encountered while drilling wells is of particular importance in today's oil industry. One indication of its importance is the fact that regulatory bodies in the United States and other countries require that personnel either be trained in well control or be tested to prove their competence. Unquestionably, training plays an important role in successful well control, for a drilling crew that knows and understands the principles and technical procedures of well control is a crew that is less likely to experience a well blowing out of control.

Interestingly, a recent study* of blowout statistics covering the last 36 years in the U.S. Gulf Coast, revealed that "…blowouts continue to occur at about a constant rate…." The authors of the study deduced that most blowouts happened because (1) a formation with higher-than-expected pressure was penetrated, (2) formation fluids were swabbed in during a trip, or (3) formation breakdown and lost circulation occurred because the mud weight was too high or the crew lowered the drill string into the hole too rapidly. The wells then blew out because, as the authors stated, "Once the first pressure barrier was lost,…the kicks could have been controlled, if the BOPs had been closed, would have worked correctly, or had been installed as planned."

It is therefore just as vital as it ever has been to continue training drilling crews in the principles of well control. This manual is intended to be a training aid for all personnel who are concerned with well control—rotary helpers, drillers, toolpushers, company representatives, or anyone whose job takes him or her directly onto a rig location. The book presents a practical approach to well control in that it emphasizes the things a rig crew should know and be able to do to control a well. It is also used as a basic textbook by those who attend well-control courses conducted by many training organizations.

The original edition of *Practical Well Control* was commissioned by Bill Butler of CS Inc. and was written by Bill Rehm, a recognized well-control authority. Jim Fitzpatrick, who appended additional text and included offshore floating drilling operations, helped in updating the second edition. The third edition included more material to reflect changes in federal rules concerning well control in the U.S. This fourth edition adds still more material and updated information and is designed to assist trainers whose courses are accredited by the International Association of Drilling Contractor's (IADC's) WellCAP program.

PETEX wishes to extend its thanks to those in the industry who graciously reviewed the material and lent support in the updating effort. In particular, Paul Sonnemann, an independent well-control specialist, must be recognized for his extensive review and helpful comments. Also, Tom Thomas, the manager of modular training at SEDCO FOREX Schlumberger, offered many helpful suggestions and comments, and provided answers to many technical questions. Glenn Shurtz of Well Control School provided much helpful information, as well, especially on horizontal drilling concerns. Further, Steve Kropla, IADC's director of accreditation and certification programs, lent his support and assistance in preparing this edition. The excellent work of the PETEX staff must also be recognized, for without the dedication of the writers, editors, layout and design personnel, proofreaders, and others on the staff, this manual could not have been revised.

In spite of the input received from many individuals in the petroleum industry, PETEX is solely responsible for the manual's content. While every effort was made to ensure accuracy, the manual is intended to be only a training aid, and nothing in it should be considered approval or disapproval of any specific product or practice.

Ron Baker, Director
Petroleum Extension Service

---

*Skalle and Podio, "Trends Extracted from 1,200 Gulf Coast Blowouts During 1960–1996," *World Oil*, June 1998: 67–72.

Chapter 1

Pressure Concepts

In well control, the two pressures of primary concern are formation pressure and hydrostatic pressure. *Formation pressure* is the force exerted by fluids in a formation. It is measured at the depth of the formation with the well shut in. It is also called reservoir pressure or, since it is usually measured at the bottom of the hole with the well shut in, shut-in bottomhole pressure.

In drilling, *hydrostatic pressure* is the force exerted by drilling fluid in the wellbore. When formation pressure is greater than hydrostatic pressure, formation fluids may enter the wellbore. If formation fluids enter the wellbore because formation pressure is higher than hydrostatic pressure, a kick has occurred. If prompt action is not taken to control the kick, or kill the well, a blowout may occur. To control a well, a proper balance between pressure in the formation and pressure in the wellbore must be maintained; hydrostatic pressure should be equal to or slightly higher than formation pressure.

ORIGIN OF FORMATION PRESSURE

One generally accepted theory of how pressures originate in subsurface formations relates to how sedimentary basins are formed. As layer upon layer of sediments are deposited, overburden pressure on the layers increases, and compaction occurs. *Overburden pressure* is the pressure exerted at any given depth by the weight of the sediments, or rocks, and the weight of the fluids that fill pore spaces in the rock. Overburden pressure is generally considered to be 1 pound (lb) per square inch per foot (psi/ft). Overburden pressure can vary in different areas because the amount of pore space and the density of rocks vary from place to place. In deepwater formations just below the seafloor, the overburden is almost entirely seawater. Overburden pressure is therefore about the same as the pressure caused by the weight of seawater—about 0.45 psi/ft depending on its salinity. Regardless of the actual value of

overburden pressure, as it increases, compaction occurs, and the porosity of the rock layer decreases.

As compaction occurs, any fluids in the formation are squeezed into permeable layers, such as sandstone. If the permeable layer into which the fluids are squeezed is continuous to the surface—that is, if the layer eventually outcrops on the surface—pressure higher than normal cannot form (fig. 1.1). If, however, a layer's fluid is trapped because of faulting or some other anomaly, pressure higher than normal can form; the formation can become overpressured.

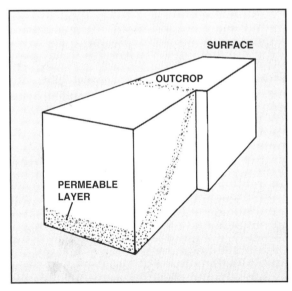

Figure 1.1   Pressure higher than normal cannot form if the layer outcrops on the surface.

Higher-than-normal formation pressure can result from several geological conditions. In some cases, the same conditions that trap hydrocarbons can also cause higher-than-normal pressure. Examples of such geological conditions are faults, large structures, massive shale beds, massive salt beds, and charged sands.

Faults

Formation pressure normally increases with depth, but deep rocks that have been faulted may have higher-than-normal pressures. The fault may trap fluids in the formation and allow abnormally high pressure to develop. Since a *fault* is a sudden break in a formation, when a faulted formation is drilled, the bit may encounter abnormally high pressure within a short interval; that is, it is possible to go from normal pressure to abnormally high pressure within a short time. Therefore, when faulted zones are being drilled, the crew must be alert to the possibility of encountering abnormally high pressures with very little warning. The high pressures that appear at different depths in the Lake Arthur field in South Louisiana are the result of a highly faulted structure. High pressures encountered in drilling next to salt domes are often the result of local faulting around the dome. High pressures related to faulting can also be found in mountainous country.

Large Structures

Any structure such as an anticline or dome may have abnormally high pressures above the oil-water or gas-water contact in the oil or gas zone because hydrocarbons are less dense than water. If the anticline or dome is large, abnormal pressures may be quite high. Since anticlines and domes sometimes serve as traps for hydrocarbons, drilling often takes place on such structures. Thus, drilling crews should be alert to the possibility of abnormally high pressures in such situations.

High pressures may be expected when drilling into the reservoir beds—usually sandstone, limestone, or dolomite—of any structure. The high pressures that were experienced in the early days of the East Texas field came from an anticlinal structure. Since large structures are often first drilled by the crew on a wildcat well, the crew should be aware of the possibility of high pressure.

Massive Shale Beds

Transition zones—formations in which pressures begin to depart from normal—and abnormally high pressure may develop within massive shale beds because thick, impermeable shale restricts the movement of fluid. As sediments are laid down on the surface and then sink deeper, they support the considerable weight of the overburden. Fluids trapped within the shale cannot escape fast enough, and they also support the weight of the overburden. Confined liquids supporting such massive weight are under higher-than-normal pressure for the depth.

When thick shales are encountered, therefore, pressure should be expected to increase abnormally with depth. Shale-related pressures can occur at any depth, from near the surface to very deep. High pressures in the U.S. Gulf Coast, the North Sea, the South China Sea, and in other deep basins of the world are often related to massive shale beds.

Massive Salt Beds

Since salt beds are plastic, they transmit all overburden weight to the rock below. Therefore, high pressures should be expected in and below thick salt beds. High pressures usually are not found in thin and erratic salt beds, however. Thick, plastic salt beds cause high pressures in the Middle East in formations below the Farrs salt, and in the United States in beds below the Louann salt. Pressures in the Zechstein salt in the North Sea and in northern Germany are also related to the fact that salt transmits the rock weight above it to the formation below it. Drilling mud with densities from 16 pounds per gallon (ppg) to 19 ppg may be necessary to control the pressures within and just below massive salt beds.

Charged Sands

Abnormally high formation pressure can be found in relatively shallow sands that have become charged from an underground blowout. A shallow sand can become charged when a well is shut in on a kick originating in a zone deeper than the sand. Pressure from the lower zone then enters the wellbore and escapes to the upper sand. As a result, the upper sand becomes overpressured by the fluids from the lower zone. Later, when another well is drilled into the charged sand, the drilling crew may be caught off guard when the charged sand kicks.

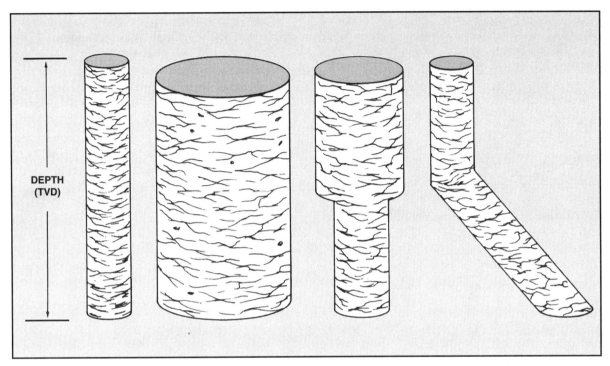

Figure 1.2   Hole geometry has no effect on hydrostatic pressure. Pressure exerted at bottom is the static pressure. Pressure exerted at bottom is the same for all containers because fluid density and depth are the same.

HYDROSTATIC PRESSURE

The term "hydrostatic" is derived from *hydro*, meaning water or liquid, and *static*, meaning at rest. Both fluid in the formation and fluid in the wellbore are under hydrostatic pressure, but in most well-control discussions, formation pressure refers to fluid pressure in the formation; hydrostatic pressure refers to the pressure of drilling fluid in the wellbore. Hydrostatic pressure increases directly with the density and depth of the fluid in the wellbore. *Hole geometry*—the diameter and shape of the fluid column—has no effect on hydrostatic pressure (fig. 1.2). In the wellbore, hydrostatic pressure is the result of the weight, or density, of the drilling fluid and the true vertical depth of the column of fluid.

True vertical depth is the length of a straight vertical line from the surface to the bottom of the hole. *Measured,* or *total, depth* is the length of the well as measured along the actual course of the hole. Thus, true vertical depth and measured depth can differ, especially in directionally drilled holes (fig. 1.3). In determining hydrostatic pressure, true vertical depth is the measurement used.

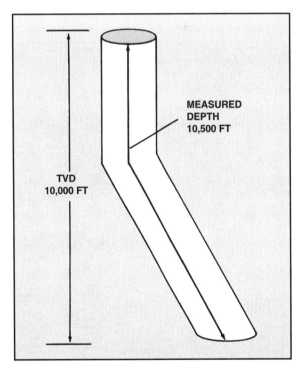

Figure 1.3   In this case, true vertical depth is different from measured depth, because the hole is directionally drilled.

Figure 1.4   Simple mud balance

Hydrostatic pressure can be calculated mathematically as—

$$HP = C \times MW \times TVD \quad \text{(Eq. 1)}$$

where

HP = hydrostatic pressure, psi

C = a constant (value depends on unit used to express mud weight)

MW = mud weight, ppg or other units

TVD = true vertical depth, ft.

Mud weight can be expressed in ppg, pounds per cubic foot (pcf), specific gravity, or other units. In the United States, mud weight is usually expressed in ppg, except on the Pacific Coast, where it is usually expressed in pcf. If mud weight is measured in ppg, the value of C in equation 1 is 0.052. If mud weight is measured in pcf, the value of C in equation 1 is 0.00694. As an example of using the equation, find the hydrostatic pressure in a wellbore if the mud weight is 12 ppg and the true vertical depth (TVD) is 11,325 ft.

$$HP = 0.052 \times 12 \times 11,325$$
$$= 7,066.8$$
$$HP = 7,067 \text{ psi.}$$

To find hydrostatic pressure if the mud weight is 90 pcf and TVD is 11,325 ft, the calculation is—

$$HP = 0.00694 \times 90 \times 11,325$$
$$= 7,073.6$$
$$HP = 7,074 \text{ psi.}$$

MEASURING MUD WEIGHT

Mud weight is usually measured with a simple mud balance (fig. 1.4). It can also be measured, however, with a special pressurized mud balance (fig. 1.5A and

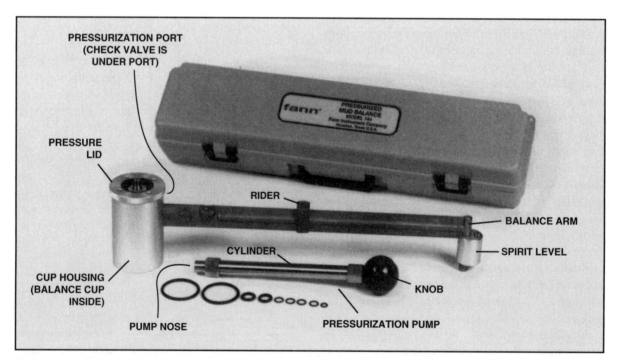

Figure 1.5A   Pressurized mud balance *(Courtesy Fann Instrument Company)*

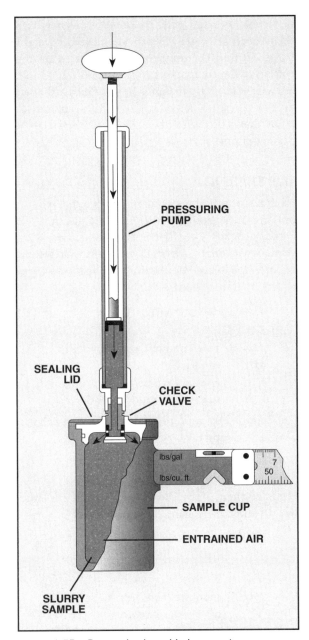

Figure 1.5B   Pressurized mud balance cutaway

mud to be tested. Put the perforated cap on the cup and rotate it until it is firmly seated. Be sure that some of the mud comes out through the hole in the cap to free any trapped air or gas. Hold the cap firmly on the cup and, while keeping the hole covered, wash or wipe the outside of the cup until it is clean and dry. Place the balance arm (beam) on the fulcrum or base support and balance it by sliding the weight along the beam's graduated scale. If the mud balance has a bubble, the bubble will be directly under a center line etched on the metal around the bubble when balance occurs. Read the mud weight at the edge of the sliding weight.

Using a pressurized mud balance is very much like using a regular mud balance except that the mud sample is put under pressure to approximate downhole conditions. Depending on the manufacturer of the balance, some type of pressure device is secured to the sample cup to pressure up the sample. A procedure from a manufacturer of a pressurized mud balance follows. (Refer to fig. 1.5.)

1. Collect a fluid sample.
2. Place the balance's stand or carrying case on a flat, level surface.
3. Measure and record the temperature of the sample, then transfer the sample to the balance cup, filling to between ¼ and ⅛ in. of the top. Tap the side of the cup several times to break up any entrained air or gases.
4. Place the lid on the cup with the check valve in the down or open position. (Some of the test sample may be expelled through the valve.) After fitting the lid onto the cup, pull the check valve up into the closed position.
5. Rinse the pressurization port and balance with water, base oil, or solvent; then dry it off.
6. Slide the cup housing over the balance cup from the bottom, aligning the slot with the balance arm. Screw the closure over the pressure lid and tighten as tight as possible by hand to insure that the pressure lid is completely seated.
7. Fill the pressurization pump with the test sample.
8. Push the nose of the pump onto the pressure port of the lid.
9. Pressure the sample cup by maintaining a downward force on the pump cylinder housing. At the same time, force the pump knob down with about 50 to 70 lb of force and release the cylinder housing. Remove the pump. (The

1.5 B). Crew members may be instructed to use a pressurized balance to obtain mud weight when air or gas is entrained in the mud. Pressurizing the balance's sample cup decreases the volume of any entrained air or gas in the mud to a negligible amount. As a result, the mud density measurement will more accurately reflect downhole conditions.

To use a regular mud balance, place the instrument on a flat, level surface. Fill the clean, dry cup with the check valve in the lid is pressure-actuated. When pressure builds in the cup, the check valve is pushed closed.)

10. Clean the mud from the outside of the cup housing and lid. Wipe off any excess water, base oil, or solvent.

11. Fit the knife edge of the balance into the fulcrum (not shown in the figure) and balance the assembly by moving the rider along the arm. Make sure the balance is level; check the spirit level.

12. Record the mud density from the side of the rider nearest the cup. Report the measurement to the nearest 0.1 ppg, 1 pcf, 0.01 grams per cubic centimetre, or 10.0 psi/1,000 ft.

13. Reconnect the empty pump assembly and push downward on the cylinder housing to release the pressure inside the cup.

14. Remove the pressure lid being careful not to spill the sample, then pour out the sample. Clean and dry all parts of the balance as soon as possible.

It is important to periodically check the accuracy of either type of mud balance. One procedure for a regular balance is to check the weight of distilled or deionized water, ideally at a temperature of 39°F (the temperature at which water is at its densest). The water should weigh 8.33 ppg or 62.31 pcf. If it does not, then the discrepancy can be noted and compensated for when mud is weighed. If the balance is very inaccurate, it should be discarded and a new, calibrated one obtained.

MUD WEIGHT CONSTANTS

Where mud weight is measured in ppg, the value for the constant C, which is 0.052, is derived from the fact that 1 cubic foot (ft^3) contains 7.48 U.S. gallons (gal). If a weightless container measuring 1 ft on each side is filled with a substance weighing 1 ppg, the substance occupies 1 ft^3, or 7.48 gal, and weighs 7.48 lb, because 7.48 gal × 1 ppg = 7.48 lb. To find the pressure in psi exerted on the bottom of the container, 7.48 lb is divided by 144 square inches (in.^{2}), because 144 in.^{2} are contained in 1 square foot (ft^2). Since 7.48 ÷144 = 0.05194, or 0.052, a column of liquid that is 1 ft high and weighs 1 ppg exerts 0.052 psi on bottom.

Where mud weight is measured in pcf, the value for C, which is 0.00694, is also derived from the fact that a weightless container measuring 1 ft on each side contains 1 ft^3. If the container is filled with a substance that weighs 1 lb, then the substance weighs 1 pcf. To find the pressure in psi exerted on the bottom of the container, 1 lb is divided by 144 in.^{2}, because 144 in.^{2} are contained in 1 ft^2. Since 1 divided by 144 equals 0.0069444, or 0.00694, a column of liquid that is 1 ft high and weighs 1 pcf exerts 0.00694 psi on bottom.

MUD GRADIENT

Hydrostatic pressure can also be calculated by multiplying the pressure gradient of the fluid by *TVD*. *Pressure*, or *mud*, *gradient* is the amount pressure changes with depth. It is obtained (where mud weight is expressed in ppg) by multiplying the mud weight in ppg by the constant 0.052:

$$MG = MW \times C \qquad \text{(Eq. 2)}$$

where
    MG = mud gradient, psi/ft of depth
    MW = mud weight, ppg
       C = constant (value depends on unit used to expresss mud weight).

For example, the mud gradient of a mud that has a weight of 12 ppg is—

    $MG = 12 \times 0.052$
    $MG = 0.624$ psi/ft of depth.

Mud gradient, then, can be substituted for mud weight and the constant C in equation 1:

$$HP = MG \times TVD \qquad \text{(Eq. 3)}$$

where
    HP = hydrostatic pressure, psi
    MG = mud gradient, psi/ft
    TVD = true vertical depth, ft.

The example problem in which the mud weight is 12 ppg and *TVD* is 11,325 ft can now be solved by using equation 3:

    $HP = 0.624 \times 11,325$
    $HP = 7,067$ psi.

If mud gradient is known, then mud weight can be calculated by transposing the terms of equation 2:

$$MW = MG \div 0.052 \qquad \text{(Eq. 4)}$$

where
    MW = mud weight, ppg
    MG = mud gradient, psi/ft.

For example, to find mud weight when the mud gradient is 0.936 psi/ft:

$$MW = 0.936 \div 0.052$$
$$MW = 18 \text{ ppg.}$$

This calculation is useful when the pressure at a certain depth is known and it is necessary to calculate the mud weight required to balance that pressure.

FORMATION PRESSURE GRADIENT

Just as mud pressure changes with depth, depending on the weight or density of the mud, formation pressure changes with depth. Formation pressure changing with depth is called *formation pressure gradient*. Normally, the hydrostatic pressure of the fluids in the formation causes formation pressure. Under abnormal conditions, in completely closed formations (formations that do not eventually connect to a surface outcrop), formation pressure is caused by the fluid bearing some or almost all of the weight of the overlying rocks.

The greater the depth of the formation, the greater is its pressure. Fluid in a porous and permeable formation has higher pressure at greater depth in the same way that pressure increases with depth underwater. The vertical distance causes fluid pressure to be greater with increased depth whether in a lake, borehole, or underground formation.

Formation fluid pressures encountered in drilling fall into three categories: normal, abnormally high, and subnormal. Normal formation pressure is generally assumed to be 0.465 psi per foot, which is the pressure gradient of a column of salt water normally found on the Gulf Coast of the United States. This value varies from area to area, depending on the amount of salt and other minerals dissolved in the salt water and other factors. Abnormally high pressures theoretically range from 0.465 psi per foot to 1 psi per foot. Subnormal pressures are formation pressures lower than 0.433 psi per foot. They may be found in partially or completely depleted reservoir formations, formations located at high elevations, and formations that reach the surface on hillsides. Normally pressured and subnormally pressured formations can be present in the same borehole. Abnormally pressured formations are frequently found in boreholes where normal pressures are exposed.

Using the 0.465-psi-per-foot gradient as normal formation pressure, a formation at 10,000 feet exerts a pressure of 4,650 psi (10,000 feet × 0.465 psi per foot). A well drilled with 10-ppg mud, which has a gradient of 0.52 psi per foot, exerts a hydrostatic pressure of 5,200 psi at 10,000 feet. The pressure overbalance from the wellbore to the formation is, in this case, 550 psi (5,200 – 4,650), and formation fluids would not enter the wellbore.

OTHER PRESSURE CONCEPTS

In well control, not only are hydrostatic and formation pressures of concern, but also of importance are differential pressure, surface pressure, pressure losses through the circulating system, and trapped pressure.

Differential Pressure

Differential pressure refers to a difference in pressure between two areas. For example, if hydrostatic pressure in the borehole is higher than pressure in the formation, a pressure differential exists between the two. In this case, the differential pressure is positive because hydrostatic pressure is higher than formation pressure. On the other hand, if formation pressure is higher than hydrostatic pressure, differential pressure is negative.

Surface Pressure

Surface pressure is pressure exerted on the well and the well equipment at or very near the surface. Examples of surface pressure include the pressure developed by the mud pump as it moves drilling mud down the drill string, and the pressure that develops on the drill pipe pressure and casing pressure gauges at the surface when a kicking well is shut in. Surface pressure is also the pressure developed by a gas kick as it nears the surface. When a well is shut in and the kick circulated out of the well, the gas moves up the annulus. As it goes up, the hydrostatic pressure on the gas becomes less and less. As a result, the gas expands. As the gas expands, it increases pressure on whatever it is in contact with. The gas exerts this pressure in all directions. On the surface, the pressure increase can be read on the casing pressure gauge. Once the gas goes out of the choke, it rapidly expands and casing pressure drops.

Since gas pressure is exerted in all directions, it also affects downhole pressure. The expanding gas pressure increases on any uncased (open) formations uphole from the depth at which the kick entered the well. If the pressure increases too much, it exceeds the strength of a formation, breaks it down (fractures it), and fluids flow into the fractured formation. Such flow is an underground blowout, which can be very difficult to control.

Usually, the formation that fractures is immediately below the shoe of the last string of casing set in the well, because the weakest uncased formation lies at that depth. (Generally, the shallower a formation, the easier it fractures.) If the last string is surface casing, the underground blowout can work its way around the casing string and erupt to the surface. In such cases, the well often craters—that is, the blowout creates a large opening (crater) around the wellhead.

Pressure Losses

Another important pressure concept is pressure losses through the circulating system. Drilling fluid exits the pump under high pressure. In most cases, it then goes through the pump's outlet piping, up the standpipe, into the rotary hose, through the swivel and kelly (or top drive), down the drill pipe and drill collars, out the bit, and back to the surface through the annulus. By the time the mud leaves the bell nipple under the rig floor, the original pressure on it has been used up. It can only flow back to the shakers by gravity; it has no pressure.

Each element in the circulating system after the mud pumps causes the mud to lose pressure. As the mud flows through each part of the circulating system, it rubs against the walls of the pipe and equipment. This friction of the mud against the equipment, as well as internal friction within the flowing mud itself, absorbs pressure. Typically, at least 25 percent of the pressure the mud started with at the pump is used up by the time the mud reaches the bit. At the bit, anywhere from 50 to 70 percent of the pressure is expended as the mud exits the bit's nozzles or passageways. Thus, only about 5 to 25 percent of the pressure the mud began with is left to move it up the annulus. As an example, assume that a pump is moving mud at a standpipe pressure of 2,400 psi. As it goes through the surface equipment and down the drill stem, 600 psi of pressure is lost because of friction. It arrives at the bit under 1,800 psi. It exits the bit nozzles and in so doing loses an additional 1,700 psi. Only 100 psi is available to move the mud up the annulus, which, by the time the mud reaches the surface, is also lost.

In general, pressure losses increase when pump speed is increased. Pressure losses also increase as mud weight is increased. What is more, pressure losses are generally higher with thick, viscous muds than with thin, less viscous muds.

Trapped Pressure

Trapped pressure is surface pressure in excess of the amount needed to balance formation pressure. It can occur for several reasons: improper handling of migrating gas, poor use of pressure gauges during pump startup in kill operations, and thermal expansion of drilling well fluids.

To check for and eliminate trapped pressure, one suggested procedure is to (1) note *SIDPP* and *SICP*, (2) open the choke to reduce *SICP* a safe amount (typically 100 psi), (3) close the choke completely and wait a minute or two to see if *SIDPP* goes down and seems to stabilize the same amount (100 psi) lower. If it does, repeat steps 2 and 3. (4) If *SIDPP* does not go down, or goes back up, trapped pressure has been removed. Any further bleeding lets in more influx.

Annular Pressure Loss

Annular pressure, the pump pressure required to move mud up the annulus, must be considered in well-control operations. Even though this pressure is often less than 200 psi at the reduced pump speeds used in killing a well, it can be significant because it contributes extra back-pressure on the bottom of the hole. Since annular pressure contributes extra back-pressure, it therefore increases bottomhole pressure. Put another way, when mud is being circulated, bottomhole pressure equals hydrostatic pressure plus the pressure required to move mud up the annulus. Annular pressure loss exists only when mud is being circulated, and it is caused by the friction that resists the flow of fluid in the annulus. When the pump is stopped, annular pressure loss ceases. When mud is moving up the hole, however, the friction of the mud against the drill pipe and sides of the wellbore and the internal resistance of the mud to flow produce friction losses that must be overcome.

Equivalent Circulating Density

Another way to look at the bottomhole pressure increase caused by friction losses in the annulus is in terms of equivalent circulating density. *Equivalent circulating density* is a combination of the original mud weight plus the equivalent mud weight increase due to pressure loss in the annulus. The following equation can be used to calculate it:

$$ECD = MW + (APL \div 0.052 \div TVD) \quad \text{(Eq. 5)}$$

where

ECD = equivalent circulating density, ppg

MW = mud weight, ppg

APL = annular pressure loss, psi

TVD = true vertical depth, ft.

As an example calculation, suppose that—

MW = 13.0 ppg

APL = 100 psi

TVD = 8, 000 ft,

then

ECD = 13 + (100 ÷ 0.052 ÷ 8,000)

= 13 + 0.24

ECD = 13.24 ppg.

It cannot be overemphasized that, because of annular pressure loss, greater pressure exists at the bottom of the hole when mud is being circulated than when the pump is off. In the example, 0.24 ppg exerts an additional 100 psi on bottom. When the pump is shut down, this 100 psi is lost, and a kick may result.

Chapter 2

Causes and Warning Signs of Kicks

Drilling personnel should know the causes and warning signs of kicks and be able to identify them readily. Since the well and the mud-circulating equipment are a closed system, any formation fluid that intrudes into the system usually shows up as a change in the flow rate of fluid returning from the well and as a change in the total volume of fluid in the pits. Exceptions can occur. For example, when oil-base drilling mud is in use, a kick composed mainly of methane and other light hydrocarbon gases may dissolve in the oil in the mud and may not be evident until the gas nears the surface, comes out of solution, and expands. While not as soluble as light hydrocarbon gases, hydrogen sulfide (H_2S) and carbon dioxide (CO_2) gases can also dissolve in oil-base mud and come out (evolve) as they near the surface. The same thing can happen with H_2S and CO_2 kicks in water-base muds, because they also dissolve in these types of mud.

Gas kicks that dissolve in mud can be difficult to detect, especially if the kick is relatively small. Since they are absorbed by the mud, little or no pit gain occurs. Also, no measurable increase in the flow rate of the mud being pumped out of the well occurs. Later, however, as the dissolved gas in the mud nears the surface, it begins evolving from the mud and rapidly expands. This rapid expansion suddenly increases the return flow rate and can create a large pit gain if the well is not rapidly shut in.

Also, when a gas kick dissolves in the mud, and crew members do detect it, they may believe that the well has taken a saltwater kick: the pit gain may be relatively small and the difference between the shut-in drill pipe pressure and the shut-in casing pressure may be small. When dissolved gases come out of solution, however, they expand rapidly. As a result, the shut-in casing pressure rises quickly. Crew members should be aware of the problem and keep close surveillance on the casing pressure. Because gases can dissolve in mud, most operators and contractors prefer to consider all kicks as gas kicks and react appropriately. It is important to anticipate the possibility of rapidly increasing *SICP* and to have a plan in place to deal with it before circulating out a kick.

Surface indications depend on the kick's size and on the temperature and pressure. A gain in pit volume is probably the most reliable indicator of a gas kick in oil-base mud. In any case, training of personnel, specific procedures, and good supervision are the most effective means of detecting kicks and preventing blowouts.

CHARACTERISTICS OF KICKS

A *well kick* is an influx of formation fluids such as oil, gas, or salt water into the wellbore from a formation that has been penetrated by the wellbore. It occurs when the pressure exerted by the column of drilling fluid in the wellbore is lower than the pore pressure in the formation, and when the formation is permeable enough to allow fluids to flow through it. Once a kick occurs, the intruded fluid further reduces the hydrostatic pressure of the mud column, since formation fluids are usually less dense than drilling muds.

As a result, formation fluid may flow into the wellbore at an increasing rate. Therefore, a well kick should be stopped or controlled as soon as possible to prevent the entry of additional formation fluids. The larger the kick, the more difficult it can be to bring the well back under control properly.

An equation can be used to illustrate how quickly a kick can grow in size if the well is not promptly shut in. From Darcy's law on rock permeability (*permeability* is the ease with which fluids flow in rocks), it can be shown that the flow rate of gas into the wellbore increases as wellbore depth through a gas sand increases:

$$Q = 0.007 \times md \times \Delta p \times L \qquad \text{(Eq. 6)}$$
$$\div \mu \times ln \, (R_e \div R_w) \, 1{,}440$$

where

Q = flow rate, barrels/minute (bbl/min)
md = permeability, millidarcys (md)
Δp = pressure differential, psi
L = length of section open to wellbore, ft
μ = viscosity of intruding gas, cp
R_e = radius of drainage, ft
R_w = radius of wellbore, ft.

As an example, assume that—

md = 200 md
Δp = 624 psi
L = 20 ft
μ = 0.3 cp
$ln \, (R_e \div R_w)$ = 2.0.

Therefore:

$$Q = 0.007 \times 200 \times 624 \times 20 \div 0.3$$
$$\times 2.0 \times 1{,}440$$
$$= 17{,}472 \div 864$$
$$Q = 20 \text{ bbl/min.}$$

The solution shows that if 20 ft of the example gas sand is drilled, the flow rate of the gas entering the hole is about 20 bbl/min. If 2 min are required to close the well in, a pit gain of 40 bbl occurs in addition to the gain incurred while drilling the 20-ft section.

Thus, it can be seen that the rate at which fluid enters the wellbore from a formation primarily depends on (1) the permeability of the formation, (2) the difference between the pressure exerted by the drilling fluid and the pressure exerted by the formation, and (3) the amount of formation drilled. In general, if the formation is not very permeable, the rate of the influx of fluid is slow; if the formation is highly permeable, the rate of fluid influx is fast. Similarly, if the differential pressure between the drilling fluid and the formation fluid is low, the influx is slow; if high differential pressure exists, the influx is fast.

Formation pressure becomes greater than mud pressure when formation, or pore, pressure increases to the point that it exceeds the hydrostatic pressure exerted by the mud column. Formation pressure can also exceed hydrostatic pressure when the height of the mud column in the hole is allowed to drop. Even though the mud weight is adequate to control formation pressure with the well full of mud, if the height of the mud column drops, formation pressure can exceed hydrostatic pressure. The drop usually occurs because insufficient mud is put into the hole to replace the volume of pipe pulled; however, lost circulation can also cause the height of the mud column to drop.

BEHAVIOR OF GAS

In vertical wells, a gas kick, because it is considerably lower in density than the drilling mud, usually migrates up the hole. As the gas migrates upward, it expands. If the well is shut in, gas expansion shows up on the casing pressure gauge as an increase in pressure. In some cases, however, gas may not migrate. For example, in very viscous (thick) muds, gas migration can be very slow or nonexistent; the gas simply cannot make its way through the thick mud. Moreover, the chemical makeup of the mud may prevent or retard gas migration. Also, in deviated wells, the gas can become trapped under ledges or in horizontal sections of the hole. It rises to the top of the borehole and cannot migrate. In such a shut in well that is not being circulated, *SICP* may remain relatively constant because the gas is not migrating up the hole. Crew members should therefore be aware that trapped gas or gas unable to migrate may expand rapidly when circulation is started. Circulation moves the gas up the hole where hydrostatic pressure is less and rapid expansion occurs with a subsequent rapid increase in shut-in casing pressure.

When hydrocarbon fluids enter the wellbore, crew members should be aware that these fluids can be gas or liquid, depending on the pressure and temperature at the depth where the fluids entered. Normally, at high pressures hydrocarbon fluids tend to be liquid; at high temperatures they tend to be gas. Conversely, at low pressures hydrocarbon fluids tend to be gas and at low temperatures they tend to be liquid. In any case, it is important to remember that a kick may enter the hole as a liquid but, as it is circulated up to the surface where the pressure is less, liquid may evolve to gas. The danger is that crew members may believe that the hydrocarbon liquid intrusion is a saltwater kick and may not be prepared for the rapid rise in casing pressure as the gas evolves from the liquid. For this reason, most operators and contractors instruct crew members to assume that all kicks are gas kicks until proven otherwise.

If a float or other type of check valve is not installed in the drill string, gas can enter the string. After the well is shut in, the mud in the drill string can U-tube into the annulus. When the choke is closed a few moments later, the top part of the drill pipe may be void of mud altogether. As bottomhole pressure increases with *SICP*, the increased pressure forces annular fluids (including kick fluids) back into the drill string, eventually filling it from the bottom before *SIDPP* starts to increase. Kick fluids entering the drill stem is undesirable with a gas kick, which is why some well-control experts recommend the use of floats, prefer to obtain *SIDPP* readings as quickly as possible, prefer to circulate the kick out before circulating kill-weight mud, and emphasize the importance of shutting in a well before large pit gains occur.

CONSEQUENCES OF NOT RESPONDING TO KICK

If the warning signs of a kick are ignored and crew members do not take prompt action, a kick can become a blowout. If formation fluids push the mud from the well and erupt to the surface, a blowout is underway. Not only is human life endangered, but also a rig and well that could be worth millions of dollars are in jeopardy. Often, since natural gas and other hydrocarbon fluids are present, the fluids catch fire, often creating a spectacular flare that endangers personnel and attracts considerable attention. What is more, in some cases, the formation fluids can contain hydrogen sulfide (H_2S), which is a highly poisonous gas that can maim or kill quickly. (See appendix A for more information about H_2S.) To add to the problem, oil, gas, and saltwater in the fluids

blowing out of the well can create ground, water, and air pollution, which is very undesirable.

Also, if the crew recognizes a kick and shuts in the well, but does not promptly start kill procedures, the kick, if it is gas or mostly gas, can migrate (rise) up the hole. Gas is very light in weight (density) and, as a result, moves to the surface even though the well is shut in. As it migrates upward, it remains at a constant pressure if not allowed to expand. Although its pressure remains constant, the migrating gas exerts pressure higher and higher up the closed in well. The resulting increase in wellbore pressure may be enough to fracture a downhole formation; the kick and mud may then flow into the fractured formation. Such flow is an underground blowout, which can be very difficult to control.

TYPES OF BLOWOUTS

A blowout is the uncontrolled flow of fluid from a wellbore. An *underground blowout* is the flow of fluid into a subsurface formation. If a kick occurs and the well is shut in, a zone located at another depth from the kicking formation can be broken down, or fractured, by the higher pressures needed to control the kick. Mud and formation fluids can then escape into the fractured zone and an underground blowout can occur. While being wasteful and potentially dangerous to existing or future wells in an area, a serious consequence to an underground blowout is the possibility that a pathway from the weaker formation to the surface may result, creating a surface blowout outside the wellbore.

Blowouts occur with equal frequency during drilling and during tripping. In general, they occur during drilling because formation pressure increases; during tripping, they occur because hydrostatic pressure decreases. Since instruments to detect a kick or to aid in the prediction of a kick are relatively inexpensive, most operators recommend that they be used at all times.

CAUSES OF KICKS

A kick occurs when the hydrostatic pressure exerted by the column of drilling fluid in the hole is less than the pressure exerted by the fluids in a porous and permeable formation penetrated by the wellbore. Therefore, the primary means of controlling a well is to maintain the correct hydrostatic pressure. In the course of drilling a well, however, conditions may occur that allow formation pressure to exceed hydrostatic pressure, and the well can kick. Personnel should be aware of these conditions and be prepared to take prompt and proper action to control the well. In general, kicks are caused by insufficient mud weight, failure to keep the hole full of mud, swabbing, surging, lost circulation, and abnormal pressure.

Insufficient Mud Weight

During drilling operations, the weight of the mud is the primary means of controlling a well. If the mud weight develops less pressure than formation pore pressure, the well is underbalanced, and fluids from a permeable formation can enter the wellbore. On the other hand, an overbalanced condition, in which mud weight develops more pressure than formation pore pressure, can create problems such as fracture of weak formations, low penetration rates, and lost circulation. Drilling in a nearly balanced condition is usually preferred, even though an underbalanced condition can improve penetration rate.

Bear in mind that mud weight can be reduced unintentionally. For example, in mud systems using mud centrifuges, it is possible for the centrifuge to remove barite or other weighting materials from the mud as it moves through the surface cleaning equipment. If too much barite is removed, the mud weight could be reduced to a level below that needed to contain formation pressure.

Mud weight can also be reduced in other ways. For instance, crew members could inadvertently dilute the mud with water or other materials to the point that the mud no longer develops enough downhole pressure to contain formation pressure. Also, mud-weighting materials can, during periods when the mud is not moving through the system, settle out of the mud in the surface tanks. If the mud in the tanks is not sufficiently agitated to suspend the weighting material until circulation is restarted, the material could settle to the bottom of the tanks. The mud would thus be considerably lower in density and not be able to develop sufficient downhole pressure.

The temperature of the mud also affects its density. Generally, the higher the temperature is the less dense the mud. In cases where the temperature is very high and where only a very small overbalance is being maintained, the mud's weight can be reduced to the point where formation fluids can enter the well.

During cementing operations, where the density of the cement is needed to balance formation pressure, it is important that the cement's weight be monitored closely. Since the cement replaces at least a portion of the mud in the hole, a plan must be made and followed to ensure that the well is not inadvertently over- or underbalanced.

Failure to Keep Hole Full of Mud

Probably the single greatest avoidable cause of well kicks is failure to keep the hole full of mud during trips into or out of it. When the drill string is pulled from the hole, the fluid level drops because of the volume of steel being removed (fig. 2.1). This drop in mud level reduces hydrostatic pressure. The hydrostatic pressure formula,

$$HP = 0.052 \times MW \times TVD, \qquad \text{(Eq. 1)}$$

where

HP = hydrostatic pressure, psi

MW = mud weight, ppg

TVD = true vertical depth, ft,

shows that, if for any reason MW or TVD is changed, HP also changes. If the mud level in the hole is allowed to drop, in effect, TVD is changed and thus HP is also changed. The amount of hydrostatic pressure lost from pipe being pulled out of the hole can be calculated. If the pipe is pulled wet (that is, if the pipe is kept full of mud), the following equation can be used:

$$PL_w = \left[MG \times (DP_c + DP_d) \right] \div \left[C_c - (DP_c + DP_d) \right] \qquad \text{(Eq. 7)}$$

where

PL_w = pressure lost for each ft of pipe pulled wet, psi

MG = mud gradient, psi/ft

DP_c = capacity of drill pipe, bbl/ft

DP_d = displacement of drill pipe, bbl/ft

C_c = capacity of casing or hole, bbl/ft.

As an example, assume that—

MG = 0.624 psi/ft

DP_c = 0.01393 bbl/ft

DP_d = 0.00648 bbl/ft

C_c = 0.07019 bbl/ft.

Figure 2.1   Just as the liquid level in a cylinder falls when a steel rod is removed from it, so does the mud level in the hole fall as pipe is removed from it.

Therefore:

$$PL_w = 0.624 \ (0.01393 + 0.00648) \div 0.07019 - (0.01393 + 0.00648)$$
$$= 0.624 \times 0.02041 \div 0.0719 - 0.02041$$
$$= 0.01274 \div 0.0478$$
$$PL_w = 0.27 \ psi.$$

In this case, hydrostatic pressure is reduced by 0.27 psi for each ft of pipe that is pulled wet from the hole. At a rate of 0.27 psi/ft, a drop of about 25 psi in hydrostatic pressure occurs for each 90-ft stand that is pulled wet. Thus, five stands reduce hydrostatic pressure by 125 psi, and ten stands reduce hydrostatic pressure by 250 psi. An equation for determining the pressure lost for each ft of pipe when the pipe is pulled dry, or empty, is also available:

$$PL_d = (MG \times DP_d) \div (C_c - DP_d) \qquad \text{(Eq. 8)}$$

where

PL_d = pressure lost for each ft of pipe pulled dry, psi

MG = mud gradient, psi/ft

DP_d = displacement of drill pipe, bbl/ft

C_c = capacity of casing or hole, bbl/ft.

As an example, assume that—

MG = 0.624 psi/ft

DP_d = 0.00648 bbl/ft

C_c = 0.07019 bbl/ft.

Therefore:

$$PL_d = (0.624 \times 0.00648) \div (0.07019 - 0.00648)$$
$$= 0.00404 \div 0.06371$$
$$PL_d = 0.06 \ psi$$

In this example, hydrostatic pressure is reduced by 0.06 psi for each ft of pipe that is pulled dry from the hole. At the rate of 0.06 psi/ft, a drop of about 5.4 psi in hydrostatic pressure occurs for each 90-ft stand that is pulled dry. Thus, five stands reduce hydrostatic pressure by 27 psi, and ten stands reduce hydrostatic pressure by 54 psi.

To prevent hydrostatic pressure from dropping as pipe is pulled, the volume of steel and mud removed from the hole must be replaced with fluid. The exact amount of fluid required to fill the hole should be known. Then, if the hole takes less fluid than calculated to fill, an influx, or feed in, of fluid from the formation into the wellbore has occurred.

Frequent or continual filling of the hole is critical in maintaining sufficient bottomhole pressure to prevent the influx of formation fluids. The hole should be filled on a regular schedule, depending on its diameter and the diameter of the pipe. In general, a small-diameter hole should be filled more often than a large-diameter hole. Under normal conditions, many operators require that the hole be filled continuously or after no more than five stands of drill pipe are pulled.

To pull pipe dry, crew members usually slug the pipe—that is, they pump a small quantity of heavy mud into the top of the drill stem. This slug forces the level of the mud in the drill stem to drop below rig floor level so that, when the crew breaks out a joint, mud does not spill on them or the rig floor. Because the slug forces the regular drilling mud down the drill stem, the mud goes out the bottom of the drill stem and into the annulus. This effect is called U-tubing (see chapter 3 for more information on U-tubing). Thus, just after the pipe is slugged and when crew members first begin pulling stands, the trip tank will indicate that the hole is taking less mud than required to replace the drill stem. To ensure that an influx of formation fluids has not also entered the hole, one procedure is to pull two or three stands and then check to see if the hole takes the correct amount of fill-up mud. If, after pulling a few stands, the hole still takes less mud than required to replace the drill stem, then an influx has probably occurred.

It is important to remember that, since drill collars are usually so much larger in diameter than drill pipe, the hole must be filled more frequently when collars are being pulled. Most operators recommend that the hole be filled after each stand of drill collars is pulled. As a rule of thumb, one stand of drill collars requires as much replacement fluid as five to ten stands of drill pipe.

The amount of fluid required to replace the volume of pipe pulled must be calculated carefully. Use of a trip tank is the most accurate method of determining the amount of fluid taken by the hole during tripping operations. This calibrated tank allows the crew to measure relatively small changes in mud volume—often in increments of ¼ bbl or ½ bbl. If the hole does not take the correct amount of mud during a trip out, the drill pipe should be run to bottom and the influx circulated out before the trip is continued.

If a kick occurs during a trip, most operators recommend that, if possible, the pipe be run back to bottom after the well is shut in. A common procedure is to shut the well in and strip the pipe back in under pressure. Most operators recommend that the pipe be stripped back to bottom, because it can be difficult, if not impossible, to kill the well with uphole methods and avoid fracturing a weak formation still exposed to the open hole.

Just as pulling the drill stem causes the level of mud in the hole to drop, running the drill stem back into the hole causes the level of mud in the hole to rise. It is important that the volume of mud coming out of the hole displaces a volume equal to the drill stem components run in. If more mud comes out than that required to replace the drill stem, then an influx may have occurred. Conversely, if less mud comes out than that required to replace the drill stem, then lost circulation may be occurring.

If a check (float) valve is installed near the bottom of the string, it prevents mud in the hole from filling the inside of the tubulars being run. It is as though a solid cylinder of steel was being lowered into the hole. On the other hand, if no check valve is in the string, then mud fills the tubulars as they are run. In this case, the volume the string displaces is less than the volume the string displaces with a check valve installed. Equations are available for determining the amount of volume displaced by the drill stem with and without a check valve. Since drill collars displace more than drill pipe, volume for each must be determined.

To determine tubular displacement with a check valve, the equations are:

$$V_{dc1} = OD_{dc1}^2 \div 1,029.4 \qquad \text{(Eq. 9)}$$

$$V_{dp1} = OD_{dp1}^2 \div 1,029.4 \qquad \text{(Eq. 10)}$$

where

V_{dc1} = volume displaced by drill collars with check valve installed, bbl/ft

OD_{dc1}^2 = outside diameter of drill collars, in.

V_{dp1} = volume displaced by drill pipe with check valve installed, bbl/ft

As an example, suppose a crew is running 12,870 ft of drill stem back into the hole, which consists of 540 ft of 7-in. OD drill collars and 12,330 ft of 4½-in. 16.60 X-95 drill pipe. A check valve is installed at the bottom of the string. How many barrels of mud will the string displace? The solution:

$$V_{dc1} = 7^2 \div 1,029.4$$

$$= 49 \div 1,029.4$$

$$V_{dc1} = 0.0476 \text{ bbl/ft.}$$

$$V_{dp1} = 4.5^2 \div 1,029.4$$

$$= 20.25 \div 1,029.4$$

$$V_{dp1} = 0.0197 \text{ bbl/ft.}$$

The 540 ft of drill collars displace 25.7 bbl (540 × 0.0476 = 25.7). Similarly, 12,330 ft of 4½-in. 16.60 X-95 drill pipe displaces 242.9 barrels (12,330 × 0.0197 = 242.9). Therefore total displacement of this particular string is 268.6 bbl (25.7 + 242.9 = 268.6) or about 268½ bbl. For clarity, this example uses a very simple drill stem configuration. In reality, drill stem components can consist not only of conventional drill pipe and drill collars, but also of special drill pipe and collars, such as heavyweight and heavy wall drill pipe, spiral drill collars, and the like. Further, drill collar strings often consist of collars of differing diameters; the same is true for drill pipe strings. Moreover, drill stem components such as stabilizers and reamers are often run. It is crucial to consider such items when making actual calculations.

If crew members do not install a check valve in the drill stem, the calculations to determine displacement of the string are slightly more complex than the calculations for determining displacement with a check valve. When no check valve is installed, mud normally fills the inside of the drill stem as the crew lowers it back into the hole. Therefore, the string displaces only the thickness of the walls of the tubulars. Wall thickness is determined by subtracting the tubular's ID from its OD. Equations that can be used to determine tubular displacement with no check valve installed are:

$$V_{dc2} = (OD_{dc2}^2 - ID_{dc2}^2) \div 1{,}029.4 \qquad \text{(Eq. 11)}$$

$$V_{dp2} = (OD_{dp2}^2 - ID_{dp2}^2) \div 1{,}029.4 \qquad \text{(Eq. 12)}$$

where

V_{dc2} = volume of drill collars without check valve installed, bbl/ft

OD_{dc2}^2 = outside diameter of drill collars, in.

ID_{dc2}^2 = inside diameter of drill collars, in.

V_{dp2} = volume of drill pipe without check valve installed, bbl/ft

OD_{dp2}^2 = outside diameter of drill pipe, in.

ID_{dp2}^2 = inside diameter of drill pipe, in.

Measurement of tubulars can be made on the rig with calipers; or various publications can be used to determine ODs and IDs. One publication in which to find the inside diameter of drill collars is API RP7G, *Recommended Practice for Drill Stem Design and Operation Limits*. Similarly, the inside diameter of drill pipe can be found in API Bulletin 5C, *Performance Properties of Casing, Tubing, and Drill Pipe*. For special tubulars, such as heavyweight drill pipe, IDs and other specifications can be obtained from the manufacturer. An example is the *Drilling Assembly Handbook* published by Smith International-Drilco.

As an example calculation, determine the displacement of a 540-ft string of 7-in. OD drill collars with an ID of 3 in. and a 12,330-ft string of 4½-in. 16.60 X-95 drill pipe with an ID of 3.826 in. A check valve has not been installed in the drill stem.

$$V_{dc2} = (7^2 - 3^2) \div 1{,}029.4$$
$$= (49 - 9) \div 1{,}029.4$$
$$= 40 \div 1{,}029.4$$
$$V_{dc2} = 0.0389 \text{ bbl/ft}$$
$$V_{dp2} = (4.5^2 - 3.826^2) \div 1{,}029.4$$
$$= (20.25 - 14.638) \div 1{,}029.4$$
$$= 5.612 \div 1{,}029.4$$
$$V_{dp2} = 0.0055 \text{ bbl/ft.}$$

The 540 ft of 7-in. OD, 3-in. ID collars without a check valve installed displace 21.006 bbl/ft (540 × 0.0389 = 21.006). The 9,350-ft string of 4½-in. 16.60 X-95 drill pipe with an ID of 3.826 in. displaces 51.425 bbls/ft (9,350 × 0.0055 = 51.425). Total displacement for this particular string without a check valve is therefore 72.431 bbl (21.006 + 72.431) or about 72½ bbl. Again, keep in mind that this example is simple. If heavyweight or heavy wall drill pipe, spiral drill collars, or other special drill stem components are in use, then displacement must also be calculated for them. Tapered strings (drill strings with tubulars of varying diameters and weights) must also be taken into account.

Swabbing

Swabbing is caused by the drill stem's dragging along mud as the stem is pulled out of the hole. Even though the hole may be full of mud of the correct weight, swabbing can reduce the pressure opposite a permeable formation and allow fluid in the formation to feed into the wellbore. The likelihood of swabbing is increased by (1) pulling the pipe too fast, (2) using mud of high viscosity and high gel strength, (3) having a balled-up bit, (4) having a plugged drill string, (5) having thick wall cake, or (6) having small clearances between the string and the hole. Most swabbing occurs when the bit is first moved off bottom.

A common practice to determine whether swabbing is likely is to make a short trip by pulling a few stands of pipe, then running back to bottom and circulating bottoms up to see if there are any signs of an influx. If swabbing is detected, before the density of the mud in the annulus is reduced enough to allow the well to flow, the pipe can be run back to bottom and the hole circulated to remove the invading fluid. Slowing the speed at which pipe is pulled can usually reduce swabbing enough to permit a trip out.

A safety, or trip, margin is sometimes added to mud weight to counterbalance swabbing effects, because the mud density used to kill a well is usually just enough to balance formation pressure and often does not include the safety margin that some operators consider necessary for normal drilling operations. This safety, or trip, margin may be necessary to offset swabbing that can occur during connections and trips and the periodic reductions in hydrostatic pressure that occur if the hole is intermittently filled as pipe is being pulled. Procedures for determining the amount of mud weight to add for a trip margin are usually based on the increase in bottomhole pressure desired. For shallow holes, many operators recommend 50 psi; for deeper drilling, 200 psi to 300 psi is recommended by some operators.

The mud-weight increase required to give the desired increase in bottomhole pressure can be calculated with the following equation:

$$MW \; = \; BHP_i \div 0.052 \div TVD \quad \text{(Eq. 13)}$$

where

    MW = mud weight, ppg

    BHP_i = desired increase in bottomhole pressure, psi

    TVD = true vertical depth, ft.

As an example, assume that a 10,000-ft well (TVD) is killed after it kicks. To determine the additional mud density required to provide a trip margin equal to an increase in bottomhole pressure of 250 psi,

    MW = $250 \div 0.052 \div 10,000$

          = 0.48

    MW = 0.5 ppg

The solution shows that the mud weight must be increased by about 0.5 ppg to provide a trip margin that causes a bottomhole pressure increase of 250 psi.

Another equation, using the mud's yield point, is available for calculating trip margin. Yield point can usually be found by referring to the mud engineer's report, or by using a viscometer. The equation is—

$$TM \; = \; YP \div 11.7 \, (D_h - D_p) \quad \text{(Eq. 14)}$$

where

    TM = trip margin, ppg

    YP = yield point, lb/100 ft^2

    D_h = diameter of the hole, in.

    D_p = outside diameter (OD) of pipe, in.

As an example, consider a situation with these values:

    YP = 8 lb/100 ft^2

    D_h = 8.5 in.

    D_p = 4.5 in.

The solution is—

    TM = $8 \div 11.7 \, (8.5 - 4.5)$

         = $8 \div 46.8$

    TM = 0.2 ppg.

The solution shows that the mud weight should be increased by 0.2 ppg to achieve an adequate trip margin with the example conditions shown for yield point, hole size, and pipe size.

Most operators recommend that the trip margin be added to drilling mud only after a well is killed. If the trip margin is added to kill-weight mud, and if the calculated kill-weight mud is less than that actually needed to kill the well, then the mud density believed to contain a trip margin may only be enough to balance formation pressure. In that case, another kick could occur when normal drilling operations are resumed. Therefore, most operators recommend that formation pressure be balanced first; mud containing the trip margin should then be circulated.

Surging

Surging is the increase in borehole pressure caused by downward movement of the drill string. The tendency of mud to adhere to drill pipe and to the wall of the hole creates friction as the pipe is moved downward. The pressure needed to overcome this friction is related to the movement of mud past the pipe; that is, the faster the fluid is forced to move relative to the pipe, the higher the surge pressure must be. Pressures on the wellbore caused by surging can cause lost circulation. To minimize surging, (1) run the pipe into the hole at a slow rate; (2) keep the mud in the system in good condition, with viscosity and gel strength at a minimum; (3) break circulation periodically while tripping into the hole; (4) be sure the volume of mud coming out of the hole equals the volume of pipe being run in; and (5) watch closely for tight spots in the hole. Since the mud's gel strength is related to swabbing and surging in that the higher the gel strength, the greater the pressure required to get the mud flowing again, a helpful formula is available for calculating the pressure necessary to overcome the mud's gel strength:

$$P_{gs} = (\gamma \div 300 \div d)\, L \qquad \text{(Eq. 15)}$$

where

P_{gs} = pressure required to break gel strength, psi

γ = 10-min gel strength of drilling fluid, lb/100 ft^2

d = inside diameter (ID) of drill pipe, in.

L = length of drill pipe, ft.

As an example, assume that 12,000 ft of drill pipe with an ID of 4.276 in. is in use and that the 10-min gel strength of the mud is 10 lb/100 ft^2). Therefore:

γ = 10 lb/100 ft^2

d = 4.276 in.

L = 12,000 ft

The solution is—

P_{gs} = (10 ÷ 300 ÷ 4.276) 12,000

= 0.007795 × 12,000

P_{gs} = 93.5 psi.

The solution shows that about 94 psi is required to break the mud's gel strength and get it flowing again. After the pump is started, speed should be built up slowly to achieve the required pressure to break circulation; otherwise, a surge could fracture a sensitive formation and cause lost circulation.

Lost Circulation

Lost circulation, a fairly common problem in rotary drilling operations, can cause the level of mud in the hole to fall as the mud flows into the zone of loss. As a result, the hydrostatic balance that provides primary control of a well can change. Since formation fracture is one of the causes of lost circulation, and since heavy mud can fracture a formation, the possibility of formation fracture should always be kept in mind when heavy mud is circulated to control formation pressures.

Formation strength—the ability of an exposed formation to support drilling fluid of a certain density without circulation being lost—is related to the weight of the overburden and the pressure of the fluid in the pore spaces of the formation. If the pressure exerted by the column of drilling fluid in the wellbore is greater than the fracture pressure of a formation, then the formation will fracture, whole mud will be lost from the hole, and fluid level in the wellbore will drop. The drop in fluid level due to lost circulation can cause bottomhole pressure to decrease below the level required to balance the pore pressure of an exposed formation, resulting in a kick or possibly an underground blowout. Lost circulation may occur even when formation fracture pressure is not exceeded. For example, when cavernous, faulted, vuggy, jointed, or fissured formations are penetrated, they can take whole mud from the well whenever formation pressure is less than hydrostatic pressure.

Abnormal Pressure

Abnormal formation pressures are those greater than the pressure exerted by a full column of formation fluid of normal weight. In most areas, the fluid considered to be of normal weight is formation salt water. Even though normal pressure is often expressed as 0.465 psi/ft, which is the pressure gradient of salt water with a weight of slightly less than 9 ppg, the value for normal pressure can vary, depending on the salinity of local formation water.

As mentioned earlier, some causes of abnormal pressures are faulted structures, salt domes, and underground blowouts that charge other formations penetrated by the wellbore. Other causes are uplift and erosion and undercompacted shales. In areas where sediment composed of sand or shale was deposited rapidly and the *connate water* (water in a rock's pore spaces that was present when the rock was deposited) could not escape from the rock, abnormal pressure can occur. Methods such as well logging and measurement-while-drilling (MWD) techniques, which are capable of measuring shale compaction, or density, can be used to predict such abnormal pressure zones. Drilling parameters studies, shale cuttings evaluation, geophysical analysis, direct pressure measurement, and use of the normalized penetration rate, or d exponent, are also used to evaluate abnormal pressures.

Annular Gas Flow After Cementing

After cement is displaced and allowed to set, chemical changes occur in the cement. These changes sometimes allow gas to enter the cement from adjacent formations. Because the cement is not fully set, it may be permeable enough to allow the gas to channel through the cement and enter the annulus above the cement. The gas in the annulus then rises, expands, and unloads the fluid in the annulus. Additional gas enters the annulus through the channels in the cement and can lead to the well's blowing out if not detected. Crew members should therefore monitor annular pressure after cementing to check for a gas influx and be prepared to shut in the annulus if flow occurs or pressures increase significantly.

WARNING SIGNS OF KICKS

Since physical laws determine the occurrence of pressures in the earth, it is often possible to detect these pressures before they can cause a blowout. The indications of pressure in the earth and the response of a rig to these indications are usually very clear and straightforward. Therefore, an alert driller or toolpusher should be able to recognize the indications and respond to them properly. Indications include well flowing with the pump shut down, changes in the drilling rate, an increase in the flow of mud from the well, a pit gain, a decrease in pump pressure and an increase in pump speed, an increase in rotary torque, an increase in drag and fill-up, a change in cutting size, an increase in string weight, an increase in different types of gas, an increase in salinity, an increase in flow-line temperature, a change in shale density, gas-cut mud, a change in the normalized drilling rate, indications from seismic analysis, indications from well logs, changes in the properties of the mud, and indications from mud logging values.

Note that many kick indicators can occur on the rig. The occurrence of only one of the many indicators may not be a sign that the hole has encountered abnormally high formation pressure. When, however, several of the indicators appear and are examined together, the chances increase that a kick is about to occur or has occurred. Also bear in mind that some of the indicators are virtually sure signs that the well has kicked. Generally, when there is a pit-level increase and/or an increase in the flow rate of the fluid returning from the well, the well has kicked (assuming that personnel have not added fluid to the active system). Before the increase in level or return flow, however, very likely there were other signs that warned that the kick was about to occur. Many of these signs can be recognized on the rig floor as the well is being drilled or pipe is being tripped. For example, a change in the drilling

rate (a drilling break) often precedes increased flow from the well. Therefore, drillers should be alert for signs on the rig floor that may happen before the kick actually enters the well and creates a return flow increase and subsequent gain in pit level.

Well Flows with Pump Shut Down

One indication of a kick is that the well flows with the pump shut down. Indeed, one procedure a driller can follow to see if a well has kicked is a flow check. To make a flow check, the driller picks up the bit off bottom, stops the mud pumps, and, after giving the pumps time to come to a complete stop and the flow to stabilize, checks the mud-return line to see if mud is flowing from the well, even though the pumps are off. If mud does flow from the well with the pumps off, it is very likely that the well has kicked. One exception to be aware of: if a slug of heavy mud has been spotted in the drill string (perhaps to prevent mud from spilling out of the drill pipe and onto the floor and crew while making a trip), the heavy slug can force drilling fluid out of the drill stem and up the annulus. In such a case, mud flows from the well with the pumps off, but a kick may not have occurred. This effect of mud moving out of the drill stem and up the annulus is known as U-tubing. (See Chapter 3.)

So, while it is not absolutely foolproof, checking for well flow is one of the best indications that an influx of formation fluids has entered the well. When the kick fluids enter the annulus they displace the mud in the annulus and force it to the surface. Thus, if the well flows with the pumps shut down, it is a good indication that the well has kicked. To be certain, the driller can shut in the well and see whether pressure appears on the drill pipe gauge and the casing gauge. If it does, then a kick has occurred.

Flow Check while Drilling

To make a flow check while drilling, the driller picks the bit up off bottom, shuts down the mud pumps, and waits long enough to ensure that the pumps have come to a complete stop and that normal flow has stabilized. Normal flow, or flow-back, is the flow that occurred from the well the last time the pumps were stopped without a kick resulting. If after a short period, however, greater than normal flow-back occurs, a kick has likely occurred. Shutting in and checking shut-in drill pipe pressure and shut-in casing pressure is the next step to confirm or deny that a kick has occurred.

Flow Check while Tripping

A flow check while tripping, unlike a flow check while drilling, is not a good kick indicator. During a tripping flow check, the pipe is usually stopped. With the pipe not moving, pressure below the drill stem is higher than it was when the pipe was being pulled because the stationary drill stem is not swabbing mud. A well is therefore less likely to flow during a tripping flow check than it is while the pipe is being pulled.

Trip Sheets

The best way to keep track of fill-up mud during a trip is to use a trip sheet (fig. 2.2). Crew members use it to precisely determine whether the hole is taking the correct amount of fill-up mud to replace the drill stem as they remove the drill stem from the hole.

Note that the trip sheet shown in figure 2.2 has places to record the displacement of the drill collars (DC 1, DC 2), heavy wall drill pipe (HWDP), regular drill pipe (DP 1, DP 2), and additional tubulars or tools (other) that may be in the hole. The driller fills in the size of the tubulars, the barrels per foot (bbl/ ft) or barrels per stand (bbl/stand) they displace, the number of feet or stands of the tubulars, and the total volume of each in barrels (Vol. bbls). In the columns below, the driller then notes the calculated amount of mud the hole should take to replace the stands of tubulars as they are pulled. Then, by checking the trip tank gauge (assuming a trip tank is being used), which tells the actual amount of mud that went into the hole to replace the stand, the driller notes the amount of hole fill. If there is a discrepancy between the calculated amount and the actual

RIG: _____				DATE: _____		
WELL: _____				TIME: _____		
DRILLER: _____		**TRIP SHEET**		DEPTH: _____		

REASON FOR THE TRIP: _____
Number of stands to have top of DC's one DP stand below BOP's: _____

		DISPLACEMENT:	DC1	DC2	OTHER	HWDP	DP1	DP2
PULL ON:	✓	Size						
EVEN		bbl/ft or						
SINGLE		bbl/stand						
DOUBLE		x ft or stands						
		= Vol. (bbls)						

STAND NO.	TRIP TANK GAUGE	CALCULATED Hole Fill (bbls) per Increment	MEASURED Hole Fill (bbls) per Increm'	MEASURED Hole Fill (bbls) Accumul.	DISCREPANCY per Increm'	DISCREPANCY Accumul.	REMARKS
0							

Figure 2.2   Trip sheet with spaces to show displacement of drill stem components, number of stands pulled, and calculated versus actual amounts of fill-up mud. *(Courtesy Sedco Forex Schlumberger)*

amount of fill-up mud, the driller notes it. During a trip out, a discrepancy can indicate that formation fluid has been swabbed into the hole.

Counting Pump Strokes

Another way to keep track of the fill-up mud required to replace the drill stem elements pulled from the hole is to use the rig's regular mud pump. By counting pump strokes, crew members can calculate the volume of mud the pump puts out. They can use pump-stroke counts to measure mud volume because a reciprocating pump, when operated properly, is like a positive-displacement meter: for each stroke, the pump puts out a given volume of mud. The volume depends on the size of the pump liners, the size of the pump piston rods, and the length of the pump's stroke. By knowing the volume of mud the pump puts out on each stroke and by counting the number of strokes, crew members can determine the volume of mud put into the hole. Pump manufacturers publish tables for each pump they manufacture that gives mud volumes pumped on each stroke for a given liner and piston size and stroke length.

Crew members should also be aware that counting pump stokes to keep track of the fill-up amount has limitations. For one thing, the pump may not be 100 percent efficient. That is, the pump may not put out precisely the volume of mud calculated for the volume of its liners and length of stroke.

Another problem with counting pump strokes to keep track of fill-up mud is that triplex pumps are often supercharged with centrifugal pumps, which can pump mud through the main pump when the main pump is putting out relatively low back-pressure against the charging pump. In this case, more mud goes into the well than is calculated by the pump-stroke count. Finally, if a rig is not equipped with a device that automatically keeps track of the number of pump strokes, it is easy for personnel to lose count and miscalculate the amount of fill-up mud actually pumped into the well.

Recirculating Tank

Still another way to keep track of fill-up volume is to use a trip tank with pump that continuously fills the hole as the pipe is pulled. Using a recirculating tank ensures that the hole is constantly being filled as the drill stem is pulled.

Changes in Drilling Rate

Observation of drilling rate changes is a direct means of detecting highly pressured shale or sand formations. Usually, when the bit enters an overpressured formation, the rate of penetration increases; however, when an oil-base mud and a diamond drilling bit are in use, the rate of penetration may decrease. A sudden increase in the drilling rate is a *drilling break*; conversely, a sudden decrease in the drilling rate is a *reverse drilling break*.

A drilling break may indicate that the bit is entering an overpressured sand section, because the bit usually drills faster when a reduction in pressure overbalance occurs; that is, when the pressure in the formation equals or exceeds the hydrostatic pressure of the mud column. A generally accepted policy is to drill no more than 2 ft to 4 ft into the break before conducting a flow check.

Of course, the rate of penetration is also affected by such factors as how well the mud cleans the bottom of the hole, bit weight, rotary speed, and the mud's fluid properties. The type of bit and its condition also influence the penetration rate. Nevertheless, when the penetration rate suddenly changes, it can indicate that the formation being drilled has changed, and the crew should be alert for the possibility of a kick.

Increase in Flow of Mud from the Well

If the well kicks, the flow rate of mud returning from the hole increases. The entrance of fluid into the wellbore from a formation must cause flow rate to increase, even though the increase may be difficult to detect. In general, the best way to detect an increase in return flow is by means of flow-measurement devices. Even if such devices are not available, however, if it is suspected that the well may be flowing, stop drilling, raise the kelly above the rotary, stop the pump, and check the return line for flow from the well. This procedure is a *flow check*. Stopping the pump stops circulation and causes a reduction in bottomhole pressure equivalent to the annular pressure drop; if the well continues to flow with the pump shut down, the well must be kicking.

If the well does not flow when the pump is shut off and remains static for a short period of time— many operators recommend waiting two or three minutes—probably no well kick is occurring. If oil-base mud is being used, however, some operators recommend longer waiting periods—up to 30 minutes if no flow is apparent.

If the well flows with the pump shut down, a gain in pit volume will occur. Sometimes, however, the pits stop gaining fluid when the pump is running. In such cases, the pump should be stopped and the well shut in to check for pressure on the drill pipe and casing pressure gauges. Stopping the pump and shutting in the well to see if pressure exists is the most important step any time it becomes necessary to kill the well. If the well flows with the pump shut down, but little or no pressure appears when the well is shut in, it is likely that the only step necessary is to increase the mud weight slightly to obtain an overbalance of hydrostatic pressure to formation pressure. If pressures appear on the drill pipe and casing when the preventer is closed and the well is shut in completely, however, well-killing procedures are generally required.

It should be noted that return-flow measuring instruments on an offshore floating rig may not provide a reliable indication of flow rates, because of drilling vessel movements. Such movements affect only the flow-line sensor; pit-volume totalizer equipment sums movement in the pits and is generally reliable. A longer time may elapse before any changes are observed, however.

Pit Gain or Loss

Unless a gain of fluid in the mud tanks, or pits, is caused by switching fluid in the tanks or some similar action, a pit gain is a positive indication that formation fluid is entering the wellbore. Many operators require every drilling and workover rig to have some type of pit-level indicating device to show gain or loss of mud quickly (fig. 2.3). For exploration and development wells, where pressures are expected to be high, many operators and contractors consider an indicating and recording pit-level instrument essential. The recorder should be placed so that the driller can see the chart while drilling and making trips, and he or she should be notified anytime mud is added to or taken from a working pit. Unscheduled pit-level drills should also be conducted to train the driller and crew to be alert to fluid level changes in the pits. A pit gain is sure evidence that fluid in the hole is being displaced by formation fluid entering the well.

A pit loss (a decrease in the volume of mud in the tanks) can also be a sign of a potential kick. One condition that can cause the volume of mud in the tanks to fall is lost circulation. As mentioned previously, lost circulation occurs when quantities of whole mud flow from the wellbore and into a formation. As mud is lost to the formation, the length of the mud column in the hole drops. If enough mud is lost, the hydrostatic pressure developed by the mud column can be reduced to a point that fluids from a downhole permeable zone could flow into the well and create a kick.

Lost circulation is not the only event that causes a loss in pit volume. For example, routing mud through the surface mud cleaning equipment can temporarily cause a decrease as the mud in the tanks moves into the cleaners. Also, dumping or moving mud from the active tanks to a reserve pit or tank can trigger a pit-loss alarm. In these two cases, the well is probably not kicking as long as all the loss in volume can be accounted for by the dumping or rerouting of the mud.

When a kick occurs, the amount of annular pressure required to contain it depends largely on how quickly the well is closed in. Quick closure retains more mud in the well than slow closure. When large amounts of mud are unloaded from the well, higher pressure is needed at the surface to contain formation pressure, because of the shortened mud column remaining in the well. With higher annular surface pressure required, the risk of formation fracture and a possible underground blowout may be greater. Therefore, the crew should be able to recognize a pit gain immediately, perform a flow check, and shut the well in.

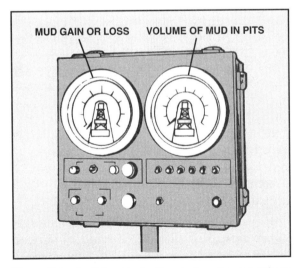

Figure 2.3 A pit-level indicating device shows a gain or loss in pit volume.

When mud is not being circulated in a well, the levels in each pit of the mud system are essentially the same. When circulation starts, the level in the suction pit drops more than the levels in the other pits, with the pit nearest the return being the highest. Because of this difference in pit levels, most pit-level recording devices are designed to measure and average the level in each pit and combine, or totalize, the average on the recorder as a single indication. Most pit-level recording devices also provide a constant readout of barrels gained or lost. A gain in pit volume is a signal for the driller to perform a flow check immediately. If the well flows, it should be shut in.

The rate of pit gain can be an indication of the permeability of the kicking formation. For example, a high-permeability kick is indicated by rapid flow of mud from the well and fast pit-level rise. Flow may start at almost the same time that the high-pressure zone is encountered. A kick from a high-permeability formation can be sudden and very dangerous.

If a high-permeability formation is only slightly underbalanced by mud pressure, the kick may be difficult to detect quickly. The flow rate is initially slow, and the pit gain may be very gradual until the gas is near the surface. When near-surface expansion begins, the well unloads rapidly, bottomhole pressure is reduced, and fluid flow into the well increases rapidly. A drilling break may be associated with the pit gain, but the pit may have only a small rise until gas is circulated far up the hole.

If the kick originates from a low-permeability, or tight, formation underbalanced by mud pressure, only a slow pit gain may occur. If the underbalance is small, only gas-cut mud may appear. Further, a change in the rate of penetration may not occur. Because the influx of fluids from tight formations is slow, the crew has ample time to react. It can be dangerous, however, to assume that a kick is coming from a tight formation without considerable experience in the area. Kicks from high-permeability formations that are only slightly underbalanced also begin with a slow pit gain.

Decrease in Pump Pressure and Increase in Pump Speed

Circulating pressure is related to fluid friction losses in surface piping, drill pipe and drill collars, bit nozzles, and in the annulus. In addition, any imbalance in hydrostatic pressure between the inside and the outside of the drill pipe affects circulating pressure. For example, if gas is encountered during drilling, it rises and expands in the annulus. Gas expansion displaces some of the fluid in the annulus so that a lighter column of fluid is present in the annulus than in the drill pipe. Unless the preventer is closed, the condition is unbalanced between the fluid in the drill pipe and the mixed column of mud and gas in the annulus. With the unbalanced condition of lighter fluids in the annulus, circulating pressure gradually decreases and, unless the pump throttle is changed, pump speed may increase. If much gas is involved, flow from the well increases, a rapid pit gain occurs, and a blowout is under way unless the preventer is closed.

Increase in Rotary Torque

Torque increases with depth in normally pressured zones but shows a greater increase in a transition zone, where formation pressures are becoming abnormal. In a transition zone, large amounts of shale cuttings can enter the wellbore and impede drill string rotation. The bit also takes larger bites into the formation as it turns. As a result, cuttings pile up around the collars and increase rotary torque. Thus, an increase in rotary torque can be a good indicator of increasing formation pressure and a potential well kick.

Increase in Drag and Fill-Up

If formation pressure is greater than hydrostatic pressure during the time circulation is stopped for a connection or trip, the formation may close in around the drill pipe or collars (fill-up). This closing in causes the pipe to drag as it is moved. If the pipe is off bottom, the formation may slough or cave into the hole, preventing the bit from returning to its previous depth because of hole fill-up. Of course, some shales are sensitive to water in the mud—they expand and slough into the hole when exposed to water in the drilling fluid, causing fill-up and drag. When fill-up and drag increases on trips and connections, however, it may be a sign of increasing formation pressure, and not simply of shale problems.

Change in Cutting Size and Volume

An increase in cutting size may occur as mud overbalance is lost. On the other hand, cuttings may decrease in size or disappear altogether in soft coastal and marine sediments. Or sometimes the shaker screen may become completely covered with long slivers of shale—in oilfield parlance, the shaker becomes *blinded*.

A blinded shaker may be an indication of higher downhole pressure. In any case, a change in the character and size of the cuttings on the shaker is a warning of a downhole change that may be leading to higher pressure. Also, an increase in volume of cuttings may occur as pore pressure increases because drilling underbalanced yields higher penetration rates and possible hole enlargement.

Increase in String Weight

The drill stem weighs less in a mud-filled borehole than it does in air because of the buoyant effect of mud. Just as a ship floats in water because the ship is buoyant, so does the drill stem float in a hole that contains drilling mud. The denser the mud the greater is its buoyant effect on the drill stem. Therefore, when a kick occurs and the formation fluids are less dense than the drilling mud, the buoyant force of the mud is reduced, causing an increase in string weight. This increase can sometimes be observed on the weight indicator at the surface, but only after large amounts of mud have been displaced. By that time, outflow from the well can reduce string weight or even blow the string out of the hole.

Increase in Different Types of Gas

Some kick indicators require special instrumentation for detection and analysis. One such indicator is an increase in gas levels. Using a gas detection device, a base, or normal, trend line for hydrocarbon gases returning from the hole can be established. This normal trend can be compared to subsequent data. If certain patterns of gas readings are seen, it is likely that pore pressures have increased. To understand these patterns it is necessary to distinguish between background, connection, and trip gas.

Background Gas

During drilling operations, gas contained in the cuttings and arriving at the surface can be measured. Measuring these gas quantities establishes a base-, or trend, line. By monitoring gas levels, an increase from the baseline can be detected, and this increase may indicate that a kick is occurring or is about to occur.

Connection Gas

When the kelly or top drive is raised to add a joint of pipe to the drill string, swabbing can occur; when the pumps are shut down, bottomhole pressure is decreased by the reduction in the amount of circulating friction pressure in the annulus (annular pressure loss). If either of these causes bottomhole pressure to fall below formation pressure, then formation fluids (including gas) can enter, or feed into, the hole. These small feed ins may not be detected at the flow line, but a gas detector can show a short term increase above a base line once the influx has been circulated to the surface.

Trip Gas

An influx of gas into the wellbore during a trip can occur because of the swabbing action of pipe being pulled from the hole, and because of the lack of annular pressure loss, since the pump is shut down. If, following a trip back into the hole, trip gas is circulated to the surface, it will show up as an increase in the gas readings. In circumstances where trip gas may go undetected while tripping out and remain in the hole until after the trip in, as in a hole being drilled with oil-base mud or in a highly deviated well, crew members should be alert for trip gas, which can rapidly unload the hole as it expands near the surface.

Increase in Salinity, or Chloride

Saltwater (brine) in a formation is generally saltier than the water used to make up the drilling mud. A significant increase in salinity could therefore mean that formation saltwater has entered the hole. Salinity, or chloride, checks of the mud filtrate can show the salinity increase (fig. 2.4).

Figure 2.4 Mud filtrate being checked for chloride, or salinity, content. After potassium chromate is added to the filtrate, silver nitrate is added until an orange-red color persists for about thirty seconds.

A salinity change in the mud filtrate indicates change and, if it occurs, the source of the change should be determined. First, the makeup water for the drilling mud should be checked. If it does not show a change in salinity, the problem must come from the hole.

Usually, chloride, or salinity, changes in the mud are first detected by the mud engineer, who routinely performs chloride tests on the mud. While the usual procedure is for the mud engineer to report findings to rig supervisors, it is important for the driller to recognize the significance of the chloride test and to be aware of the type of makeup water in the drilling mud in use on the well. Should a chloride increase occur, it may very well be an indication that pore pressures are also increasing.

Increase in Flow-Line Temperature

Since temperature and pressure are related, the thermal gradient may drop sharply just above a transition zone, and it may increase rapidly in an abnormally pressured zone. If flow-line temperature can be measured accurately, the abnormal zone may be predicted. Because of the variables that affect flow-line temperature, it is necessary to use an end-to-end plot that shows all the variables. An end-to-end plot is useful for confirming that the cause of a flow-line temperature change is a change in one of the variables rather than a change in formation pressure. The cause and magnitude of the change is noted on the plot, and that amount is added to or subtracted from the actual reading to produce a continuous plot. A normal trend line can be established, and excursions from the normal trend can be recognized readily. The high and low points could have a 5° F to 6° F difference. Accurate plotting is difficult because of the changeable nature of a circulating system. Some of the variables that affect flow-line temperature are (1) mud weight, (2) solids content, (3) flow properties, (4) circulation rates, (5) hole geometry, and (6) water depth and temperature when drilling offshore.

Change in Shale Density

Normally, shale density increases with depth, because shale's unit weight is greater when free water is squeezed out of it and into sand formations by the process of compaction. When density decreases below a normal trend line, increased formation pressure can be suspected. The use of this method to detect formation pressure is difficult, however, because of problems in selecting representative particles of shale, and in making precise measurements of density.

Shale density changes as a transition zone of higher pressure is penetrated. The drilling rate and size of cuttings may also increase. Direct shale density measurements can be made, but interpretation must be done with care, because such measurements are often masked by changes in mineralogy.

Gas-Cut Mud

Often, gas-cut drilling mud is not a sign of a well kick because gas cutting may result from gas transported in the cuttings up the hole when drilling a gas-bearing formation. Gas cutting usually does not cause a large reduction in bottomhole pressure because gas compresses; the weight of the mud above the gas keeps gas volume small. If the volume of gas in the mud is small, then the bottomhole pressure reduction is also small. As gas is circulated up to the surface, however, hydrostatic pressure decreases and gas volume increases. Gas expands most near the surface and thus makes the mud light at the flow line. From even a few feet below the surface to the bottom of the hole, hydrostatic pressure is sufficient to keep gas volume small, and very little reduction in bottomhole pressure occurs.

Gas-cut mud can occur for various reasons. For example, when a gas-bearing formation is drilled, gas gets into the mud from the rock that is destroyed by the bit. In this case, gas cutting is usually an indication that a reservoir rock or a gas-bearing shale has been penetrated; it is not a result of the mud weight's being too low to overbalance formation pressure. If any doubt exists, however, the pump should be stopped and a flow check of the well made.

Gas cutting also occurs during drilling into a low-permeability formation that contains gas under higher pressure than the pressure developed by the mud column. Since the formation is not very permeable, the influx of gas is slow. Gas from such a source can cause an increase in background, trip, or connection gas, and is a warning that pressures may be building in the wellbore.

When mud pressure is close to formation pressure, trip or connection gas can also enter the wellbore and appear as gas-cut mud when circulated to the surface. As stated earlier, when the pump is turned off, bottomhole pressure is reduced by the amount of friction loss in the annulus. When pipe is pulled up, bottomhole pressure is further reduced by the friction loss of the mud falling back down around the bit. Therefore, between drilling with the pump on and pulling pipe with the pump off, a relatively large difference in bottomhole pressure can exist. This difference may be enough to cause the well to be momentarily underbalanced, thereby allowing connection or trip gas to enter the hole. Therefore, the reduction in bottomhole pressure can be calculated from the pit volume increase. By developing a pressure-per-volume value for the annulus—psi/bbl in this case—and multiplying this value by pit-volume increase in bbl, the reduction in bottomhole pressure can be determined:

$$P = (MG \div AV) \, PVI \qquad \text{(Eq. 16)}$$

where

P = reduction in bottomhole pressure, psi

MG = mud gradient, psi/ft

AV = annular volume, bbl/ft

PVI = pit volume increase, bbl

As an example, suppose in a 20,000-ft well that the mud is gas-cut from 18 ppg to 9 ppg, the mud gradient is 0.936 psi/ft (determined from 0.052×18 ppg), the annular volume is 0.0896 bbl/ft, and the pit volume increase is 10 bbl. How much has bottomhole pressure been reduced?

$$P = (0.936 \div 0.0896) \, 10$$
$$= 10.4 \times 10$$
$$P = 104 \text{ psi.}$$

From the example, it can be seen that, even though the mud weight at the surface has been cut to 9 ppg, half that of the 18 ppg mud weight at bottom, bottomhole pressure is reduced by only 104 psi. Even though 104 psi is a relatively small value in view of how much the mud weight has been reduced at the surface, it may, of course, be sufficient to cause the well to flow. It is worth noting that, even when gas cutting is as extreme as in the example, it may have less effect on bottomhole pressure than when the pumps are shut down during drilling.

In summary, gas cutting does not cause a large reduction in bottomhole pressure; therefore, mud weight should never be increased only because of gas-cut mud. If doubt exists about whether gas cutting is reducing bottomhole pressure by an amount large enough to allow formation fluids to enter the hole, however, the pump should be shut down and a flow check made. Finally, if the well does kick, pit volume increase is one way to determine how much bottomhole pressure was reduced.

Change in Normalized Drilling Rate (d Exponent)

The d exponent as a value for predicting abnormally pressured formations in the Gulf Coast was devised in 1966 by Jorden and Shirley of Shell Oil Company. They postulated that the drilling rate, or rate of penetration (ROP), could be used to identify overpressured formations because ROP was dependent on the differential between wellbore pressure and formation pressure; that is, ROP increased as the pressure differential decreased. By studying data from several wells drilled in southern Louisiana, Jorden and Shirley concluded that "drilling performance data can be used to detect the top of overpressured sediments in areas where the approximate depth of overpressuring is known. A plot of normalized rate of penetration will show a trend of continually decreasing penetration rates with depth and a reversal of this trend as overpressures are penetrated by the drill bit. This technique can be used as a means to avoid taking a kick and to identify overpressures prior to logging." They derived the d exponent from the general drilling equation:

$$R \div N = a \, (W^d \div D) \qquad \text{(Eq. 17)}$$

where

R = penetration rate, ft/hr

N = rotary speed, rpm

a = a constant, dimensionless

W = weight on bit, lb

d = exponent in general drilling equation, dimensionless

D = hole diameter or bit size, in.

Jorden and Shirley found that, if the constant a were dropped, the equation could be rearranged and solved for the exponent d. Following is the rearranged and solved equation:

$$d = \log (R \div 60N) \div \log \qquad \text{(Eq. 18)}$$
$$(12W \div 1,000D)$$

where

d = d exponent, dimensionless

R = penetration rate, ft/hr

N = rotary speed, rpm

W = weight on bit, 1,000 lb

D = bit size, in.

As an example of solving the equation, assume that—

$$R = 30 \text{ ft/hr}$$
$$N = 120 \text{ rpm}$$
$$W = 35$$
$$D = 8.5 \text{ in.}$$

Therefore:

$$
\begin{aligned}
d &= \log [30 \div (60 \times 120)] \div \log \\
&\quad [(12 \times 35) \div (1,000 \times 8.5)] \\
&= \log (30 \div 7,200) \div \log (420 \div 8,500) \\
&= \log 0.0042 \div \log 0.0494 \\
&= -2.377 \div -1.306 \\
d &= 1.82.
\end{aligned}
$$

Normally, values for d increase with depth until a transition or overpressured zone is penetrated by the bit; values for d then decrease. Since the d exponent should be plotted for each 5 ft to 10 ft of hole when the top of the abnormally pressured zone is being looked

for, Jorden and Shirley developed a nomograph for determining d exponents quickly (fig. 2.5). To use the nomograph, first determine the rate of penetration (R) and rotary speed (N). As an example, assume that R = 20 ft/hr and that N = 100 rpm. Place a straight edge between the two points on the R and N lines on the nomograph, 20 and 100 in this example. Note where the straight edge falls on the $R \div 60N$ line and mark it; in this case, it is between 0.003 and 0.004. Next, determine weight on bit (W) and bit size (D). As an example, assume that W = 25,000 lb and that D = 9⅞ in. Place a straight edge between the two points on the W and D lines on the nomograph and mark where the straight edge falls on the $12W \div 1,000D$ line—0.030 in this case. Finally, place a straight edge between the mark on the $R \div 60N$ line and the mark on the $12W \div 1,000D$ line, and read the value for the d exponent where the straight edge crosses the d line on the nomograph. In this example, d = 1.64. The important fact to

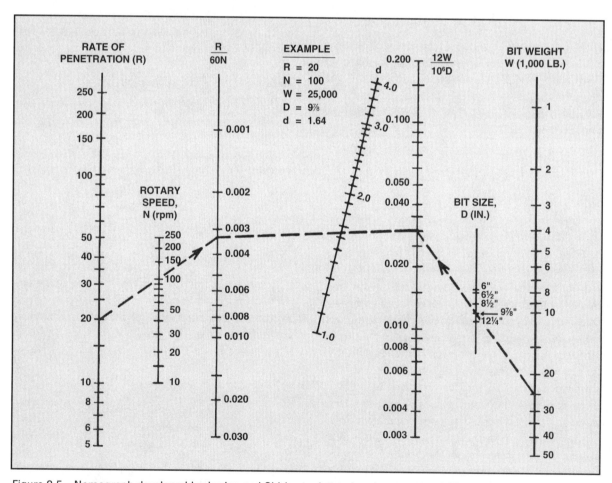

Figure 2.5   Nomograph developed by Jorden and Shirley to determine d exponent quickly

remember about use of the d exponent is that its value should increase with depth; if it decreases, it is possible that a transition or overpressured zone has been penetrated. Since the d exponent is influenced by mud-weight variations, modifications have been made to correct it for changes in mud weight. The following is one formula:

$$d_c = d \, (MW_1 \div MW_2) \qquad \text{(Eq. 19)}$$

where

d_c = corrected d exponent

MW_1 = normal mud weight, 9.0 ppg

MW_2 = actual mud weight used, ppg.

Some operators and drilling contractors prefer to modify equation 19 by assigning a different value to MW_1. Instead of assuming a normal mud weight of 9.0 ppg, MW_1 is given the value of the mud weight in use before the weight was changed. MW_2 is then given the value of the mud weight in use after the weight was changed. Thus:

MW_1 = previous mud weight, ppg

MW_2 = current mud weight, ppg.

For example, if—

d = 1.64

MW_1 = 12.1 ppg

MW_2 = 12.7 ppg,

then

d_c = 1.64 (12.1 ÷ 12.7)

= 1.64 (0.95)

d_c = 1.56

The corrected d exponent is 1.56.

Indications from Seismic Analysis

Large structures, massive shales, faults, and other possible sources of abnormally high formation pressure often can be identified from seismic records. It is from such records, or sections, that the first indication of possible abnormal pressures in an area can be obtained. Because interpreters can see structure sizes, fault movements, depths and thickness of salt beds, and other structures on seismic sections, they can often predict the presence of abnormal pressure. Further, since measurement of pressures in the earth is a normal part of seismic interpretation, the pressures that develop in massive shales can be identified and measured. In short, modern seismic techniques and interpretation methods can provide operators and contractors with information to prepare them for abnormally pressured zones.

Indications from Well Logging Data

Formation logs can be very helpful in obtaining knowledge of high formation pore pressure after the formation is drilled. While such logs may be of no assistance in predicting high-pressure zones on a current well, they may be of great help on subsequent wells drilled in the area. Shale resistivity, acoustic travel time, conductivity, and radioactivity are some of the characteristics recorded by logs to confirm that abnormal pressures have been penetrated. One method used to identity abnormally pressured formations during drilling is to compare the actual log measurements of shales with trends established in normally compacted shale beds. Normal shale compaction results in decreasing porosity with depth, increasing resistivity with depth, and decreasing sonic travel time with depth. If deviations from these trends are observed, abnormal pore pressures will probably be present.

Change in Properties of the Mud

Properties of the drilling mud may change as formation pressure increases. For example, when pressure in the formation increases faster than the pressure of the mud column, more cuttings and cavings can dissolve into the mud and increase its viscosity. If an anhydrite caprock on a salt dome or the dome itself is drilled, mud viscosity usually increases. An increase in viscosity shows up as increased water loss and, in the case of salt, increased salinity in the mud filtrate. Changes in mud properties can be the result of downhole conditions and should be checked as soon as they are noticed.

Indications from Mud Logging

Mud loggers can occasionally find and interpret other indicators of abnormal pressure, but these indicators may be difficult for the crew to recognize. For example, mud loggers can detect subtle reactions between drilling mud, the bit, and the formation that may give indications that abnormally pressured formations are being penetrated. Still other indicators that a mud logger may be able to detect are changes in the type of shales or clays being penetrated by the bit.

FALSE KICK INDICATORS

While a main concern of anyone involved in well control is the recognition of kicks, it is also helpful to know that sometimes a situation can occur in which an indication of a kick exists, but, upon further examination, can prove to be false.

Kelly Cut

When a mud-saver valve is not installed on the end of the kelly and the kelly is broken out of the drill stem during a connection, mud in the kelly empties onto the rig floor. When the crew makes up the kelly in the new joint of pipe, a quantity of air exists inside the kelly, and in the joint added to the string. This air is then pumped down the drill stem along with mud and, when the air gets to bottom, bottomhole pressure causes its volume to shrink. Conversely, as the air and mud move up the annulus, the air's volume expands as pressure is reduced. This expanding air can cause an increase in the return rate of mud flowing to the shaker and a slight pit gain. An alert crew will notice these increases and may believe that the well is kicking, even though it is not.

Background Gas

An increase in background gas can be a sign that the well has kicked. Increases in background gas, however, are often not a sign of a kick. Even though the gas level increases above a baseline, gas concentrations in formations can change drastically, even though there is no increase in formation pore pressure.

Bottoms Up with Oil-Base Mud

Because gas from a formation can dissolve in an oil-base mud, a drilling crew not used to drilling with oil-base mud can make the mistake of thinking the well they just tripped out is gas free. However, when they get back to bottom and the driller circulates bottoms up, the trip gas can break out, expand, unload the hole above it, and cause any gas dissolved in the oil mud below to unload its gas because of the hydrostatic reduction. This unzipping of gas down the hole has lead to the reputation of oil muds unloading massive amounts of mud when approaching bottoms up after trips. Many companies require the first bottoms up circulation to go through the choke, with the crew ready to restrict flow if large pit gains require it. Bear in mind that even though flow and pit gain while circulating bottoms up with an oil-base mud may be false indications of a kick, such flow and gain can result in real kicks.

Fluid Transfers

One of the more important things that crew members can do on any rig is maintain good communications between the driller, derrickman, and other crew members whose job involves handling drilling fluid. If they do not, then a transfer of drilling mud from one tank to another, may create a pit-level alarm. If they move mud out of the tanks, a low-level alarm may sound and the driller may think that lost circulation has occurred. Conversely, if they move mud into the tanks, a high-level alarm may sound and the driller may think that a kick has occurred. Such a false kick indicator should never occur.

Chapter 3

Shut-In Procedures and Shut-In Pressures

If a well's casing is set and adequately cemented in competent formation, usually the well can be shut in safely when a kick occurs. Although shut-in procedures are basically the same for every well, well conditions and company policies often dictate variations. Therefore, many operators post their shut-in procedures for each well on the rig floor and require crew members to learn them. Since land rigs and offshore bottom-supported rigs use surface-mounted blowout preventers (BOPs), basic shut-in procedures for these types of rigs are the same. Special shut-in procedures are required for floating rigs, however, because the BOP stack is usually mounted on the seafloor. Regardless of whether a surface stack or a subsea stack is in use, different shut-in procedures are required for a kick that occurs while drilling and for a kick that occurs while tripping.

SHUT-IN PROCEDURES WITH SURFACE STACKS WHILE DRILLING

A surface stack is normally used on land rigs and on bottom-supported offshore rigs like jackups and platforms. Since there are several variations in the procedures used to shut in a well with a surface stack while drilling, it is important that everyone be familiar with and adhere to the procedures on the rig. For instructional purposes, an example shut-in procedure (a soft shut-in) that has been successfully used follows:

1. Stop the rotary and sound the alarm.
2. Pick up the drill string until the kelly saver sub is above the rotary table. (Prior spaceout should be made to ensure that a tool joint is not in a ram BOP when the string is picked up.) On rigs with top drives, raise the string to the first space out point that ensures the bit is at least a few feet off bottom.
3. Stop the pumps.
4. Check for flow.
5. If the well flows, open the line from a BOP outlet to the choke manifold.
6. Close the BOP (usually the annular preventer).
7. Close the choke. (The choke to be used, as well as other valves in the intended flow path through the manifold, should initially be in the open position.)
8. Confirm that all flow from the well has stopped. No flow should occur from the choke manifold, the bell nipple, or back through the drill stem.
9. With the well fully shut in, allow a few minutes for pressures to stabilize; then, record shut-in drill pipe pressure (SIDPP).
10. Record shut-in casing pressure (SICP).
11. Record the pit-level increase.
12. Notify supervisor.

SHUT-IN PROCEDURES WITH SURFACE STACKS WHILE TRIPPING

Just as many variations in procedure are available to shut in a surface stack while drilling, so are many variations available to shut in a surface stack while a trip is being made. Again, it is important for everyone to know and to follow the procedures on the rig. The following is one procedure that has been used successfully:

1. Immediately set the drill pipe on slips; sound the alarm.

2. Install and make up a full-opening drill pipe safety valve in the drill pipe. The valve should be in the open position.

3. Close the drill pipe safety valve.

4. Open the HCR valve to a remote-controlled or manually adjustable choke.

5. Close the BOP (usually the annular preventer).

6. Close the choke.

7. Confirm that all flow from the well is stopped. No flow should occur from the choke manifold, the bell nipple, or back through the drill stem.

8. Pick up and make up the kelly or top drive (insert a pup joint or a single joint between the safety valve and top drive to allow easy rig floor access to top-drive unit).

9. Record *SIDPP*.

10. Record *SICP*.

11. Record the pit-level increase.

12. Notify supervisor.

SHUT-IN PROCEDURES WITH SUBSEA STACKS WHILE DRILLING

Since a floating drilling rig such as a drill ship or semisubmersible is in motion during drilling operations, and since the BOP stack is usually located on the seafloor at the mud line, care must be taken to ensure that a tool joint does not interfere with closing a ram preventer. As is the case with surface stacks, several shut-in procedures are available for use with subsea stacks. The following procedure has been used successfully to shut in a subsea stack during drilling operations:

1. Stop the rotary; sound the alarm.

2. Pick up the drill string until the kelly saver sub (or tool joint if using a top drive) clears the rotary table.

3. Stop the pumps.

4. Open the outer and inner subsea choke-line fail-safe valves.

5. Close the annular preventer.

6. Close the remote controlled choke.

7. Confirm that all flow from the well is stopped. No flow should occur from the choke manifold, the bell nipple, or back through the drill stem.

8. Record *SIDPP*.

9. Record *SICP*.

10. Record the pit-level increase.

11. Notify supervisor.

Some operators and contractors also prefer to hang off the drill stem on the ram preventers to prevent accumulation of gas below the subsea annular preventers and to prevent damage to the ram seals caused by pipe motion. Following is a hang-off procedure:

1. Conduct a spaceout to ensure that the rams will not close on a tool joint and that the lowest drill string safety valve is accessible.

2. Close the rams.

3. Slowly lower the drill string until a tool joint contacts the rams. Observe the weight indicator for a decrease in string weight.

4. If using a ram BOP that does not have automatic locking devices, actuate the ram locking devices.

SHUT-IN PROCEDURES WITH SUBSEA STACKS WHILE TRIPPING

To shut in a well with a subsea stack while tripping, the following procedure may be used:

1. Set the top drill pipe joint on the slips; sound the alarm.

2. Stab and make up a full-opening drill pipe safety valve in the drill pipe. Make sure the valve is open.

3. Close the drill pipe safety valve.

4. Open the outer and inner subsea choke-line fail-safe valves.

5. Close the annular preventer.

6. Close the remote controlled drilling choke.

7. Confirm that all flow from the well is stopped. No flow should occur from the choke manifold, the bell nipple, or back through the drill stem.

8. Record *SIDPP*.

9. Record *SICP*.

10. Record the pit-level increase.

11. Notify supervisor.

Some operators recommend that when a float is in the string, closing in and stripping back in may be done as long as the float functions properly. If the float starts to leak, they recommend that a safety valve and inside BOP be installed before stripping is continued. Under no circumstances should an attempt be made to run the drill pipe back to bottom with the preventers open and the well flowing.

SHUT-IN PROCEDURES WHILE RUNNING CASING

Shutting in a well while running casing is similar to shutting in a well when tripping drill pipe. The main differences involve the device used to stop potential flow up the casing and whether to close a BOP or a diverter, which depends on the type of casing being run. When running surface casing, the BOP stack is not usually nippled up, since there is no casinghead to nipple onto. In such cases, crew members will have to use a diverter or other procedures to close in the well. Because casing is normally run with a float shoe, once the diverter or BOPs are closed, the shoe prevents backflow through the casing. Also, the common cement circulating head can be closed to prevent flow up the casing. It is important to remember to plan for having to close in around casing. Ram BOPs will need to be properly sized to close around the casing. Further, annular closing pressure may need to be reduced to prevent collapsing the casing.

SHUT-IN PROCEDURES WHILE CEMENTING

During cementing operations, one factor to keep in mind is that joints of casing shorter than normal may be run into the well to ensure that the casing shoe hangs at the correct depth near the bottom. With joints shorter than normal, space out becomes important. Space out is the placement of the casing joints in the BOP stack in relation to the position of the casing couplings. When properly spaced out, the BOPs will be able to close on the casing's body and not on a coupling. The BOP may not be able to close properly on a coupling. If a kick is detected, the driller should first ensure that the casing is properly spaced out. Then, the cement pump is shut down, the BOP closed (usually the annular), and the supervisor notified.

HARD AND SOFT SHUT-IN PROCEDURES

Depending on the preference of the operator, the well may be shut in hard or soft. *Hard shut-in* means closing the BOP without first opening an alternate flow path through the choke. *Soft shut-in* means closing the BOP with the choke and HCR, or fail-safe, valves open. A majority of operators seem to prefer hard shut-in because it keeps the kick influx to a minimum and simplifies the shut-in procedure; however, hard shut-in may increase the likelihood of breaking down the formation during shut-in.

Hard Shut-In Procedure

If a hard shut-in is used, the choke and the HCR, or fail-safe, valves are set in closed position during normal operations. When a kick occurs, the following hard shut-in procedure may be used while drilling:

1. Stop the rotary; sound the alarm.
2. Pick up the drill string until the kelly saver sub (or tool joint if using a top drive) clears the rotary table.
3. Stop the pumps.
4. Close the annular preventer and adjust its closing pressure as required.
5. Confirm that all flow from the well is stopped. No flow should occur from the choke manifold, the bell nipple, or back through the drill stem.
6. Open the HCR, or fail-safe, valves.
7. Read and record *SIDPP* and *SICP*. Allow the pressures to stabilize before reading the gauges.
8. Read and record the pit-level increase.
9. Notify supervisor.

The primary advantage of a hard shut-in is that the kick influx is held to a small volume because the well is closed in more quickly. No documentation exists to show that hard shut-in causes *water hammer* (a pressure concussion caused by suddenly stopping the flow of liquids in a closed container) and damage to the wellbore, even if gas is in the kick fluids.

Soft Shut-In Procedure

If the soft shut-in procedure is used, the choke is set in full-open position during normal operations. When a kick is taken, the following soft shut-in procedure may be used:

1. Open the remote choke-line (HCR) valve.
2. Close the annular preventer.
3. Close the drilling choke.
4. Adjust closing pressure on the BOP.

5. Read and record *SIDPP* and *SICP* after allowing them to stabilize.

6. Read and record the pit-level increase.

The primary disadvantage of a soft shut-in is that it requires more steps and time than a hard shut-in. The result can be a larger influx of kick fluids.

VERIFICATION OF SHUT IN

Regardless of whether the well is shut in hard or soft, it is important to verify that it is actually shut in and that all equipment is holding pressure. Any leaks erode equipment and make it impossible to precisely monitor closed-in pressures. The BOP equipment shutting in the annulus should have no leaks; also, with the well completely shut in, no flow should be coming from the flow line. On surface stacks, while checking annulus shut in, crew members should ensure that the casing valve on the wellhead is not leaking. Further, there should be no broaching or flow around the wellhead.

To ensure that flow is not occurring through the drill stem, the mud pump's pressure-relief valves should be holding—that is, no flow should occur through the pump's relief valves. Also, crew members should check all connections on the stand pipe, as well as the stand pipe itself, for leaks.

Finally, crew members should check to be sure that no leakage is occurring in the choke manifold. All fittings in the manifold, the chokes, and the lines should be pressure-tight with no leakage. Also, they should check the flare line or overboard lines, if offshore, to ensure that no leaks are occurring at any point in these lines.

SPECIAL SHUT-IN PROCEDURES

When a kick occurs and the rig is not drilling or tripping drill pipe, special shut-in procedures may be required. For example, some companies require special procedures when a kick occurs during a trip and drill collars have reached the BOP stack. Other operations that require special shut-in procedures include kicks that occur during logging operations in open hole and kicks that occur when the drill string is completely removed from the hole. It is important for every crew member to know the required procedures during all phases of the drilling operation. Also, it is important for them to notify their supervisor whenever a kick or a suspected kick occurs.

Wireline Operations

When a kick occurs during wireline operations, usually the drill stem is out of the hole. If a kick is detected, the annular BOP should be closed to shut in the well, since it will seal around the wireline. Most operators do not recommend shutting in the pipe, blind, shear, or blind-shear rams on wireline. Should the blind or shear rams need to be closed to shut in the well, most operators recommend having a line cutter accessible during wireline operations that will cut the line and allow it to fall back into the hole. With the wireline out of the way, the blind or shear rams can effectively close the open hole.

Open Hole

When all the drill stem is removed from the hole and it becomes necessary to shut in the well, most operators recommend that the blind or blind-shear rams be closed. While an annular preventer can be used to shut in open hole (hole with the drill stem removed from it), doing so places a great deal of wear on the sealing element. As a result, the element will have to be replaced before it can be used again. Once the blind or shear rams are closed on the open hole, crew members should close the choke to completely shut in the well and notify their supervisor.

VERIFYING SHUT IN

As mentioned earlier, regardless of the procedure used to shut in a well, it is important to verify that the well is completely shut in; otherwise, formation fluids could continue to enter the hole, making it difficult or impossible to regain control of the well. To confirm shut in, many operators use the following procedures.

Annulus

To ensure that the annulus is closed, check to be sure that the annular BOP is completely closed on the well and is holding pressure. Then, be sure that no fluids are flowing out of the mud return (flow) line.

Drill Stem

To confirm that pressure is holding on the drill stem, check to be sure that the mud pump's pressure relief valves have not popped open and that closed-in pressure will not exceed the pressure at which the relief valves will open; then, check the standpipe manifold. Every joint should be pressure tight with no leaks.

Wellhead and BOPs

On surface stacks on land and on bottom-supported offshore units, crew members should ensure that the casing valve on the wellhead is closed and, once closed, is capable of holding the anticipated annular pressures. Further, they should check for pressure broaching to the surface outside the wellbore.

Choke and Choke Manifold

Finally, the choke and choke manifold should be checked to ensure that the choke, when completely closed, is holding pressure. If not, then well fluids could be leaking past the choke and an alternate choke may need to be employed. Crew members should also ensure that all the fittings and lines in the choke manifold are holding pressure. If not, prompt repairs should be made.

SHUT-IN DRILL PIPE PRESSURE AND SHUT-IN CASING PRESSURE

Once the well is shut in, vital well-control information can be gained from the drill pipe and casing pressure gauges located at the surface. These gauges may be thought of as having very long stems. The stem for the drill pipe gauge is the drill string; if the bit is on bottom, it goes all the way to the bottom of the hole. The stem for the casing gauge is the walls of the hole and casing. Because the gauge stems go all the way to bottom, *SIDPP* and *SICP* indicate bottomhole conditions. In addition, as long as the drill stem is full of clean mud from the pits, as it usually is when a kick occurs with the bit on bottom and drilling, *SIDPP* can be used to determine the increase in mud weight required to kill the well. The weight of the mud in use when the kick occurs plus the increase in weight calculated from *SIDPP* gives the new mud weight required to control the well. Since the annulus, or casing, mud has cuttings and kick fluids such as gas, oil, or salt water in it, using its pressure for calculations does not give an accurate value for mud weight increase.

Since kick fluids contaminate the mud in the annulus, but often do not contaminate the mud in the drill stem, *SIDPP* is usually lower than *SICP*. Under certain circumstances, however, *SIDPP* can exceed *SICP*. For example, large amounts of cuttings in the annulus can offset the tendency of kick fluids to reduce the hydrostatic pressure of the annular mud. Rather, large quantities of cuttings in the annulus can increase the density of the annular fluids above that of the mud in the drill stem to the extent that *SICP* is actually lower than *SIDPP*.

Further, in cases where low-density drilling fluids are being used—aerated mud, for example—the density of formation fluids that intrude into the annulus—saltwater, for example—can exceed the density of the drilling fluid. As a result, when the well is shut in, *SIDPP* will exceed *SICP*. Also *SIDPP* can exceed *SICP* when kick fluids enter the drill stem. Another instance in which *SIDPP* can be greater than *SICP* occurs when the annulus is partially or totally blocked so that bottomhole pressure cannot transmit to the surface outside the drill stem. The hole may collapse or become blocked with debris.

Finally, *SIDPP* can read higher than *SICP* simply because of malfunctioning pressure gauges. If any doubt exists about gauge accuracy, crew members should verify it by taking readings from more than one gauge—for example, the standpipe pressure gauge indication should agree with the drill pipe pressure gauge indication on the choke control console.

Kick Fluid Density and *SICP*

Since most kicks are composed of gas, oil, or salt water either alone or in combination, and since the density of these fluids relates to *SICP* values, it is useful to know the range of densities, or weights, that these fluids exhibit. In general, gas ranges in weight from 1.5 ppg to 3.0 ppg, oil from 5.0 ppg to 7.0 ppg, and salt water from 8.6 ppg to 10.0 ppg.

Knowing whether the kick fluids in the wellbore are composed mainly of gas or salt water is not essential to successful well control; however, if the kick fluid can be identified, problems relating to expansion, percolation, explosion, or changes in mud properties can be prepared for and thus be avoided or solved more easily. Most operators agree that the best policy is to handle all kicks as though they were composed of gas until the influx is circulated out. If a well is shut in on a gas kick, *SICP* is greater than if the well is shut in on a saltwater kick, assuming that both fluids have the same volume. An equation is available that can be used to illustrate that gas kicks induce higher *SICP*s than saltwater kicks:

$$SICP = FP - HP - P_f \qquad \text{(Eq. 20)}$$

where

$SICP$ = shut-in casing pressure, psi

FP = formation pressure, psi

HP = hydrostatic pressure from top of kick to surface, psi

P_f = hydrostatic pressure of the kick fluid, psi.

As an example, assume that—

well depth = 13,750 ft

mud weight = 12.0 ppg

mud gradient = 0.624 psi/ft

annular volume = 0.0459 bbl/ft

kick size = 30 bbl

$SIDPP$ = 600 psi

height of kick in annulus = 654 ft

FP = 9,180 psi

HP = well depth – height of kick × mud weight × .052

P_f = pressure gradient of fluid × height of kick.

If the weight of the gas is 2.15 ppg (pressure gradient is 0.112 psi/ft), then—

$SICP$ = 9,180 – 8,172 – 73

$SICP$ = 935 psi.

If the weight of the salt water is 9.25 ppg (pressure gradient is 0.481 psi), then—

$SICP$ = 9,180 – 8,172 – 315

$SICP$ = 693 psi.

Assuming that the volume of the kick is the same, the example shows that $SICP$ can vary dramatically, depending on whether the kick is gas or salt water. The difference is almost 250 psi in this case.

Another difference between a gas kick and a saltwater kick is the expansion behavior of the kick as it is circulated to the surface. If a constant bottom-hole pressure method of circulating the kick out of the well is used, then a gas kick must be allowed to expand as it moves up the annulus. Therefore, $SICP$ and pit level will increase until the gas reaches the surface. On the other hand, with a saltwater kick, little or no expansion occurs in the salt water as it is circulated up the annulus. Since little or no expansion occurs, $SICP$ and pit level remain close to their original shut-in values.

The density of the kick influx can be calculated with the following equation:

$$D_i = MW - [(SICP - SIDDP) \div (L \times 0.052)] \quad \text{(Eq. 21)}$$

where

D_i = density of influx, ppg

MW = original mud weight, ppg

$SICP$ = shut-in casing pressure, psi

$SIDPP$ = shut-in drill pipe pressure, psi

L = influx height in the annulus, ft (L = influx, or kick, size, bbl ÷ annular volume, bbl/ft).

If the density of the intruded fluid, or influx, is less than 3 ppg, it is considered to be gas. If the density of the fluid is from 6 to 9 ppg, it is considered to be a mixture of gas and liquids, such as salt water and oil. If the density is from 9 to 10 ppg, the kick is considered to be salt water. Be aware that most operators recommend that all kicks be treated as gas kicks until the kick is circulated out of the choke.

As an example of calculating the density of a kick influx, assume that the mud weight is 9.6 ppg, $SIDPP$ is 400 psi, $SICP$ is 700 psi, pit gain is 15 bbl, and annular volume is 0.0231 bbl/ft. In this case, L is about 649 ft, since 15 bbl divided by 0.0231 bbl/ft equals about 649 ft. Using these values in equation 19:

$$D_i = 9.6 - [(700 - 400) \div (649 \times 0.052)]$$
$$= 9.6 - (300 \div 34)$$
$$= 9.6 - 8.8$$
$$D_i = 0.8 \text{ ppg (gas).}$$

Hydrocarbon Fluid Considerations

When all, or nearly all, of the fluids in a kick are hydrocarbons, personnel handling the kick need to be aware of how such fluids behave in a well. For one thing, under certain circumstances hydrocarbon gases will not migrate up the hole. This lack of migration can occur because of the mud's characteristics. For instance, a very viscous mud can prevent hydrocarbon gases from migrating. In cases where gas migration does not occur, crew members circulating a kick need to be aware that when the gas nears the surface where pressure is reduced, it will expand rapidly and casing pressure will rise accordingly.

When hydrocarbon fluids enter the wellbore, they can be either liquid or gas, depending on the temperature and pressure on the fluids where they enter. Under normal wellbore conditions, the higher the pressure is, the more likely the hydrocarbons will be liquid. Conversely, the lower the pressure is, the more likely the hydrocarbons will be gas. On the other hand, the higher the temperature is, the more likely the hydrocarbons will be gas, and the lower the temperature is, the more likely the hydrocarbons will be liquid.

This behavior of hydrocarbon fluids affects well control in several ways. For example, at a particular depth, the pressure and temperature at that depth may cause an intruded hydrocarbon fluid to be liquid. Therefore, the pit level increase may be relatively small, and the crew may believe that a saltwater or oil kick has occurred. However, as the intruded liquid is circulated up the wellbore, the pressure is reduced and the liquid becomes gas. This behavior of hydrocarbon fluids is one reason many operators and contractors require their drilling personnel to handle all kicks as though they were gas kicks.

THE WELL AS U-TUBE

The pressures that appear on the drill pipe and casing pressure gauges, and thus the relationship between hydrostatic and formation pressure, may be easier to understand if the well is thought of as a large U-tube. The drill stem forms one leg of the U-tube, and the annulus between the drill stem and the hole forms the other. When mud-column pressures on both sides of the U are equal to each other and equal to or greater than formation pressure, no formation fluids can enter the wellbore. Both sides of the U stand full of mud, and no pressure appears on the drill pipe or casing pressure gauges when the pump is stopped and the well shut in (fig. 3.1).

Now assume that a kick occurs and that the well is shut in. The drill stem is still full of clean mud, but the annulus has mud and kick fluids in it. With the pumps off and the well shut in, pressure appears on the drill pipe and casing pressure gauges, because formation pressure is higher than hydrostatic pressure, and because kick fluids have entered the annulus. As an example, assume that the well is 15,000 ft deep and that 15-ppg mud is in use. If a 400-ft-long gas kick with a weight, or density, of 2.3 ppg is taken, *SIDPP* is 780 psi and *SICP* is 1,044 psi, because hydrostatic pressure is 11,700 psi, formation pressure is 12,480 psi, and the kick's hydrostatic pressure is 48 psi (fig. 3.2). The annulus side of the U shows

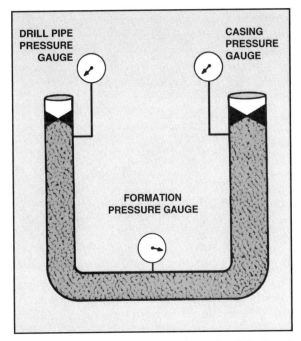

Figure 3.1 With the drill pipe and annulus full of mud equal to or greater than formation pressure, the pumps stopped, and the well shut in, no pressure appears on the drill pipe and casing pressure gauges.

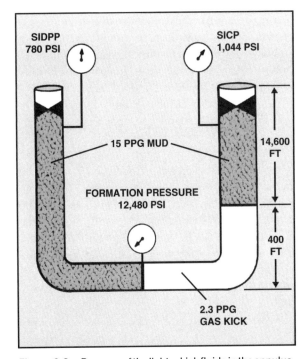

Figure 3.2 Because of the lighter kick fluids in the annulus, shut-in casing pressure is higher than shut-in drill pipe pressure.

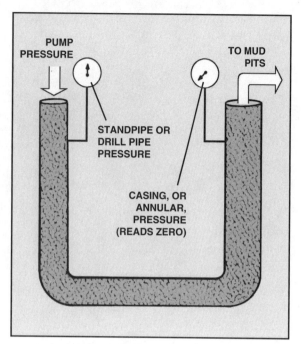

Figure 3.3 When mud is circulating, pump pressure is indicated on the standpipe or drill pipe pressure gauge. Casing, or annular, pressure reads zero because all the pump pressure is expended as the mud is circulated.

higher pressure on the surface than the drill stem side because of kick fluids in the annulus.

Imagining the well as a U-tube is also useful when mud is being circulated during normal operations. The pressure required to circulate mud through the drill stem–annulus U-tube—*pump pressure*—can be read directly on the standpipe, or drill pipe, pressure gauge. Annulus pressure can be read on the casing pressure gauge. When circulating, drill pipe pressure indicates the pressure developed by the mud pump. Casing pressure indicates zero, because all the circulating pressure is lost as the mud is pumped down the drill string, out of the bit, and up the annulus (fig. 3.3). Of course, if the circulation rate is increased by speeding up the pump, higher pressure will appear on the standpipe, or drill pipe, gauge. Also, if the mud weight is increased, higher pressure will be required to maintain the same circulation rate; in effect, the pump will have to work harder to maintain the same circulation rate with heavier mud. The higher, or new, pressure required can be estimated by equation 22:

$$NCP = OCP \times NMW \div OMW \quad \text{(Eq. 22)}$$

FP – SICP = HPT

HPT – HP2 = HP3

where

NCP = new circulating pressure, psi

OCP = old circulating pressure, psi

NMW = new mud weight, ppg

OMW = old mud weight, ppg.

As an example, assume that the old circulating pressure is 2,800 psi with 10 ppg mud. With the mud weight increased to 11 ppg, the new circulating pressure is—

$$NCP = 2,800 \times 11 \div 10$$
$$= 2,800 \times 1.1$$
$$NCP = 3,080 \text{ psi.}$$

If the well is circulated and back-pressure is held on the annulus by the choke, the new circulating pressure will be added to shut-in drill pipe pressure. In the example, if 200 psi back-pressure is being held on the well, then the total pressure will be 200 plus 3,080, which equals 3,280 psi. This pressure can be read on the standpipe gauge.

Using *SIDPP* to Calculate Formation Pressure

When mud-column pressure becomes less than formation pressure, formation fluids can intrude into the wellbore, and a gain in pit volume equal to the volume of intruding fluid usually occurs. When the well is shut in, pressure on the drill pipe is equal to the difference between formation pressure and hydrostatic pressure. Since hydrostatic pressure can be calculated from the known height and weight of the mud column inside the drill stem, the drill stem represents a pressure gauge by which formation pressure can be determined using equation 23:

$$FP = HP + SIDPP \quad \text{(Eq. 23)}$$

where

FP = formation pressure, psi

HP = hydrostatic pressure, psi

$SIDPP$ = shut-in drill pipe pressure, psi.

As an example, if hydrostatic pressure is 5,300 psi and shut-in drill pipe pressure (*SIDPP*) is 350 psi, what is the formation pressure?

$$FP = 5,300 + 350$$
$$FP = 5,650 \text{ psi.}$$

Using *SIDPP* to Calculate Mud-Weight Increase

Since *SIDPP* is an indication of formation pressure, it can be used to calculate how much the mud weight must be increased to kill the kick from the bit to the surface. Keep in mind that the drill stem must be full of clean mud from the pits for *SIDPP* to be used to determine mud-weight increase accurately. As mentioned before, *SICP* cannot be used to calculate mud-weight increase, because kick fluids contaminate the drilling mud in the annulus. As a result, *SICP* is usually higher than *SIDPP*, because the invading kick fluids reduce hydrostatic pressure in the annulus. As stated before, the denser the kick fluids are, the closer *SICP* will be to *SIDPP*. Thus, if the kick is mostly salt water, *SICP* will be lower than if the kick is mostly gas, assuming that both are of the same volume.

Because *SIDPP* is essential to well control, most operators recommend that a check be made to ensure that the drill string is full of mud before *SIDPP* is used to calculate the mud weight needed to kill the well. In the case of a large kick, mud can U-tube out of the string, because fluids in the annulus have been lightened by the kick. Unreasonably high *SIDPP* is an indication of U-tubing. Mud fill in the drill string can be checked by holding *SICP* constant with the choke while pumping 300 to 400 strokes. If the drill stem is full of mud, *SIDPP* will remain the same as before pumping. If it is partly empty, the pressure will be lower after it has been filled with mud. If *SIDPP* does go down after pumping, a second pumping should be given to be sure that the pressure is correct. When the pipe is full, values for *SIDPP* will match.

Once *SIDPP* is obtained, the following equation can be used to calculate the amount of mud-weight increase needed to balance formation pressure:

$$MWI = SIDPP \div TVD \div 0.052 \text{ (Eq. 24)}$$

where

MWI = mud-weight increase, ppg

$SIDPP$ = shut-in drill pipe pressure, psi

TVD = true vertical depth, ft.

Once the mud-weight increase is found, it should be added to the old mud weight to calculate the new, or kill-, weight mud required to kill the well; thus,

$$KWM = OMW + MWI \text{ (Eq. 25)}$$

where

KWM = kill-weight mud, ppg

OMW = old mud weight, ppg

MWI = mud-weight increase, ppg.

As an example calculation, suppose that a well kicks and is shut in. The well's *TVD* is 7,890 ft, the mud weight before the well kicked is 9 ppg, and *SIDPP* reads 400 psi. How much must the mud weight be increased to kill the well, and what is the kill-weight mud?

$$MWI = 400 \div 7,890 \div 0.052$$
$$= 0.0507 \div 0.052$$
$$= 0.97$$
$$MWI = 1 \text{ ppg.}$$

Then, to find kill-weight mud:

$$KWM = 9 + 1$$
$$KWM = 10 \text{ ppg.}$$

The solution shows that the mud weight of 9 ppg must be increased by 1 ppg to 10 ppg to kill the well.

With the advent of hand calculators, most operators and contractors use equations 22 and 23, or similar equations, to determine mud-weight increase. However, once *SIDPP* is obtained, mud-weight increase can also be found by use of a table (table B.8 in Appendix B). The value taken from the table added to the mud weight already in the hole gives the new mud weight necessary to kill the well. The mud-weight increase required can be read opposite the depth and under *SIDPP*.

Rising *SIDPP* and *SICP*

Sometimes, after a well is shut in, *SIDPP* and *SICP* slowly rise, instead of stabilizing at a steady value. The rise in pressure occurs either because gas is percolating up through the mud, or because the kicking formation has low permeability. If the slowly rising pressure is due to gas percolating up the hole, drill pipe pressure does not represent formation pressure; rather, it represents an elevated pressure because of gas at constant pressure moving up the hole. As a result, the required mud-weight increase determined from *SIDPP* will not be accurate.

If the slowly rising pressure is due to a formation with low permeability, the kick is entering the hole very slowly, and it will become evident after several circulations that *SIDPP* is reading too low and that mud weight must be increased more. It is difficult to tell the difference between percolating gas and a low-permeability formation until the influx has been circulated out of the hole. At that time, if *SIDPP* does not stabilize, the problem is one of formation fluids entering the hole because of a low-permeability formation.

Since the difference between low permeability and percolating gas is not evident, one rule of thumb is to assume that any pressure rise after the first hour of shut-in is due to percolating gas. A slow pressure increase may be allowed for the first hour unless it begins to approach the maximum allowable value. By then, however, kill operations should have begun to make sure bottomhole pressure does not fall, because more kick fluids could enter.

Low or No *SIDPP* and *SICP*

Sometimes indications of a kick exist, but when the well is shut in, little or no pressure appears on the drill pipe pressure gauge. If no *SIDPP* appears, it is possible that (l) the drill pipe and casing pressure gauges are shut off, (2) a float is in the drill pipe, (3) the well has no pressure, or (4) pressure is too low to show on the gauges. If little or no *SIDPP* appears, first open the choke to see if the well will flow. If it flows, a drill stem float may be why no pressure appears on the gauge; or the gauge could be broken or shut off. If no float is in use and if the gauges are working properly, open the kelly bypass valve, if the rig is equipped with one, and check for pressure. If doubt still exists, some operators recommend using the driller's method to circulate the hole and see if additional mud weight is required to balance formation pressure.

MAXIMUM SURFACE PRESSURE AND VOLUME WITH GAS KICKS

When a well is shut in on a gas kick and the gas is circulated to the surface, its volume will expand as it approaches the surface because of reduced hydrostatic pressure. The gas will achieve its maximum volume at the surface, where pressure on the gas is lowest. Annular surface pressure, which can be read on the casing pressure gauge, will rise as the gas nears the surface and will achieve maximum value when the gas vents to atmosphere. The maximum value can be critical if it exceeds the pressure ratings of the surface equipment or casing. The amount of annular surface pressure that develops on a well that is properly circulated out depends on several conditions:

1. The greater the imbalance between hydrostatic and formation pressures, the higher the surface pressure.

2. The larger the volume of the kick, the higher the surface pressure.

3. The lower the density of the influx, the higher the surface pressure.

4. Pressures increase as the annulus becomes smaller.

5. Pressures increase as hole depth increases.

6. Pressures increase as mud density increases.

7. Circulating mud of the weight required to kill the well at the same time the kick is circulated out produces lower surface pressures than circulating the kick out without increasing the mud weight, assuming that circulation begins with the kick on or near the bottom.

8. Note that gas percolation while the well is improperly left closed in can increase surface pressures close to formation pressure; the calculations assume that no such mistake is made.

Calculating Maximum Surface Pressure

An equation is available for calculating a value for maximum surface pressure that develops from a properly handled gas kick. Because of many unknowns, however, such as the nature of the kick material, influx fluid temperature, mud flow pattern, and distribution of the kick, maximum surface pressure cannot be calculated accurately. Since so many unknown factors exist, the equation is designed to give a worst-case answer; that is, the maximum surface pressure calculated from the equation should give a result of the highest magnitude. The equation is—

$$MSP_{gk} = 0.2 \sqrt{(P \times V \times W) \div C} \qquad \text{(Eq. 26)}$$

where

MSP_{gk} = maximum surface pressure from a gas kick, psi

P = formation pressure, psi

V = original pit gain, bbl

W = mud weight to kill well, ppg

C = annulus capacity at surface, bbl/ft.

As an example, suppose that—

P = 6, 000 psi

V = 15 bbl

W = 14 ppg

C = 0.0352 bbl/ft.

The solution is—

$$MSP_{gk} = 0.2 \sqrt{(6,000 \times 15 \times 14) \div 0.0352}$$
$$= 0.2 \sqrt{1,260,000 \div 0.0352}$$
$$= 0.2 \sqrt{35,795,455}$$
$$= 0.2 \times 5,982.93$$
$$MSP_{gk} = 1,197 \text{ psi.}$$

Keep in mind that this solution is not an accurate representation of maximum surface pressure; rather, it is an indication of the highest value that could be achieved.

Calculating Maximum Pit Gain

If the kick contains gas, the gas will expand as it nears the surface; at the surface it will achieve maximum volume. As the gas expands, it pushes more and more mud out of the hole ahead of it, so that the pits gain more and more volume until the gas is released through the choke to the surface. An idea of how much pit gain will occur can be calculated with the following formula:

$$MPG_{gk} = 4 \sqrt{(P \times V \times C) \div W} \quad \text{(Eq. 27)}$$

where

MPG_{gk} = maximum pit gain from a gas kick

P = formation pressure, psi

V = original pit gain, bbl

C = annulus capacity, bbl/ft

W = mud weight to kill well, ppg.

As an example, assume that—

P = 6,000 psi

V = 15 bbl

C = 0.0352 bbl/ft

W = 14 ppg.

The solution is—

$$MPG_{gk} = 4 \sqrt{(6,000 \times 15 \times 0.0352) \div 14}$$
$$= 4 \sqrt{3,168 \div 14}$$
$$= 4 \sqrt{226.29}$$
$$= 4 \times 15$$
$$MPG_{gk} = 60 \text{ bbl.}$$

The solution for the example problem shows that the original 15-bbl gas kick will cause the pits to show a 60-bbl gain when the gas reaches the surface and expands to its maximum volume.

MAXIMUM SAFE PRESSURES

Other considerations must also be taken into account concerning gas expansion. For one thing, the rated working pressure of the wellhead components in which the casing hangs must be known so that the rated pressure is not exceeded by the rising gas. Rated working pressures can be obtained from the manufacturer of the wellhead components. Casing burst pressures must also be known to prevent the rising gas from bursting the casing. Table B.12 on page B-15 in Appendix B is one source of casing burst pressures; another is API Bulletin 5C2, *Performance of Casing, Tubing, and Drill Pipe*. [This bulletin may be obtained from API, 1220 L Street, NW, Washington, D.C., 20005, (202)-682-8375.]

Similarly, drill pipe and tubing collapse pressures should also be known, because the gas pressure could be high enough to collapse these tubulars. API Bulletin 5C2 is a source of drill pipe and tubing collapse pressure. For drill pipe collapse pressures, the International Association of Drilling Contractors (IADC) *Drilling Manual* is another source. (Tables B.17, B.18, B.19, and B.20 in Appendix B are reproduced from the *Drilling Manual*. The manual is available from IADC, P.O. Box 4287, Houston, Texas 77210-4287.) Drill pipe collapse pressure depends on the diameter and grade of pipe in use, whether the pipe is new or used, and its used condition. For example, according to the *Drilling Manual*, 4½-in., 16.60, X-95 new drill pipe has a collapse pressure of 12,750 psi. This same pipe, in grade 3 used condition (the worst case), has a collapse pressure of only 3,930 psi. Further, biaxial loading on the pipe must be considered when using the tables. Multiple and simultaneous loads imposed on tubulars can greatly affect their rated pressure capabilities. The *Drilling Manual* is a good source for biaxial loading factors.

GAS MIGRATION AND EXPANSION

When a well is shut in on a kick that contains gas, the gas will percolate, or migrate, up the hole, even if the well is allowed to remain static. Gas migration can cause confusion during a well-control operation, because it can be overlooked, does not always visibly occur, and, if it does occur, may happen early or late in the operation. Gas or gas bubbles float, or migrate, up the hole because they are lighter than mud.

When gas bubbles rise, either they expand, or, if not allowed to expand, they cause an increase in surface pressure and wellbore pressure. Therefore, if a well is shut in for a long time, surface pressures can gradually rise until they cause lost circulation if not relieved. If the gas bubble is allowed to expand without control, however, it unloads the hole. With the hole unloaded, kick size increases, which in turn causes more unloading. This cycle of influx and unloading has caused the loss of many wells.

Gas Laws and Gas Expansion

If gas is circulated from the bottom of the well, it must be allowed to expand as it moves up the annulus to avoid excessive pressures in the wellbore. *Boyle's law* states that if the temperature of the gas is kept constant, an increase in pressure causes a decrease in the volume of the gas. *Charles's law* states that if the pressure is held constant, an increase in temperature causes an increase in the volume of gas. And if the volume of the gas is constant, an increase in the temperature of the gas results in an increase in the pressure of the gas. When Boyle's and Charles's laws are combined, the ideal gas equation is obtained:

$$P_1 V_1 \div T_1 \ = \ P_2 V_2 \div T_2 \qquad \text{(Eq. 28)}$$

where

P_1 = formation pressure, psi

P_2 = hydrostatic pressure at any depth in the wellbore, psi

V_1 = original pit gain, bbl

V_2 = gas volume at surface, bbl

T_1 = temperature of formation fluid, degrees Rankine ($^\circ$R = $^\circ$F + 460)

T_2 = temperature at the surface, $^\circ$R.

In an oilwell, the drilled gases are a complex mixture of hydrocarbon gases. Hydrocarbon gases depart from the ideal gas law by an amount equal to the compressibility factor, Z, of the gases. The following equation can be used for hydrocarbon gases:

$$P_1 V_1 \div T_1 Z_1 \ = \ P_2 V_2 \div T_2 Z_2 \qquad \text{(Eq. 29)}$$

where

Z_1 = compressibility factor under pressure in formation, dimensionless

Z_2 = compressibility factor at the surface, dimensionless.

This equation shows that if the gas does not expand as it is circulated or as it migrates to the surface,

bottomhole pressure increases and may exceed formation fracture gradient, causing formation breakdown. The equation can also be used to calculate how much gas expands as it reaches the surface. For example, suppose that a 12,000-ft well takes a 10-bbl kick with a mud weight of 13 ppg. Bottomhole temperature is 220° F, bottomhole pressure is 8,112 psi, and the bottomhole compressibility factor is 1.40. Finally, surface pressure is 14.7 psi, surface temperature is 120° F, and the surface compressibility factor is 1.0. Therefore:

P_1 = 8,112 psi

P_2 = 14.7 psi

V_1 = 10 bbl

V_2 = unknown

T_1 = 220° F + 460 = 680°R

Z_1 = 1.40

T_2 = 120° F + 460 = 580° R

Z_2 = 1.0.

The solution is—

$$(8,112 \times 10) \div (680 \times 1.40) \ = \ (14.7 \times V_2) \div (580 \times 1.0)$$
$$81,120 \div 952 \ = \ 14.7 \times V_2 \div 580$$
$$85.2 \ = \ 0.02535 \times V_2$$
$$V_2 \ = \ 85.2 \div 0.02535$$
$$V_2 \ = \ 3,361 \text{ bbl at surface.}$$

The solution shows that in this case a 10-bbl kick at bottom can expand to 3,361 bbl by the time it reaches the surface if left uncontrolled.

Shortened Gas Expansion Equation

The compressibility factor of hydrocarbon gases is usually determined experimentally, and formation temperatures are not always available. Therefore, the following shortened equation for gas expansion is used for calculations by some well-control operators. In general, it states that if the volume of gas doubles, the pressure is reduced by half.

$$P_1 V_1 \ = \ P_2 V_2 \qquad \text{(Eq. 30)}$$

where

P_1 = formation pressure, psi

P_2 = hydrostatic pressure, psi

V_1 = original pit gain, bbl

V_2 = gas volume at surface, bbl.

As an example problem using the shortened formula, assume—

$$P_1 = 5,200 \text{ psi}$$
$$P_2 = 14.7 \text{ psi}$$
$$V_1 = 10 \text{ bbl}$$
$$V_2 = \text{unknown.}$$

The solution is—

$$5,200 \times 1 = 14.7 \times V_2$$
$$V_2 = 5,200 \times 10 \div 14.7$$
$$V_2 = 3,537 \text{ bbl.}$$

The solution shows that a gas kick that displaced 10 bbl of mud from the bottom will occupy 3,537 bbl by the time it is vented to the atmosphere.

As the gas is circulated up the annulus, the hydrostatic pressure above the gas column decreases, and the gas expands rapidly as it nears the surface. If the gas column is not allowed to expand as it is circulated or as it migrates to the surface, bottom-hole pressure increases for every increment the gas column moves up the hole.

Gas Migration Equation

The rate at which gas migrates up the hole can be calculated. One way in which to calculate the rate is to observe the rise in *SICP* for a 1-hr period after the well is shut in. (If *SICP* is rising rapidly, the observation period should be reduced, depending on how fast the rise occurs.) After the observation period, the increase in *SICP* is recorded and used in the gas migration equation:

$$R_{gm} = \Delta SICP \div MG \qquad \text{(Eq. 31)}$$

where

R_{gm} = rate of gas migration, ft/h
$\Delta SICP$ = change in *SICP* after 1 hr, psi
MG = mud gradient, psi/ft.

As an example, assume that the original *SICP* is 500 psi, that *SICP* after 1 hr is 1,100 psi, and that the mud gradient is 0.572 psi/ft. What is the gas migration rate?

$$R_{gm} = (1,100 - 500) \div 0.572$$
$$R_{gm} = 1,049 \text{ ft/hr, or}$$
$$R_{gm} = 17.5 \text{ ft/min.}$$

Gas Migration When Gas Is in Solution

Well-control problems can result in blowouts because of the solubility of certain gases in specific types of mud. For example, methane gas dissolves in oil-base muds and H_2S dissolves in water-base muds. This fact can sometimes make it more difficult to detect a kick. A large gas influx entering the wellbore may change the pit volume very little if the gas dissolves in the mud. The influx is then circulated up the wellbore in the mud column until the hydrostatic pressure on top of the gas decreases to a certain point; then the gas flashes and comes out of solution.

Detecting the kick by observing the flow line or mud pits can be very difficult until the kick is fairly close to the surface and expands rapidly. Moreover, gas dispersed in wellbore fluids does not migrate up the hole, and a flow check may not show a noticeable flow. If a kick is suspected, some operators recommend close monitoring of the flow line for a long period to determine if an influx has invaded the wellbore. Well-control procedures for handling gas-soluble kicks are essentially the same as for any other kick, although little pit gain may occur until the gas is near the surface. For safety's sake, a rotating head should be considered to prevent gas from reaching the rig floor; crews should also be alerted to the need for extra watchfulness when the potential for gas-soluble kicks is present.

PRACTICE CALCULATIONS FOR MUD-WEIGHT INCREASE

Determining the mud-weight increase necessary to kill a kick is a basic part of well-control calculations. The following problems may be solved by first using either table B.8 or equation 24:

$$MWI = SIDPP \div TVD \div 0.052 \quad \text{(Eq. 24)}$$

where

MWI = mud-weight increase, ppg
$SIDPP$ = shut-in drill pipe pressure, psi
TVD = true vertical depth, ft.

Then, equation 25 should be used:

$$KWM = OMW + MWI \qquad \text{(Eq. 25)}$$

where

KWM = kill-weight mud, ppg
OMW = old mud weight, ppg
MWI = mud-weight increase, ppg.

1.
$$TVD = 11,250 \text{ ft}$$
$$SIDPP = 300 \text{ psi}$$
$$\text{Mud weight} = 14.7 \text{ ppg}$$
$$MWI = \underline{\hspace{1cm}}$$
$$\text{Required mud weight} = \underline{\hspace{1cm}}$$

2.
$$TVD = 14,250 \text{ ft}$$
$$SIDPP = 720 \text{ psi}$$
$$\text{Mud weight} = 16.2 \text{ ppg}$$
$$MWI = \underline{\hspace{1cm}}$$
$$\text{Required mud weight} = \underline{\hspace{1cm}}$$

3.
$$TVD = 5,500 \text{ ft}$$
$$SIDPP = 150 \text{ psi}$$
$$\text{Mud weight} = 10.3 \text{ ppg}$$
$$MWI = \underline{\hspace{1cm}}$$
$$\text{Required mud weight} = \underline{\hspace{1cm}}$$

4.
$$TVD = 8,820 \text{ ft}$$
$$SIDPP = 780 \text{ psi}$$
$$\text{Mud weight} = 12.6 \text{ ppg}$$
$$MWI = \underline{\hspace{1cm}}$$
$$\text{Required mud weight} = \underline{\hspace{1cm}}$$

5.
$$\text{Measured depth} = 11,640 \text{ ft}$$
$$TVD = 9,820 \text{ ft}$$
$$SIDPP = 250 \text{ psi}$$
$$\text{Mud weight} = 11.7 \text{ ppg}$$
$$MWI = \underline{\hspace{1cm}}$$
$$\text{Required mud weight} = \underline{\hspace{1cm}}$$

HIGH-PRESSURE, HIGH-TEMPERATURE CONSIDERATIONS

Generally speaking, a high-pressure well is a well in which wellhead pressure could reach or exceed 10,000 psi when the well is shut in on a full column of gas originating from the zone of highest pressure in the well. A high-temperature well is a well in which wellhead temperature could reach or exceed 300° F under the conditions created by an uncontrolled flow from the zone of highest pressure through an open choke manifold. A high-pressure, high-temperature (HP-HT) well combines the two extremes. Drilling HP-HT wells requires special planning, operating procedures, and equipment, particularly when an oil-base mud is being used.

Planning

Many contractors and operators recommend that a well simulator or computer model be used to estimate the maximum gas and fluid flow rates and the maximum temperature that could result from an uncontrolled flow from the highest pressure zone through an open choke manifold. While simulators and computer models may not give perfect results, they can certainly help crew members in knowing the pressures and temperatures to anticipate. Also, since some well-control procedures for handling HP-HT wells are different from normal wells, crew members should be well briefed and prepared before spudding an HP-HT well. Some contractors and operators also recommend that the well's casing program include plans for running an extra, contingency casing string. Finally, many contractors and operators recommend that the ram BOPs be fitted with rams that can close around production casing being run into the well.

Operating

It is considered good practice for the driller to make a flow check of an HP-HT well every time crew members make a connection. Also, a drop-in sub for a pump-down check valve should be installed in the string. In many cases, it is good practice to drop the dart and let it seat in the sub before tripping out. (It may not be necessary for short trips.) Many well-control operators also recommend that the well be circulated bottoms up through the choke if it is suspected that fluid has been swabbed in. When making connections using a top drive, it is good practice to install a full-opening safety valve on the bottom of the stand before adding it to the drill stem hanging in the rotary. The safety valve makes it easier to disconnect and install additional drill stem valves at the rig floor level if it becomes necessary.

Kill Procedures

Select a pump kill-rate speed that does not exceed the capacity of the surface equipment. Pay particular attention to situations where the capacity of the mud-gas separator is reached or is likely to be reached. Further, take into consideration the temperature limits of the choke through which the kick is circulated. It is possible for gas expansion downstream of the choke to be so great that the choke manifold or mud-gas separator could become blocked with hydrates or ice.

Also, if, after the well is closed in, it is determined that MASP or other pressure limitations will be exceeded, it may be better to attempt to bullhead the influx back into the formation. (See the section on bullheading in Chapter 7.)

Equipment

Be sure that the temperature rating of all the BOP elastomers (seals) exposed to well fluids are higher than the maximum anticipated temperature at the wellhead and BOP stack. The elastomers should also be rated to withstand the anticipated peak temperature and pressure for at least 1 hour. Also, make sure that all the well-control equipment is trimmed for H_2S service. What is more, it is recommended that an antifreeze injection system be installed in the choke and kill manifolds. Further, these manifolds should also have mud temperature measurement probes upstream from the chokes to help determine wellhead temperature. Or, the temperature probes can be installed in the BOP stack. These probes will help determine whether hydrates will form. Finally, the flare line from the choke manifold should be rated for at least 5,000 psi and be equipped with valves that can be remotely operated to close the mud-gas separator line and open the flare line.

On HP-HT wells, it is important that the size of the mud-gas separator be large enough to handle the anticipated volumes. In some instances, two separators may be required. Also, the gas vent lines from the separator may need to be 8 to 10 in. in diameter and may be required to maintain a 15 to 20-ft mud seal. Most operators recommend that a low-pressure differential sensor (typically capable of measuring a 20-psi differential), that sends its signal to the choke control panel on the rig floor, be installed on the mud-gas separator. When the differential drops below the sensor's set point, it is a warning that the mud-gas separator is being overloaded because full separation of mud and gas cannot occur at low differential pressure. Also, some operators install heaters and injectors to warm and inject low-pressure mud into the separator. Such equipment helps break up and prevent hydrate formation.

Chapter 4

Circulation and Well Control

Because the mud pump is used to circulate kick fluids out of the hole and to circulate kill-weight mud into the hole, it is one of the basic tools of well control. Also, because the rate, or speed, at which the pump is run affects pressures, the pump rate is one of the basic values in well control. It is usually measured in strokes per minute (spm). Pump rate plays a vital role in the successful control of a well, because even small changes in pump speed can cause large changes in pressure at the bottom of the hole, where constant pressure is especially critical. Since the fundamental goal of most well-control procedures is to maintain a constant bottomhole pressure equal to or only slightly higher than formation pressure, accurate control of the pump's speed is necessary.

KILL RATE

To circulate a well that has kicked, most operators require that the pump rate be reduced to a speed below that used for normal drilling. Called the *kill rate,* this reduced pump rate affords several advantages to any well-control method: (1) it reduces circulating friction losses so that circulating pressures are less likely to cause excessive pressure on exposed formations; (2) it gives the crew more time to add barite or other weighting materials to the mud; (3) it reduces strain on the pump; (4) it allows more time for the crew to react to problems; and (5) it allows adjustable chokes to work within proper orifice ranges. Often, only one kill-rate speed will be required during a well-killing operation; however, most operators recommend that several kill rates be selected. For example, one recommendation is to select kill rates of ½, ⅓, and ¼ the normal drilling pump rate. Sometimes an even slower rate may be recommended—for example, two to four barrels per minute—to reduce pressure further and allow more time for weighting up.

KILL-RATE PRESSURE

When the pump rate is reduced to the kill rate, circulating, or pump, pressure is also reduced. This reduced pump pressure is *kill-rate pressure.* It is obtained by pumping down the drill pipe and drill collars, out of the bit nozzles, and up the annulus. Running the pump at a reduced speed and breaking circulation will produce the kill-rate pressure. While kill-rate pressure can be easily read on the standpipe gauge or the drill pipe pressure gauge, the relationship between pump rate and pump pressure can also be estimated by the following equation:

$$P_2 = P_1 \times SPM_2^2 \div SPM_1^2 \quad \text{(Eq. 32)}$$

where

P_1 = original pump pressure at SPM_1, psi

P_2 = reduced or increased pump pressure at SPM_2, psi

SPM_1 = original pump rate, spm

SPM_2 = reduced or increased pump rate, spm.

As an example of working with the equation, calculate P_2 if—

$$P_1 = 1,200 \text{ psi}$$
$$SPM_1 = 60 \text{ spm}$$
$$SPM_2 = 30 \text{ spm}.$$

The solution is—

$$P_2 = 1,200 \times 30^2 \div 60^2$$
$$= 1,200 \times 900 \div 3,600$$
$$= 1,200 \times 0.25$$
$$P_2 = 300 \text{ psi}.$$

The solution shows that when the pump speed is reduced by half—from 60 spm to 30 spm, in this case—the circulating pressure drops by a factor of four, from 1,200 psi to about 300 psi.

Another example also shows the relationship of pump speed to pump pressure. Assume that—

$$P_1 = 3,000 \text{ psi}$$
$$SPM_1 = 90 \text{ spm}$$
$$SPM_2 = 30 \text{ spm}.$$

The solution is—

$$P_2 = 3,000 \times 30^2 \div 90^2$$
$$= 3,000 \times 900 \div 8,100$$
$$P_2 = 333 \text{ psi}.$$

In this case, the speed is cut by two-thirds rather than half; therefore, pump pressure is reduced more than four times—in this case, from 3,000 psi to about 333 psi, a drop by a factor of nine.

Although kill-rate circulating pressures can be calculated, most operators and contractors prefer that the pump actually be slowed to each kill rate, and that the standpipe pressure at each rate be read and recorded. Since it is easier to determine a kill rate and kill-rate pressure (KRP) before a well kicks, both should be established prior to a kick. Most operators and contractors require that their crews routinely determine and record the kill rate and KRP at least once each tour, usually during the grease-up and maintenance period after the new crew comes on tour. The kill rates and pressures for all pumps should be determined and entered on the drilling report. If the mud weight or bit nozzle sizes are changed during the tour, or if a large amount of footage is drilled, then new KRPs should be taken.

If a KRP has not been recorded prior to shutting in the well on a kick, it is easy to determine one. For example, assume that the well has been closed in and SIDPP reads 300 psi. One procedure to determine

KRP is to bring the pump up to a speed that is reasonably below the normal drilling rate. While bringing up the pump speed, SICP is held constant by adjusting the choke. In this example, let's say the pump rate is adjusted to 30 spm and that this speed puts 800 psi on the drill pipe pressure gauge. Thus KRP at 30 spm is 500 psi because 300 psi (SIDPP) – 800 psi (total drill pipe pressure) = 500 psi (KRP). It is also important to select a speed that does not create so much pressure that SICP reaches MASP.

Bear in mind that the kill-rate pressure recorded prior to a kick may not be the correct KRP at a later time. Not only do changes in depth, mud weight, and nozzle sizes change KRP, but so can other factors that the crew may not be aware of. For example, a plugged or partially plugged bit nozzle increases KRP. Changes in annular pressure loss also change KRP. Because KRP can change without the crew's knowledge, many operators require that it be verified immediately before the kick is circulated.

INITIAL CIRCULATING PRESSURE

When a well kicks and is completely shut in, pressure normally appears on the drill pipe pressure gauge. Shut-in drill pipe pressure (SIDPP) represents the amount of hydrostatic underbalance in the well; that is, SIDPP shows how much higher formation pressure is than the hydrostatic pressure in the drill stem. Therefore, when it is time for the kick to be circulated out of the well, SIDPP must be added to KRP to maintain constant bottomhole pressure. When SIDPP and KRP are added, initial circulating pressure (ICP) is obtained. ICP is the pump pressure required to begin circulating a well that has kicked. The following is an equation for determining ICP:

$$ICP = KRP + SIDPP \qquad \text{(Eq. 33)}$$

where

$$ICP = \text{initial circulating pressure, psi}$$
$$KRP = \text{kill-rate pressure, psi}$$
$$SIDPP = \text{shut-in drill pipe pressure, psi}.$$

As an example, assume that KRP is 750 psi and SIDPP is 200 psi. To determine ICP—

$$ICP = 750 + 200$$
$$ICP = 950 \text{ psi}.$$

The solution shows that 950 psi is the pressure required to begin circulating mud into the well and the kick out. Depending on the well-control method used, *ICP* can be merely the initial, or beginning, circulating pressure required, or it can be the circulating pressure to be maintained until the kick is circulated out of the well. For example, if the kick is circulated out at the same time that kill-weight mud is circulated in (as in the wait-and-weight method), then the value for *ICP* must be allowed to decrease as heavier kill-weight mud fills the drill stem. When kill-weight mud completely fills the drill stem, hydrostatic pressure in the stem should balance formation pressure, and a final circulating pressure with a lower value than *ICP* is achieved. On the other hand, if the kick is circulated out with mud of the weight that was in the well when it kicked (as in the driller's method), then the value for *ICP* should be used until the kick is circulated out of the hole. The lighter-weight mud will not balance formation pressure.

By working example problems, it can be seen how *KRP* and *SIDPP* relate in obtaining *ICP*: simply add *KRP* and *SIDPP* to obtain *ICP*.

1. KRP = 900 psi
   $SIDPP$ = 210 psi
   ICP = _____
2. KRP = 1,000 psi
   $SIDPP$ = 540 psi
   ICP = _____
3. KRP = 610 psi
   $SIDPP$ = 100 psi
   ICP = _____
4. KRP = 900 psi
   $SIDPP$ = 820 psi
   ICP = _____
5. KRP = 820 psi
   $SIDPP$ = 150 psi
   ICP = _____

SURFACE-TO-BIT PUMP STROKES AND TIME

The number of pump strokes and the time it takes for mud from the surface to be pumped down the drill stem to the bit are also required for successful well control. Since drill pipe pressure decreases as kill-weight mud is pumped down the drill stem, surface-to-bit strokes and time are needed for charting the rate at which drill pipe pressure decreases. Keeping track of the decrease in drill pipe pressure allows well-control personnel to maintain constant bottom-hole pressure while kill-weight mud is being circulated down the drill stem.

To calculate the surface-to-bit time or the number of pump strokes required to pump mud from the surface to the bit, pump displacement, or output, in barrels per stroke (bbl/s) and the capacity of the drill pipe and drill collars in barrels per foot (bbl/ft) must be known. Pump output can be found in Appendix B, in the pump manual supplied by the manufacturer, or it can be calculated. Drill pipe and drill collar capacity may also be found in Appendix B, in cementing tables published by major cementing service companies, or it can be calculated. In any case, once values are found for pump output and for drill pipe and drill collar capacity, equations can be used to calculate surface-to-bit strokes and surface-to-bit time:

$$SBS_{dp} = C_{dp} \times L_{dp} \div PD \qquad \text{(Eq. 34)}$$

where

SBS_{dp} = number of surface-to-bit strokes to displace mud in drill pipe
C_{dp} = capacity of drill pipe, bbl/ft
L_{dp} = length of drill pipe, ft
PD = pump displacement, bbl/s,

and—

$$SBS_{dc} = C_{dc} \times L_{dc} \div PD \qquad \text{(Eq. 35)}$$

where

SBS_{dc} = number of surface-to-bit strokes to displace mud in drill collars
C_{dc} = capacity of drill collars, bbl/ft
L_{dc} = length of drill collars, ft
PD = pump displacement, bbl/s.

For the time calculation the equation is—

$$SBT = SBS \div SPM \qquad \text{(Eq. 36)}$$

where

SBT = surface-to-bit time, min
SBS = number of surface-to-bit strokes
SPM = pump rate, spm.

Using well data, an example problem can be solved. For example, if—

well depth = 14,000 ft
drill pipe = 13,400 ft, 5-in. outside diameter (OD), 19.5 XH
drill collars = 600 ft, 6-in. OD, $2^{13}/_{16}$ in. inside diameter (ID)
SPM = 45
PD = 0.123 bbl/s
C_{dp} = 0.01776 bbl/ft
C_{dc} = 0.00491 bbl/ft,

what are the number of surface-to-bit strokes, and how much time does it take for the mud to be displaced in the drill stem?

$$SBS_{dp} = 0.01776 \times 13,400 \div 0.123$$
$$= 1,935$$
$$= 0.00491 \times 600 \div 0.123$$
$$SBS_{dc} = 24$$
$$\text{Total } SBS = 1,935 + 24$$
$$\text{Total } SBS = 1,959.$$

To determine surface-to-bit time:

$$SBT = 1,959 \div 45$$
$$SBT = 43.5 \text{ min.}$$

As mentioned earlier, capacities for both duplex and triplex pumps can be calculated. The diameter of the pistons and the length of the stroke must be known for both duplex and triplex pumps; for duplex pumps, the diameter of the rods must also be known. The equation for a duplex pump at 90% efficiency follows:

$$PD = 0.000162 \times L\,[(2 \times D^2) - d^2]\,0.90 \quad \text{(Eq. 37)}$$

where

PD = pump displacement, bbl/s

L = stroke length, in.

D = piston diameter, in.

d = rod diameter, in.

As an example, find the displacement of a duplex pump at 90% efficiency if the pump has a 14-in. stroke, pistons of 5¾ in. in diameter, and a rod diameter of 2 in.:

$$PD = 0.000162 \times 14\,[(2 \times 5.75^2) - 2^2]\,0.90$$
$$= 0.000162 \times 14 \times 62.125 \times 0.90$$
$$PD = 0.1268 \text{ bbl/s.}$$

The solution shows that the displacement of the example duplex pump is 0.1268 bbl/s.

The equation for finding the displacement of a triplex pump at 100% efficiency follows:

$$PD = 0.000243 D^2 L \quad \text{(Eq. 38)}$$

where

PD = pump displacement, bbl/s

D = piston diameter, in.

L = stroke length, in.

As an example, the displacement of a triplex pump at 100% efficiency with pistons of 7 in. in diameter and with a stroke length of 10 in. is—

$$PD = 0.000243 \times 7^2 \times 10$$
$$PD = 0.119 \text{ bbl/s.}$$

The solution shows that the displacement of the example triplex pump is 0.119 bbl/s.

If desired, the capacity of drill pipe, drill collars, cased hole, or open hole can also be calculated. The ID of the pipe, collars, or casing must be known, and can be found in manufacturers' or other tables such as *API Bul 5C2, Bulletin on Performance Properties of Casing, Tubing, and Drill Pipe*. Once the appropriate ID is known, the following equation for determining capacity may be used:

$$C_h = ID^2 \div 1,029.4 \quad \text{(Eq. 39)}$$

where

C_h = capacity of drill pipe, drill collars, or cased or open hole, bbl/ft

ID = ID of drill pipe, drill collars, or cased or open hole, in.

As an example, suppose that 5-in., 19.5 XH drill pipe is being used. By checking an appropriate table, the ID of this pipe is found to be 4.276 in. Therefore:

$$C_h = 4.276^2 \div 1,029.4$$
$$C_h = 0.01776 \text{ bbl/ft.}$$

The solution shows that 5-in., 19.5 XH drill pipe has a capacity of 0.01776 bbl/ft.

EXAMPLE PROBLEMS FOR SURFACE-TO-BIT STROKES AND TIME

Using example pump and well data, determine surface-to-bit pump strokes and surface-to-bit travel time. Use tables in Appendix B for drill pipe, drill collar, and pump capacities, or calculate the values using formulas.

1.   Total depth (TD) = 8,590 ft

     True vertical depth

               (TVD) = 8,590 ft

       Drill pipe (DP) = 8,390 ft, 4½ in., 16.6 IF

       Drill collars (DC) = 200 ft, 6-in. OD, 2¼-in. ID

             Pump = 6½ × 16 duplex, 2½-in. rod OD

         Kill rate = 30 spm

     Surface-to-bit

     strokes (SBS) = 660-93

     Surface-to-bit time

             (SBT) = 22.08

2. TD = 11,240 ft

           TVD = 11,240 ft

DP = 10,790 ft, 4½ in., 16.6 XH

DC = 450 ft, 7-in. OD, 2¼-in. ID

Pump = 6 × 16 duplex, 2½-in. rod OD

Kill rate = 38 spm

SBS = *2998.66*

SBT = *'79*

3. TD = 5,201 ft

TVD = 5,201 ft

DP = 5,051 ft, 4½ in., 16.6 IF

DC = 150 ft, 6¼-in. OD, 2¼-in. ID

Pump = 6½ × 18 duplex, 2¼-in. rod OD

Kill rate = 33 spm

SBS = *1043.36*

SBT = *32 32*

4. TD = 14,450 ft

TVD = 14,450 ft

DP = 13,950 ft, 4½ in., 16.6 IF

DC = 500 ft, 7-in. OD, 2¼-in. ID

Pump = 6 × 16 duplex, 2¼-in. rod OD

Kill rate = 40 spm

SBS = *1321.5*

SBT = *33*

5. TD = 8,410 ft

TVD = 8,410 ft

DP = 8,310 ft, 4½-in., 16.6 XH

DC = 100 ft, 6-in. OD, 2¼-in. ID

Pump = 6 × 16 duplex, 2¼-in. rod OD

Kill rate = 30 spm

SBS = *775.5*

SBT = *25.8*

6. TD = 12,450 ft

TVD = 10,800 ft

DP = 11,820 ft, 5-in., 19.5 XH

DC = 630 ft, 8-in. OD, 3-in. ID

Pump = 6 × 12 triplex

Kill rate = 45 spm

SBS = *1805*

SBT = *40*

7. TD = 11,450 ft

TVD = 9,220 ft

DP = 11,210 ft, 5-in., 19.5 XH

DC = 240 ft, 7¾-in. OD, 2¼-in. ID

Pump = 6 × 10 triplex

Kill rate = 40 spm

SBS = _____

SBT = _____

8. TD = 14,200 ft

TVD = 10,244 ft

DP = 14,000 ft, 5-in., 19.5 XH

DC = 200 ft, 9-in. OD, 3-in. ID

Pump = 5 × 12 triplex

Kill rate = 50 spm

SBS = _____

SBT = _____

9. TD = 8,850 ft

TVD = 6,210 ft

DP = 8,700 ft, 5-in., 19.5 XH

DC = 150 ft, 7¾-in. OD, 2¼-in. ID

Pump = 6 × 12 triplex

Kill rate = 47 spm

SBS = _____

SBT = _____

10. TD = 9,955 ft

TVD = 7,850 ft

DP = 9,745 ft, 5-in., 19.5 XH

DC = 210 ft, 6¼-in. OD, 2¼-in. ID

Pump = 5½ × 12 triplex

Kill rate = 45 spm

SBS = _____

SBT = _____

ANNULAR VOLUME

When circulating during well-control operations, it can be useful to know the annular volume between drill pipe or drill collars and cased or open hole. Tables or the following equation can be used for finding the volume:

$$AV = (ID_h^2 - OD_{dp}^2) \div 1{,}029.4 \quad \text{(Eq. 40)}$$

where

AV = annular volume, bbl/ft

ID_h = ID of open or cased hole, in.

OD_{dp} = OD of drill pipe or drill collars, in.

As an example, suppose that 5-in., 19.5 XH drill pipe is in use, and that the diameter of the open hole is 8½ in. Also assume that 9⅝-in. OD, 53.5 pounds per foot (ppf) casing is in a hole that has an ID of 8.535 in. Thus:

$$AV_{open\,hole} = (8.5^2 - 5^2) \div 1{,}029.4$$
$$= 0.0459 \text{ bbl/ft}$$
$$AV_{cased\,hole} = (8.535^2 - 5^2) \div 1{,}029.4$$
$$= 0.0465 \text{ bbl/ft.}$$

It may also be useful to know the volume, or capacity, of the open or cased hole without drill pipe or drill collars in it:

$$C_h = ID^2 \div 1,029.4 \qquad \text{(Eq. 39)}$$

where

C_h = hole capacity, bbl/ft

ID = inside diameter of the hole (either open or cased), in.

As an example, calculate the capacity of an 8½-in. open hole. Assume that the hole does not have any washed-out areas:

$$\begin{aligned} C_h &= 8.5^2 \div 1,029.4 \\ &= 72.25 \div 1,029.4 \\ C_h &= 0.0702 \text{ bbl/ft.} \end{aligned}$$

RISER VOLUME

When drilling from floating offshore rigs with subsea blowout preventer stacks and riser systems, it may be useful to know the volume of the riser with and without drill stem elements in it. Riser volume is determined the same way as annular volume, except riser IDs are used instead of hole or casing IDs. The riser volume equation is:

$$V_r = (ID_r^2 - OD_{dp}^2) \div 1,029.4 \qquad \text{(Eq. 41)}$$

where

V_r = riser volume, bbl/ft

ID_r = ID of riser, in.

OD_{dp} = OD of drill pipe or drill collars, in.

As an example, calculate the volume of an 18⅜-in. riser with 5-in. drill pipe inside it:

$$\begin{aligned} V_r &= (18.375^2 - 5^2) \div 1,029.4 \\ &= (337.64 - 25) \div 1,029.4 \\ &= 312.64 \div 1,029.4 \\ V_r &= 0.3037 \text{ bbl/ft.} \end{aligned}$$

Thus, if the riser was 1,100 ft. long, its total volume (with 5-in. drill pipe inside it) would be 1,100 × 0.3037 = 334 bbl.

In some cases, the riser may not have drill pipe or drill collars in it and it may be necessary to know the riser's empty volume. In this case:

$$V_r = ID^2 \div 1,029.4 \qquad \text{(Eq. 42)}$$

where

V_r = riser volume, bbl/ft

ID^2 = riser ID, in.

As an example, calculate the volume of an 18⅜-in. ID riser that is 1,100 ft long.

$$\begin{aligned} V_r &= 18.375^2 \div 1,029.4 \\ &= 337.64 \div 1,029.4 \\ &= 0.3279 \text{ bbl/ft.} \\ &= 0.3279 \times 1,100 \text{ ft} \\ V_r &= 361 \text{ bbl.} \end{aligned}$$

CHOKE-AND-KILL-LINE VOLUMES

Another volume calculation of importance where subsea BOP stacks are in use is that of the choke and kill lines running from the surface to the subsea BOPs. The same calculation is used to determine choke and kill line volumes as is used to determine annular and riser volumes.

$$V_{ckl} = ID^2 \div 1,029.4 \qquad \text{(Eq. 43)}$$

where

V_{ckl} = choke line or kill line volume, bbl/ft

ID^2 = choke line or kill line ID, in.

As an example calculation, determine the volume of a 1,233-ft, 3-in. ID choke line.

$$\begin{aligned} V_{ckl} &= 3^2 \div 1,029.4 \\ &= 9 \div 1,029.4 \\ V_{ckl} &= 0.0087 \text{ bbl/ft.} \end{aligned}$$

Next, multiply the volume (in bbl) per ft by the length of the choke line. Thus, 0.0087 × 1,233 ft = 10.73 bbl.

BIT-TO-SURFACE PUMP STROKES AND TIMES

It is important for well-control personnel to be aware of *bottoms-up time*—the time it takes for mud to travel from the bottom of the hole to the surface—because they should know when to expect the kick fluids and kill-weight mud at the surface. When kill-weight mud reaches the surface, the crew can stop the pump, check the well for flow, and determine whether the well is dead. If the bottoms-up time is known for the normal pump rate when drilling and if, during a well-control operation, the kill rate is half the rate used during drilling, then bottoms-up time is twice that of the normal drilling rate. Of course, if the number of pump strokes instead of time is used to determine bottoms up, the number is the same as during drilling.

The time or number of pump strokes required to displace the heavy mud to the surface can be calculated by using equations similar to those used in surface-to-bit calculations. While the calculations are not required for well killing, they do give the crew an estimate of when kill-weight mud will arrive at the surface. For bit-to-surface pump strokes required for bottoms up, use the following equation:

$$BSS = [(D_h^2 - D_{dc}^2) \div 1{,}029.4 \times (L_{dc} \div PD)]$$
$$+$$
$$[(D_h^2 - D_{dp}^2) \div 1{,}029.4 \times (L_{dp} \div PD)] \quad \text{(Eq. 44)}$$

where

BSS = number of bit-to-surface pump strokes

D_h = diameter of hole, in.

D_{dc} = OD of drill collars, in.

L_{dc} = length of drill collars, ft

PD = pump displacement, bbl/s

D_{dp} = OD of drill pipe, in.

L_{dp} = length of drill pipe, ft

For the time required for bottoms up, use equation 45:

$$BST = BSS \div SPM \quad \text{(Eq. 45)}$$

where

BST = bit-to-surface time, min

BSS = number of bit-to-surface strokes

SPM = number of pump strokes, spm.

An example problem requiring the use of the equations is to calculate the number of bit-to-surface strokes and time if—

D_{dp} = 5-in., 19.5 XH

L_{dp} = 13,400 ft

D_{dc} = 6-in. OD (2¼-in. ID)

L_{dc} = 600 ft

D_h = 8½-in.

PD = 0.123 bbl/s

SPM = 30 spm.

The solution is—

BSS = $[(8.5^2 - 6^2) \div 1{,}029.4 \times (600 \div 0.123)]$ + $[(8.5^2 - 5^2) \div 1{,}029.4 \times (13{,}400 \div 0.123)]$

BSS = 172 + 5,001

BSS = 5,173

To calculate the time—

BST = 5,173 ÷ 30

BST = 172 min

BST = 2 h, 52 min.

Keep in mind that if bit-to-surface strokes are being computed inside casing instead of in open hole, use casing ID in place of hole diameter.

EXAMPLE PROBLEMS, BIT-TO-SURFACE STROKES AND TIME

Using the given well data, determine the number of strokes and the time required to pump mud from the bottom of the hole to the surface.

1. D_h = 8¾ in.
   D_{dp} = 4½ in.
   D_{dc} = 6 in.
   L_{dc} = 540 ft
   L_{dp} = 8,050 ft
   PD = 0.182 bbl/s
   SPM = 30 spm
   BSS = *2536.29*
   BST = *7h 40m 1h 29m*

2. D_h = 7⅞ in.
   D_{dp} = 4½ in.
   D_{dc} = 5 in.
   L_{dc} = 360 ft
   L_{dp} = 10,880 ft
   PD = 0.153 bbl/s
   SPM = 33 spm
   BSS = *2969*
   BST = *1h 30m*

3. D_h = 9⅞ in.
   D_{dp} = 4½ in.
   D_{dc} = 4 in.
   L_{dc} = 200 ft
   L_{dp} = 5,001 ft
   PD = 0.205 bbl/s
   SPM = 25 spm
   BSS = *12318 3071.8*
   BST = *2h 2m*

4. D_h = 8¾ in.
   D_{dp} = 4½ in.
   D_{dc} = 6 in.
   L_{dc} = 400 ft
   L_{dp} = 14,050 ft
   PD = 0.153 bbl/s
   SPM = 30 spm
   BSS = *5189*
   BST = *173m*

5. D_h = 12¼ in.
   D_{dp} = 5 in.
   D_{dc} = 8 in.
   L_{dc} = 500 ft
   L_{dp} = 11,950 ft
   PD = 0.123 bbl/s
   SPM = 45 spm
   BSS = _____
   BST = _____

TOTAL CIRCULATING STROKES AND TIMES

To determine the total strokes and time that it takes for the mud to circulate from the surface, down the drill stem, and back up to the surface, simply determine surface-to-bit strokes and time and bit-to-surface strokes and time and add the two together. For example, in a 14,000-ft well drilling with a 13,400-ft string of 5-in., 19.5 XH drill pipe and 600 ft of 6-in. OD and 2¹³⁄₁₆-in. ID drill collars, in 8½-in. hole, use equations 34, 35, and 36 to determine the numbers of strokes and the time. In this case, the number of surface-to-bit strokes is 1,959 and the surface-to-bit time is 43.5 min. Similarly, using equations 44 and 45 and the example just given, the number of bit-to-surface strokes is 5,173 and the time is 172 min or 2 hr, 52 min. To determine the total circulating strokes, simply add surface-to-bit strokes to bit-to-surface strokes. In this example, the answer is 1,959 + 5,173 = 7,132 total strokes. The total circulating time is 43.5 min + 172 min = 215.5 min or 3 hr, 35.5 min.

BIT-TO-CASING SHOE STROKES AND TIMES

When closed in on a gas kick and circulating it out of the well, it is helpful to know how many strokes and the time it will take for the kick to reach the casing shoe set above the open hole. To determine the time and strokes to the casing shoe, it is simply a matter of taking into account the depth of the casing shoe in relation to the drill stem diameters, hole diameter, total hole depth, and pump displacement data as used to determine bit-to-surface strokes and times. For instance, using the data given in determining bit-to-surface strokes and times in the example given before, assume that an intermediate casing shoe exists at 11,050 ft. Since total well depth is 14,000 ft, the kick will travel 2,950 ft to reach the shoe (14,000 – 11,050 = 2,950). In this case, the drill pipe length in this portion of the hole is 2,350 ft and the drill collar length is 600 ft (2,950 – 600 = 2,350). Using equation 44, the gas kick will reach the casing shoe at 1,049 strokes and in 35 min.

CHOKE-AND-KILL-LINE STROKES AND TIMES

In floating operations that use subsea BOP stacks, it is useful to know the number of pump strokes required to displace fluid in the choke or kill lines and the time is takes to displace fluid. To determine the number of strokes required:

$$S_{ckl} = ID^2 \div 1,029.4 \times (L_{ckl} \div PD) \quad \text{(Eq. 46)}$$

where

S_{ckl} = number of strokes to displace the choke line or the kill line

ID^2 = inside diameter of the choke line or the kill line, in.

L_{ckl} = length of the choke line or the kill line, ft

PD = pump displacement, bbl/s

For the time required to displace the choke line or the kill line, use equation 47:

$$T_{ckl} = S_{ckl} \div SPM \quad \text{(Eq. 47)}$$

where

T_{ckl} = time to displace fluid from the choke line or the kill line, min

S_{ckl} = number of strokes to displace the choke line or the kill line

SPM = number of pump stokes, spm.

As an example calculation, determine the number of strokes and the time required to displace the fluid from a 3-in., 833-ft kill line. The pump displaces 0.205 bbl/s at 30 spm:

S_{ckl} = 3² ÷ 1,029.4 × (833 ÷ 0.205)
S_{ckl} = 0.0087 × 4,063.4
S_{ckl} = 35

To calculate the time required to displace:

T_{ckl} = 35 ÷ 30
T_{ckl} = slightly more than 1 min.

FINAL CIRCULATING PRESSURE

Pump pressure changes with mud weight. As mud weight increases, more pump pressure is required to pump the mud at the same rate as before. In well control, mud weight is increased to control the kicking formation. Therefore, the kill-rate circulating pressure must also be increased to allow for the heavier mud weight. To calculate the increase, KRP is multiplied by the kill-weight mud (new mud weight) divided by the old mud weight (the mud in the hole when the kick occurred). The corrected pump pressure is final circulating pressure (FCP). To calculate FCP, which can be read on the drill pipe pressure gauge once new kill-weight mud fills the drill stem, the following equation can be used:

$$FCP = KRP \times (KWM \div OMW) \quad \text{(Eq. 48)}$$

where

FCP = final circulating pressure, psi

KRP = kill-rate pressure, psi

OMW = old mud weight, ppg

KWM = kill-weight mud, ppg.

As an example, assume that KRP is 600 psi, the old mud weight is 11 ppg, and the kill-mud weight is 12 ppg. What is FCP?

$$FCP = 600 (12 \div 11)$$
$$FCP = 655 \text{ psi.}$$

FCP in this example is 655 psi; that is, when kill mud completely fills the drill stem, pump pressure should read 655 psi. Even though FCP is higher than KRP because it requires more pressure to pump heavier mud at the same rate, the value for FCP will be lower than ICP, because ICP is determined by adding KRP and SIDPP. As an example, suppose SIDPP is 500 psi. If ICP is 1,100 psi, KRP is 600 psi, the old mud weight is 11 ppg, and the kill-weight mud is 12 ppg, FCP is about 655 psi, or 445 psi lower than ICP.

EXAMPLE PROBLEMS FOR FINAL CIRCULATING PRESSURE

As an aid in understanding and calculating FCP, work the following problems using equation 48.

1. KRP = 750 psi
   OMW = 11.5 ppg
   KWM = 12.5 ppg
   FCP = _____

2. KRP = 650 psi
   OMW = 13.0 ppg
   KWM = 13.3 ppg
   FCP = _____

3. KRP = 650 psi
   OMW = 11.0 ppg
   KWM = 12.2 ppg
   FCP = _____

4. KRP = 1,200 psi
   OMW = 14.5 ppg
   KWM = 15.2 ppg
   FCP = _____

5. KRP = 1,050 psi
   OMW = 16.2 ppg
   KWM = 17.0 ppg
   FCP = _____

6. KRP = 950 psi
   OMW = 14.5 ppg
   KWM = 15.0 ppg
   FCP = _____

7. KRP = 750 psi
   OMW = 17.0 ppg
   KWM = 17.8 ppg
   FCP = _____

Chapter 5

Formation Fracture Gradient

Formation fracture pressure is the amount of pressure that causes a formation to break down, or fracture. In well control, the fracture pressure of the weakest formation exposed to the wellbore must be known, because the pressures developed during well-control procedures may exceed the fracture pressure of the formation. Should fracture pressure be exceeded, the formation fractures, and lost circulation and an underground blowout or broaching could result.

Formation fracture pressure can be expressed in psi, equivalent mud weight, or fracture gradient. If fracture gradient is used, it is usually expressed in psi/ft. For example, assume that a formation 5,000 ft deep fractures when 3,640 psi is exerted on it. Thus, the fracture pressure of this formation is 3,640 psi. Its fracture gradient is 0.728 psi/ft, because 3,640 psi divided by 5,000 ft equals 0.728 psi/ft. Further, in terms of equivalent mud weight, the pressure developed by 14.0-ppg mud will fracture this example formation because 14.0 ppg × 0.052 × 5,000 ft = 3,640 psi.

FRACTURE DATA

The point at which a formation fractures has been the subject of considerable study and research since the 1960s. Large amounts of data from breakdown pressures on squeeze cementing jobs and from wells in which lost circulation occurred have been used to develop correlations between fracture pressure, well depth, and pore pressure for a given area.

One such relationship has been developed for the Louisiana Gulf Coast (fig. 5.1). Fracture data for normal pore pressure—in this example, the pressure developed by mud with a weight of 9.0 ppg—versus depth are shown as the heavy black line at the extreme left of the group of curves on the graph. This 9.0-ppg curve indicates that at 4,000 ft on the Louisiana

Gulf Coast, the fracture pressure could be expected to be equivalent to about 14.4-ppg mud. An equivalent mud weight of 14.4 ppg corresponds to a fracture pressure of 2,995 psi at 4,000 ft, or to a fracture gradient of 0.749 psi/ft. The remaining curves show how fracture pressure increases at any given depth when abnormal pore pressures are encountered. A higher pore pressure at a given depth results in a correspondingly higher fracture pressure at that depth.

KICK TOLERANCE

Kick tolerance is a calculated estimate of the size of potential kick that could fracture an exposed formation and lead to serious well-control problems. Kick tolerance is expressed in terms of ppg, volume of influx, or a combination of both. It is a way in which to express the fact that, when a well kicks and is shut in, *SICP* will be greater if a larger volume of influx is allowed to enter the wellbore. *SICP* is also greater if the kicking formation's pressure is a great deal higher than hydrostatic pressure. The worst case occurs when both formation pressure is relatively higher and a large kick is taken. With a reasonable understanding of wellbore strength, it is possible to estimate the combinations of increasing formation pressure and influx volume that will likely cause formation fracture. When calculations suggest that only a relatively small kick could be

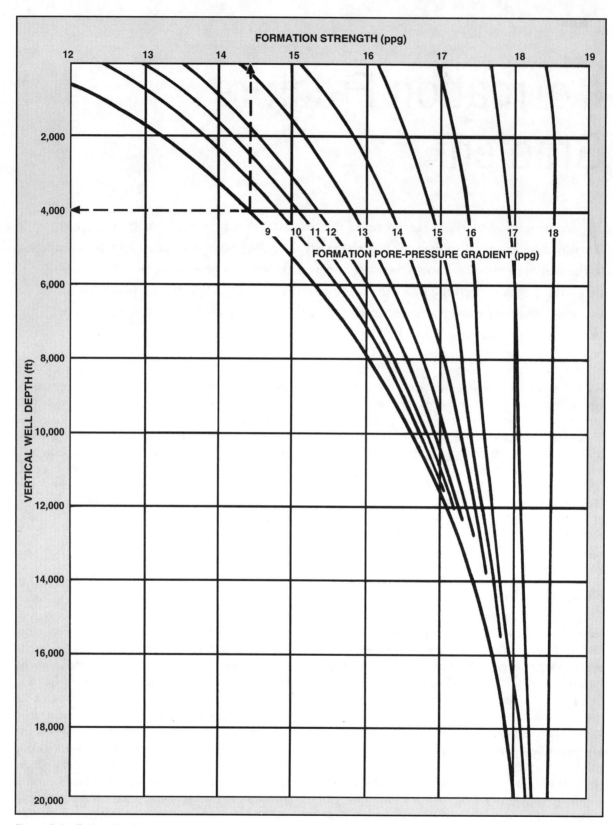

Figure 5.1   Estimating formation strength in the Louisiana Gulf Coast area

safely contained, drilling supervisors consider such options as running an additional casing string to reduce the risk of wellbore failure.

DETERMINING FRACTURE PRESSURE

In a section of open hole, the formation that is immediately below the casing shoe is almost always the weakest formation, because it is the most shallow exposed formation and usually has a pore pressure lower than the formations that lie below it. In general, formations with lower pore pressures fracture more easily than formations with higher pore pressures. Because the formation most likely to fracture is at the casing shoe, and because the resulting mud loss can cause severe difficulties in well control, the amount of pressure it takes to fracture the formation at the shoe should be determined each time a string of casing is run and cemented in a well. API defines fracture pressure as the maximum surface pressure that can be applied to a casing that is full of drilling fluid without fracturing the formation.

Over the years, a number of equations have been developed that allow estimated fracture gradient to be calculated, but many operators prefer to perform leak-off or formation-competency tests. (One equation for example, is fracture pressure in psi = 0.052 × casing TVD in ft × (fracture drilling fluid density in ppg – present drilling fluid density in ppg.) *Leak-off tests* show the pressure at which a formation actually begins to take fluid and fracture. In other words, the condition measured in a leak-off test is the pressure above a full column of mud that is required to cause the formation actually to fracture and to start to accept whole mud. Formation-competency tests are similar to leak-off tests but fluid is pumped into the well until a predetermined pressure is reached to confirm that the formation is at least as strong as expected or required.

A leak-off test can also provide data for determining maximum allowable surface pressure (MASP) and the maximum equivalent mud weight that can be used to avoid formation fracture at the casing shoe. Since bottomhole pressure depends on mud weight, a change in mud weight changes MASP: the higher the mud weight, the lower MASP will be. Further, the test can indicate the competency of the cement bond at the shoe and whether remedial cementing or another string of casing may be required.

CONSIDERATIONS BEFORE CONDUCTING LEAK-OFF TESTS

When the casing shoe is being drilled out prior to running a leak-off test, no more than about 5 ft to 50 ft of hole should be drilled out below the shoe. Generally, 5 ft to 10 ft is acceptable, but the depth depends on the operating company's requirements and, on the Outer Continental Shelf of the United States, on Minerals Management Service (MMS) requirements.

Since the mud's gel strength, yield point, and viscosity affect the amount of pressure required to circulate it, it should be conditioned to reduce these values to a minimum. In particular, gel strength should be kept as low as possible, because it affects the amount of pressure needed to break circulation, and the pressure required to break circulation must be subtracted from casing shoe fracture pressure.

Formulas are available for determining how much pressure must be subtracted because of gel strength. Since determining fracture pressure involves pumping drilling mud down the wellbore, and since the mud can be pumped either down the drill stem or down the annulus, one of two formulas is used, depending on the pumping method. When fracture pressure is determined by pumping drilling mud down the drill stem, the following formula may be used:

$$P = L(\gamma \div 300d) \qquad \text{(Eq. 49)}$$

where

P = pressure required to overcome mud's gel strength, psi

L = length of drill stem, ft

γ = 10-min gel strength of mud, lb/100 ft^2

d = ID of drill pipe, in.

As an example, assume that—

L = 12,000 ft

γ = 10 lb/100 ft^2

d = 4.276 in.

Therefore:

P = 12,000 [10 ÷ (300 × 4.276)]

    = 12,000 (10 ÷ 1,282.8)

    = 12,000 × 0.0078

P = 94 psi.

In this example, 94 psi must be subtracted from the pressure value found by the leak-off test, because it represents the amount of pressure required to break the gel strength of the mud. When pumping down

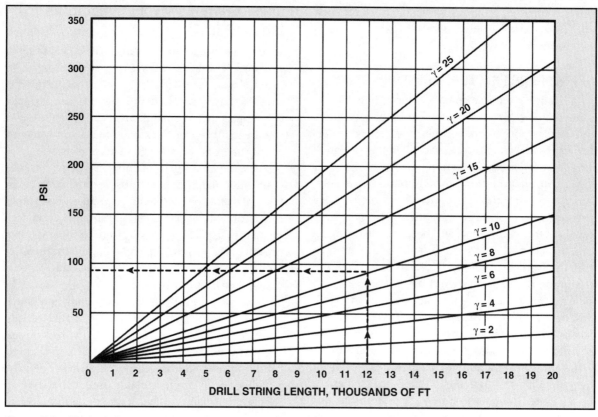

Figure 5.2   Mud gelation pressure losses versus thousands of feet of 5-in., 19.5-ppf drill pipe (developed from equation 50).

the drill stem, a graph can be helpful in determining gel-strength pressure losses (fig. 5.2). Using the example of a 12,000-ft drill string and mud with a gel strength of 10 lb/100 ft², the graph shows that the mud gelation pressure loss is about 94 psi. Other gel-strength values are also shown on the graph.

In cases where leak-off tests are conducted by pumping down the annulus, equation 49 is altered slightly:

$$P = L [\gamma \div 300 (D_h - D_p)] \qquad \text{(Eq. 50)}$$

where

P = pressure required to overcome mud's gel strength, psi
L = length of drill stem, ft
γ = 10-min gel strength of mud, lb/100 ft²
D_h = diameter of hole, in.
D_p = OD of drill pipe, in.

As an example, assume that—

L = 12,000 ft
γ = 10 lb/100 ft²
D_h = 8½ in.
D_p = 5 in.

Therefore:

P = 12,000 [10 ÷ 300 (8½ – 5)]
  = 12,000 (10 ÷1,050)
  = 12,000 × 0.0095
P = 114 psi.

In this example, 114 psi must be subtracted from the pressure value found by the leak-off test when the test is conducted by pumping down the annulus.

Conducting Leak-Off Tests

Many different techniques are available for running leak-off tests, but regardless of the method used, most operators agree that certain general points apply to all. For example, most operators recommend that an accurate pressure gauge be used, that a cementing pump (instead of the mud pump) be used, that the pump rate not exceed ½ bbl per minute, and, as previously mentioned, that the shoe be drilled out no deeper than required. To conduct a leak-off test, the following is an accepted procedure:

1. Circulate the hole to establish a uniform mud density throughout the system.
2. Pull the bit to the casing shoe.
3. Close the blowout preventer.
4. Using the cementing pump, pump down the drill stem at a rate no higher than ½ bbl/min,

and preferably at about ¼ bbl/min.

5. On a graph, plot the fluid pumped in ¼-bbl increments versus drill pipe pressure until the formation at the shoe starts to take fluid; at this point, the pressure will continue to rise, but at a slower rate (fig. 5.3).

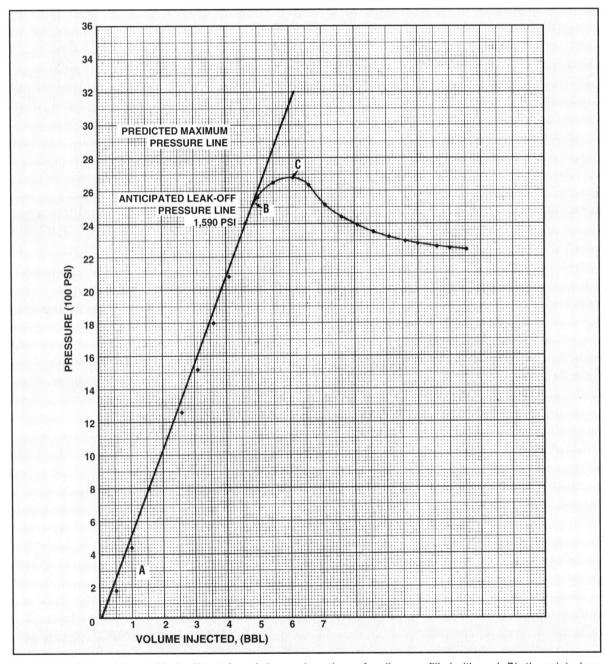

Figure 5.3 A plot of a typical leak-off test. Area *A* shows where the surface lines are filled with mud; *B* is the point where the formation first starts to take fluid. In many cases, this is the point defined as the leak-off pressure. *C* is the point where the formation fractured.

6. Repeat the test to verify the point at which the formation just starts to take fluid; this point is leak-off pressure.

By modifying leak-off pressure as necessary—such as by subtracting the amount of pressure required to break circulation—maximum allowable surface pressure can be determined. As stated earlier, a leak-off test can also reveal a poor cement job in which pressure channels behind the casing. A plot of pressure versus the volume of fluid injected during a leak-off test involving a poor cement job shows that pressure drops well before the anticipated leak-off point (fig. 5.4).

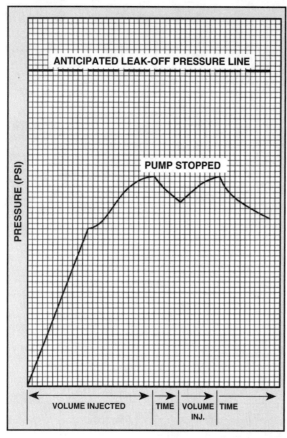

Figure 5.4 Pressure drops well before the anticipated leak-off pressure if the cement job was poorly done.

Performing Formation-Competency Tests

In cases in which it is not desirable or allowable to pump into the wellbore to the leak-off point, formation-competency tests can be run. In a formation-competency test, fluid is pumped into the shut-in well until a predetermined surface test pressure is reached that is believed to be below the pressure that actually causes the formation to break down and take fluid. The value for surface test pressure is usually obtained by assessing data from other wells drilled nearby.

One way in which to run a formation-competency test is to follow the same steps as in a leak-off test except, after the preventer is closed, slowly pump mud into the wellbore until the predetermined surface test pressure is reached. If the predetermined pressure is reached before the formation actually breaks down and fluid begins to leak off, the pressure can be used in much the same way as leak-off pressure. If, during the test, the pressure causes the formation to break down and fluid to leak off, then the formation-competency test simply reverts to a leak-off test and the results are treated accordingly.

It is common practice to express the results of a leak-off or formation integrity test as an equivalent mud weight. By calculating a mud weight that would create a hydrostatic pressure at the shoe equal to pressure exerted at the shoe during the test, the greatest mud weight the shoe can support can be calculated. This calculation is often referred to as maximum allowable mud weight or leak-off test equivalent mud weight.

When the value of maximum allowable mud weight is available from nearby wells at comparable shoe depths, it can be used to calculate a formation integrity test surface pressure limit as follows:

$$STP = (EMW - MWH)\,0.052 \times TVD_{cs} \quad \text{(Eq. 51)}$$

where

STP = surface test pressure, psi
EMW = equivalent mud weight, ppg
MWH = mud weight in hole, ppg
TVD_{cs} = true vertical depth of casing shoe, ft.

As an example problem, assume that—

EMW = 13 ppg
MWH = 10 ppg
TVD_{cs} = 5,000 ft.

Therefore:

STP = (13 3– 10) 0.052 × 5,000
STP = 780 psi.

The predetermined surface test pressure in this example is 780 psi; in other words, it might be reasonable to limit formation integrity test pressure to 780 psi, assuming comparable well geometry and operational requirements.

With the equivalent mud weight that it is believed the formation at the shoe can withstand obtained from nearby well data, and with the mud weight used to drill out the shoe known, another set of formulas for determining surface test pressure can also be used:

$$HP_1 = 0.052 \times EMW \times TVD_{cs} \quad \text{(Eq. 52)}$$

$$HP_2 = 0.052 \times MWH \times TVD_{cs} \quad \text{(Eq. 53)}$$

$$STP = HP_1 \times HP_2 \quad \text{(Eq. 54)}$$

where

HP_1 = hydrostatic pressure, psi
EMW = equivalent mud weight, ppg
TVD_{cs} = true vertical depth of casing shoe, ft
HP_2 = hydrostatic pressure, psi
MWH = mud weight in hole, ppg
STP = surface test pressure, psi.

As an example, assume that—

EMW = 16.2 ppg
TVD_{cs} = 12,450 ft
MWH = 11.5 ppg.

Therefore:

HP_1 = $0.052 \times 16.2 \times 12{,}450$
HP_1 = 10,488 psi
HP_2 = $0.052 \times 11.5 \times 12{,}450$
HP_2 = 7,445 psi
STP = $10{,}488 - 7{,}445$
STP = 3,043 psi.

In this case, the surface test pressure limit will be 3,043 psi.

Well-Control Methods

The goal of any well-control method is to kill the kick and bring the well under control. To accomplish this, the well-control method must allow personnel to (1) remove kick fluids from the hole and (2) fill the hole with mud of sufficient weight to exert pressure equal to or greater than formation pressure. Many well-control methods are available, including the driller's, wait-and-weight, concurrent, volumetric, top-kill, and low choke pressure; however, the three most often used are the driller's, wait-and-weight, and concurrent. In any event, shutting in the well stops the entry of additional formation fluids into the well. By closing in the well, bottomhole pressure becomes equal to formation pressure.

Although differences exist between the three methods, they have several similarities. For example, all three share the basic principle that constant bottomhole pressure must be maintained throughout the well-control operation, regardless of the nature of the influx. Constant bottomhole pressure is maintained by circulating the well at a constant pump rate through a choke orifice, and by changing the size of the choke orifice when necessary to adjust the back-pressure held throughout the wellbore. Further, all three methods make it possible for personnel to stop the pump, close the choke completely, and analyze a problem without jeopardizing the well at any time during the procedure. All three also require that a final circulating pressure be maintained after kill-weight mud reaches the bit.

Regardless of which of the three methods is used, after a well kicks and is shut in, $SIDPP$ and $SICP$ are given time to stabilize (usually a matter of a few minutes) and are read and recorded. KRP, which is the pressure indicated on the drill pipe or standpipe gauge when the pump is run at a reduced rate, is usually determined and recorded prior to the kick. When a kick occurs, $SIDPP$ is added to KRP to obtain ICP. During circulation, constant bottomhole pressure is maintained by keeping the pump speed constant and by adjusting the choke as required. Well-control worksheets for all methods are helpful because $SIDPP$, $SICP$, KRP, well depth, casing data, and other information needed to kill a well successfully can be recorded on the sheet and used as a reference during kill procedures.

When noting readings from the various gauges used to indicate $SIDPP$, $SICP$, pump speed, and the like, crew members should check for errors in the readings. For example, $SIDPP$ can be checked by noting the readings on the drill pipe pressure gauge in the remote choke control panel and the standpipe pressure gauge. They should both read very close to the same. Similarly, $SICP$ can be checked by a casing pressure gauge on the choke panel and a gauge installed on the wellhead that reads annular pressure.

The main difference between the three methods lies in how and when kill-weight mud is pumped down the drill stem and to the bit. In the driller's method, the kick is circulated out with the same mud that was in the hole when the kick occurred. Kill-weight mud is then circulated to control the well. In the wait-and-weight method, the kick is circulated out at the same time kill-weight mud is pumped in. In the concurrent method, instead of increasing mud weight to kill weight all at once, it is increased in steps, usually a point at a time. Each time mud weight is increased, the new mud is immediately circulated down the drill stem and circulating pressures recalculated. This process of increasing the mud weight in steps and circulating continues until kill weight is achieved.

DRILLER'S METHOD

The driller's method is the most basic of all methods and can be employed in a number of well-control situations. Because it involves the use of many techniques common to other well-control methods, the driller's method can be studied to learn basic well-control procedures.

Procedure for Driller's Method

The following is one step-by-step procedure for the driller's method of well control:

1. When the kick occurs, shut in the well in compliance with the operator's and the contractor's procedures.

2. Record SIDPP, SICP, and pit gain.

3. Fill in the well-control work sheet (fig. 6.1). Note that some information, such as KRP, is usually recorded before the kick occurs.

4. To start circulating, open the choke, slowly bring the pump up to kill rate, and hold SICP at a constant value (usually the original shut-in reading) by adjusting the choke. Keeping SICP constant for this short period of time maintains constant bottomhole pressure.

5. When the pump is at kill-rate speed, observe the drill pipe gauge; it shows ICP. Hold ICP constant by opening or closing the choke. Maintain a constant pump rate. *Do not allow the pump to change speed.*

6. Circulate the influx out, holding SIDPP constant at ICP. (Remember that ICP is equal to KRP plus SIDPP.) Keep in mind that a lag time exists—that is, a choke adjustment does not immediately appear on the drill pipe pressure gauge. The actual amount of time depends on the drill stem's length (depth). The longer the drill stem is, the longer it takes for a choke adjustment to appear on the gauge. A rule of thumb is to expect 1 second of delay per 1,000 ft of depth. For example, in a 10,000-ft well, the choke operator would expect a 20-second delay before a choke adjustment appears on the drill pipe gauge (10 seconds down the pipe and 10 seconds up the annulus). SICP response time, on the other hand, is quite rapid because annular pressure changes occur at the surface where casing pressure is read. (The rapid casing gauge response can be used to adjust SIDPP. For example, say it is necessary to decrease SIDPP by 100 psi. The choke operator can observe the casing pressure gauge while opening the choke until SICP drops by 100 psi. In the example 10,000-ft well, SIDPP should also drop by 100 psi about 20 seconds later.)

When *all* of the kick influx has been pumped out and the well is shut in, both SIDPP and SICP should be equal to the SIDPP noted when the kick occurred.

7. Stop the pump, close the choke, and mix kill-weight mud. The mud-weight increase can be calculated by using the following formula:

$$MWI \ = \ SIDPP \div TVD \div 0.052 \quad \text{(Eq. 24)}$$

where

MWI = mud-weight increase, ppg

$SIDPP$ = shut-in drill pipe pressure, psi

TVD = true vertical depth, ft.

If preferred, a table may be used (such as table B.8 in Appendix B).

8. When the pits are full of kill-weight mud, open the choke and slowly bring the pump up to the kill rate, holding casing pressure constant at the shut-in reading until new mud gets to the bit. As new mud fills the drill stem, ICP will slowly decrease toward FCP. Since it must be known when the drill stem is full of new mud in order to tell when FCP is achieved, a few calculations are needed. The number of pump strokes needed for the new mud to get to the bit (surface-to-bit strokes) can be calculated with the following equation:

$$SBS_{dp} \ = \ C_{dp} \times L_{dp} \div PD \quad \text{(Eq. 34)}$$

where

SBS_{dp} = number of surface-to-bit strokes

C_{dp} = capacity of drill pipe, bbl

L_{dp} = length of drill pipe, ft

PD = pump displacement, bbl/s.

The surface-to-bit time can be calculated with the following equation:

$$SBT \ = \ SBS \div SPM \quad \text{(Eq. 36)}$$

where

SBT = surface-to-bit time, min

SBS = number of surface-to-bit strokes

SPM = pump rate, spm.

The crew should continue mixing kill-weight mud in the pits as it is pumped into the well.

WELL-KILLING WORK SHEET
DRILLER'S METHOD

1. **Record information.**

   a. Casing size __13 3/8__ Depth __3,500__ ft

   b. Rated casing burst __3,090__ psi

   Maximum allowable surface pressure (MASP) __2,472__ psi

   c. Normal circulating pressure __3,000__ psi

   Pumping rate __60__ spm
   (Normal drilling circulating rate)

   d. Reduced circulating pressure

   (1) __1,000__ psi Rate __35__ spm

   (2) _____ psi Rate _____ spm

   e. Time of shut-in _____ am/pm

2. **Stop pump and close well completely. Allow pressure to stabilize.**

   Do not let casing pressure exceed MASP. If pressure builds to this value, circulate at highest allowed casing pressure and use low-choke pressure method to kill the well.

   a. Shut-in drill-pipe pressure (SIDPP) __260__ psi

   b. Shut-in casing pressure (SICP) __400__ psi

   c. Mud weight __12__ ppg

   d. True vertical depth __10,000__ ft

   e. Pit gain _____ bbl

   f. Circulating time, surface to bit _____ min

3. **Set circulating rate and pressures to clean well.**

   a. Start pump and open choke, as required. Pressure on the choke initially should be the shut-in casing pressure, and this pressure should be maintained while the pump speed comes up to the desired strokes per minute.

   b. Adjust choke to obtain SICP or to obtain SIDPP plus reduced circulating rate psi.

   c. Record ciruclating SIDPP __1,260__ psi

   Rate __35__ spm

   d. Maintain pump rate constant at the selected reduced speed and maintain *constant* SIDPP. If drill-pipe pressure increases, open choke; if it decreases, close the choke slightly. When the choke is adjusted, observe change on casing gauge to forecast magnitude of pressure on drill-pipe gauge.

   e. When well is free of gas, oil, or salt water, stop pump and close well. At this time, the annulus and drill-pipe pressure should be the same as original SIDPP.

   f. Record new SICP

   __260__ psi

4. **Calculate mud density to kill well.**

   The mud density increase needed is calculated from the information recorded in step 2.

   $$MWI = SIDPP \div TVD \div 0.052$$

5. **Increase surface mud system to required density.**

   If mud weighting can be done in separate pit, it should be started at step 3.

6. **Set circulating rate and pressures to kill well.**

   a. Start pump and open choke as required. Pressure on the choke initially should be the shut-in casing pressure 3f, and this pressure should be maintained while the pump speed comes up to the desired strokes per minute.

   b. Adjust choke to hold the new annulus pressure and hold constant until the drill pipe is full of the required density mud.

   c. After drill pipe is full of the required density mud, record drill-pipe pressure and hold pump rate and *drill-pipe pressure constant* by varying choke size until the annulus is filled with new mud.

   d. When required weight reaches surface, choke pressure, if any, is bled off. Stop circulating and check for flow.

Figure 6.1   Well-killing work sheet, driller's method

9. When kill-weight mud reaches the bit, stop observing casing pressure and begin observing drill pipe pressure. Hold drill pipe pressure constant at the new indicated value—*FCP*—until kill-weight mud appears at the flow line or choke. *Keep the pump rate constant.* If *KRP* is known, final circulating pressure can be estimated with the following formula:

$$FCP = KRP \times (KWM \div OMW) \text{ (Eq. 48)}$$

where
FCP = final circulating pressure, psi
KRP = kill-rate pressure, psi
KWM = kill-weight mud, ppg
OMW = old mud weight, ppg.

10. Stop the pumps and perform a flow check or shut the well in to ensure that it is dead. Consider additional circulation to condition the mud and possibly to add a safety, or trip, margin before drilling ahead.

Common Techniques in the Driller's Method

Since many of the techniques used in the driller's method are also used in the wait-and-weight and concurrent methods, well-control specialists often recommend that the driller's method be studied carefully. Driller's method techniques common to the other two include (1) shutting in the well promptly and properly, (2) holding casing pressure constant as the pump rate is brought up to kill rate to determine *ICP* properly, (3) holding the pump rate constant at the kill rate, (4) holding drill pipe pressure constant, (5) calculating drill stem capacity and number of strokes to the bit, and (6) calculating kill-weight mud. Regardless of the method, being able to shut the well in promptly and properly keeps the influx small. Being able to bring the pump rate up while holding casing pressure constant maintains sufficient bottomhole pressure to keep additional kick fluids from entering the hole. Once the pump is up to kill-rate speed, being able to keep it at a constant speed ensures that bottomhole pressure can be kept constant by adjusting the choke. Finally, being able to hold drill pipe pressure stable by adjusting the choke keeps bottomhole pressure stable prior to the time kill-weight mud is circulated. Once kill-weight mud is circulated down the drill stem to the bit, drill pipe pressure is again held constant at *FCP* to keep bottomhole pressure constant. Even though the driller's method sometimes increases the likelihood of lost circulation while killing gas kicks, certain aspects of the method can be advantageous. For example, it is easy to learn because it requires only a few calculations and circulation can begin immediately, which lessens the chance of the drill stem getting stuck, the circulating path becoming plugged, or migrating gas complicating the well-control problem.

Limitations of the Driller's Method

In spite of its advantages, the driller's method has a few limitations. For example, surface casing pressure rises to a maximum value with this method if the kick influx is gas. Since the well is circulated with mud of the weight present when the kick occurred, all of the additional pressure needed to prevent further intrusions of kick fluids into the well must be maintained by holding back-pressure with the choke. No heavier kill-weight mud is pumped until the hole is clean. Since gas in a kick expands as it is circulated up the hole, it pushes mud out of the annulus and reduces annular hydrostatic pressure; thus, the choke must hold more and more back-pressure to maintain bottomhole pressure at the correct value. As a result, casing pressure at the surface can attain a high value when the kick is mostly or all gas. High casing pressure can increase the chance of formation fracture, lost returns, and an underground blowout or broaching, depending on the relative locations of the influx and the casing shoe.

Further, the driller's method requires a relatively long time to kill the well, since two circulations are needed—one to circulate the influx out, and one to circulate kill-weight mud into and up the hole. Since pressure on the BOP stack is maintained during the two circulations, the possibility of other trouble increases.

WAIT-AND-WEIGHT METHOD

The wait-and-weight method is so named because the crew first shuts the well in, waits for kill-weight mud to be prepared, and then circulates the new, weighted-up mud into the hole. At the same time new mud is pumped in, old-weight mud and kick fluids are removed through the choke. Pumping in new mud while removing old mud and kick fluids may result in lower surface, or casing, pressure than first circulating the kick out with old mud and then circulating in new mud.

Surface pressure may be lower with the wait-and-weight method, because new, heavier mud may be present at the bottom of the annulus before a gas influx is removed through the choke; the increased hydrostatic pressure of the kill-weight mud reduces the surface pressure needed to maintain constant bottomhole pressure. If surface pressure is limited, the chances of formation breakdown are reduced; however, the chance of a reduction in formation breakdown occurs only when kill-weight mud enters the annulus before a gas influx moves up past the casing shoe, a fairly rare event.

Most operators recommend that the potential advantage of the wait-and-weight method be considered in marine and offshore drilling, on locations where lost circulation or equipment failure is a good possibility, and where reasonably good mud-mixing facilities permit kill-weight mud to be prepared quickly. The advantage of reduced surface casing pressure must be weighed against the increased likelihood of such problems as gas migration, stuck pipe, or downhole plugging of the circulating system during the waiting period. Such problems get worse as the time before circulation grows longer.

Characteristics of the Wait-and-Weight Method

Two important characteristics of the wait-and-weight method should be noted. First, pump pressure, as shown on the drill pipe gauge, decreases when new, kill-weight mud is circulated down the drill stem, even though the pump rate is kept constant. Second, casing pressure should not be held constant as new mud fills the drill stem; it must be allowed to increase if bottomhole pressure is to remain constant while circulating a gas kick of up the annulus.

To understand why drill pipe pressure decreases as new mud is pumped down the drill stem, keep in mind that (1) the drill pipe pressure gauge shows pump pressure when the well is being circulated, and (2) *ICP* includes the formation pressure in excess of the hydrostatic pressure of the old mud in the drill stem. Thus, after the new mud weight is calculated from *SIDPP* and new mud is circulated down the drill pipe, it balances the pressure exerted by the formation on the mud in the pipe, and drill pipe circulating pressure decreases.

In the driller's method, one way to keep bottomhole pressure constant as new mud fills the drill stem is to keep casing pressure constant by adjusting the choke. This procedure is acceptable as long as all gas has been circulated out of the well. With the wait-and-weight method, casing pressure must not be held constant as new mud fills the drill stem, because the kick is being circulated up the annulus at the same time. Since the kick is being circulated up the hole, any gas in the kick must be allowed to expand to avoid excessive pressures, and when gas is allowed to expand, casing pressure increases. With the wait-and-weight method, therefore, bottomhole pressure must be kept constant by using a different technique.

One popular technique is first to determine *ICP*, *FCP*, and the amount and rate at which *SIDPP* will decrease. These values are then recorded on a graph and observed as new mud fills the drill stem (fig. 6.2). If drill pipe pressure strays from the calculated value, the choke is adjusted to correct it, and bottomhole pressure is kept constant until *FCP* is reached.

Procedure for the Wait-and-Weight Method

The following is one procedure for the wait-and-weight method:

1. When the kick occurs, shut the well in.
2. Fill out the work sheet (kill sheet) and increase the mud weight.
3. When the new mud is ready, open the choke and slowly bring the pump up to kill rate while keeping casing pressure constant using the choke. (With subsea BOPs, decrease casing pressure by the amount of the choke-line friction pressure, if known.) When the pump is up to the kill rate, observe the drill pipe pressure gauge; it should indicate *ICP*.
4. While keeping the pump rate constant, continue pumping new mud down the pipe. Drill pipe pressure will decrease as new mud slugs the pipe. The decrease in drill pipe pressure can be monitored on the kill sheet graph, and the choke opened or closed to correct small fluctuations.
5. When kill-weight mud arrives at the bit, maintain *FCP* and keep the pump strokes constant until new mud reaches the surface.
6. Stop the pump, close the well in, and check for flow.

Filling in the Wait-and-Weight Method Kill Sheet

With the wait-and-weight method, it is essential to use a work, or kill, sheet that not only provides spaces for prerecorded information such as *KRP*, but also has spaces for the required calculations and a graph with divisions for plotting the drop-in drill pipe pressure as new mud is circulated down the drill stem (fig. 6.3).

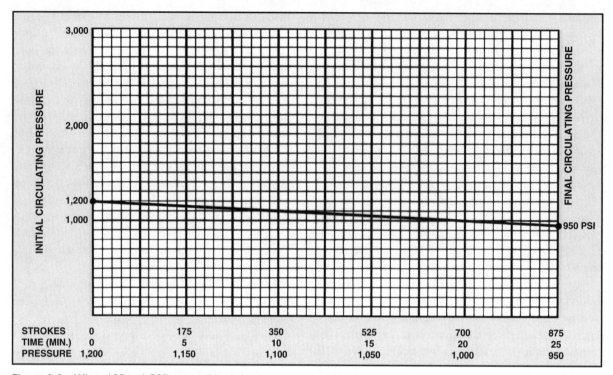

Figure 6.2   When *ICP* and *FCP* are predetermined, a graph can be referred to for the correct drill pipe pressure. In this example, *ICP* is 1,200 psi and *FCP* is 950 psi. At 15 min. and 525 strokes into the circulation, drill pipe pressure should be 1,050 psi.

A plot of the *SIDPP* reduction, when used with the number of surface-to-bit strokes and surface-to-bit time, provides a clear schedule that the well-control operator can refer to at any point during the time the drill stem is filling with new mud. By referring to the plot of *SIDPP*, the operator can ensure that proper pressure is being maintained on the drill pipe pressure gauge and thus on the bottom of the hole. A kill sheet can be filled in in the following manner:

1. Make sure that the prerecorded information is correct.

2. When a kick occurs, record *SIDPP*, *SICP*, and pit-volume increase.

3. Using the formula on the kill sheet, calculate the mud-weight increase and record it.

4. Now turn to the graph of *SIDPP*. To plot the decrease in *SIDPP*, determine *ICP* using the formula on the kill sheet.

5. Enter *ICP* in the blank on the kill sheet and at the appropriate psi value on the left vertical line of the graph.

6. Use the kill-sheet formula to determine *FCP* and enter it in the blank and on the right vertical line of the graph.

7. Place a straight edge between the point for *ICP* and the point for *FCP* and draw a line to connect the two points (fig. 6.4).

8. Determine the pressures between *ICP* and *FCP* and enter them in the divisions at the bottom of the graph. Intermediate pressures between *ICP* and *FCP* can be calculated by subtracting *FCP* from *ICP*, dividing the answer by the number of divisions, and progressively subtracting this answer from *ICP*. For example, if *ICP* is 1,500 psi, *FCP* is 1,200 psi, and 11 divisions are provided at the bottom of the graph on the kill sheet, record 1,500 at the first division. This leaves 10 divisions to get to the point where drill pipe pressure reaches 1,200 psi. Therefore, subtract 1,200 from 1,500, divide the result by 10. The answer shows that *SIDPP* drops 30 psi at each division after the *ICP* of 1,500 psi. Thus, in the first division after the one where *ICP* is recorded, the pressure value is 1,470 psi; at the second, it is 1,440 psi, and so on until the *FCP* of 1,200 psi is reached (fig. 6.5).

9. Determine the number of surface-to-bit strokes (*SBS*) using the kill-sheet formula and enter this

WELL-KILL WORKSHEET
WAIT-AND-WEIGHT METHOD

1. **PRERECORDED INFORMATION**

    kill-rate pressure at _____ strokes per minute . . . . _____ psi

    time of shut-in _____ am/pm

2. **RECORD AT TIME OF SHUT-IN**

    circulating time, surface to bit _____ min., _____ pump strokes

    shut-in drill pipe pressure (SIDPP) . . . . _____ psi

    shut-in casing pressure (SICP) . . . . _____ psi

3. **DETERMINE INITIAL CIRCULATING PRESSURE**

    kill-rate pressure + SIDPP . . . . _____ psi

4. **CALCULATING MUD WEIGHT INCREASE**

    MWI = SIDPP + TVD + 0.052 . . . . _____ ppg

    original mud weight . . . . _____ ppg

5. **NEW MUD WEIGHT REQUIRED** . . . . _____ ppg

6. **DETERMINE FINAL CIRCULATING PRESSURE**

    kill-rate pressure × new mud weight + old mud weight . . . . _____ psi

GRAPHICAL ANALYSIS

1. Plot initial circulating pressure at left edge of graph.
2. Plot final circulating pressure at right edge of graph.
3. Connect the points with a straight line.

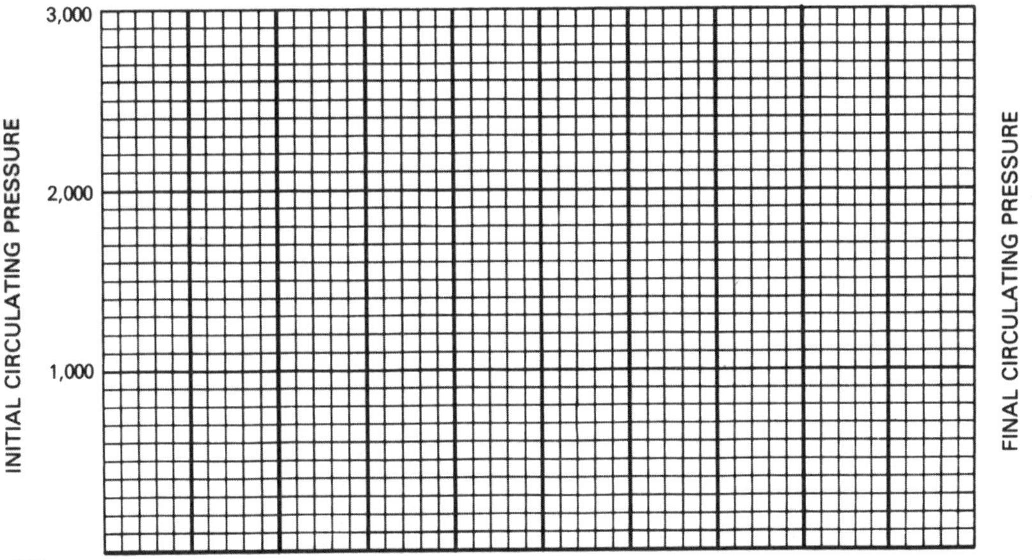

ICP
SBS
SBT

Figure 6.3   A typical kill sheet for the wait-and-weight method

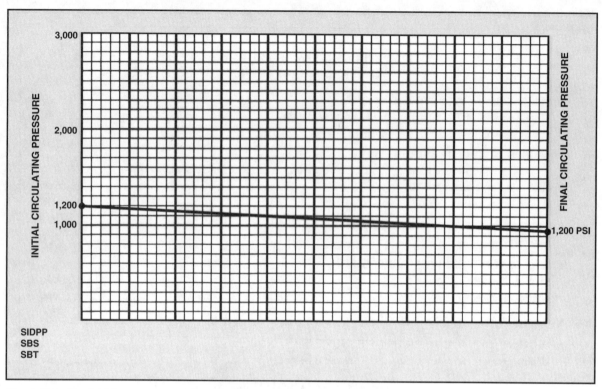

Figure 6.4   Straight line connects *ICP* and *FCP*, wait-and-weight method

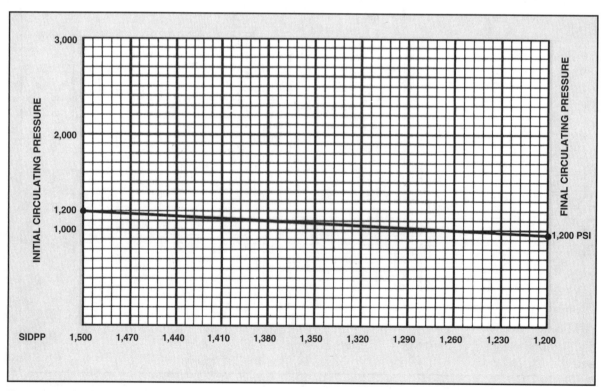

Figure 6.5   *SIDPP* drops about 30 psi at each division of the graph between *ICP* and *FCP*, wait-and-weight method.

value in the blanks on the kill sheet and at the division at the far right on the kill sheet's graph.

10. Enter 0 (zero) at the far left. Then, record the proper number of strokes at the remaining divisions. As an example, assume that SBS is 800 and that 11 divisions are marked on the graph. Divide 800 by 10 to obtain an answer of 80. Enter 80 in the first division after 0, 160 in the second, 240 in the third, and so on until the total of 800 strokes is reached (fig. 6.6).

11. To obtain surface-to-bit time (SBT), use the kill-sheet formula and record the value in the blank on the kill sheet and at the division at the far right on the kill sheet's graph. Then, enter 0 (zero) at the left and the proper times at the remaining divisions. As an example, assume that SBT is 50 min. Fifty divided by 10 is 5. Enter 5 in the first division after 0 and then progressively add 5 additional min to each division so that the times read 0, 5, 10, 15, and so on, until 50 min is reached (fig. 6.7).

12. With the pressure, number of strokes, and the time recorded, the operator can now refer to the graph and observe the calculated value for drill pipe pressure at any point during the interval when the drill stem is being slugged with new mud. For example, at 25 min into the circulating procedure, and at 400 strokes, SIDPP should be about 1,350 psi. If it is not, the choke can be adjusted to bring the pressure to the correct value or, should a large discrepancy exist, troubleshooting procedures can be implemented.

Although many operators prefer to use a graph to plot the drop-in drill pipe pressure as kill-weight mud fills the drill string, it is not essential to do so. In fact, some operators feel that a graph may complicate the procedure; they, therefore, prefer to use a kill sheet without a graph (fig. 6.8). With a kill sheet that does not have a graph, it is not necessary to draw lines with a straight edge; otherwise, it is filled out very much like any other kill sheet.

EXAMPLE PROBLEMS FOR WAIT-AND-WEIGHT METHOD

To learn how to fill in a kill sheet, practice is essential. Use the information given in each of the following problems along with the appropriate equations to fill in a wait-and-weight kill sheet.

1. Hole: TVD 10,500 ft; size 12½ in.
   Drill pipe (DP): 5 in., 19.5 ppf, XH
   Casing: 13⅜ in.; 72.0 ppf; TVD 8,000 ft
   KRP: 800 psi at 45 spm
   Pump: 6½ × 12 P-160 triplex
   SIDPP: 300 psi
   SICP: 420 psi
   Pit increase: 40 bbl
   Mud weight: 11.8 ppg

2. Hole: TVD 5,800 ft; size 12¼ in.
   DP: 5 in., 19.5 ppf, XH
   Casing: 13⅜ in.; 72.0 ppf; TVD 4,000 ft
   KRP: 700 psi at 45 spm
   Pump: 6 × 12 PT-1700 triplex
   SIDPP: 220 psi
   SICP: 300 psi
   Pit increase: 25 bbl
   Mud weight: 10.2 ppg

3. Hole: TVD 12,550 ft; size 8¾ in.
   DP: 4½ in., 16.6 ppf, XH
   Casing: 9⅝ in.; 53.5 ppf; TVD 9,500 ft
   KRP: 850 psi at 30 spm
   Pump: 5¾ × 12 F-1600 triplex
   SIDPP: 520 psi
   SICP: 750 psi
   Pit increase: 38 bbl
   Mud weight: 14.5 ppg

4. Hole: TVD 8,820 ft; measured depth (MD) 10,255 ft; size 9⅞ in.
   DP: 5 in., 19.5 ppf, XH
   Casing: 7⅝ in.; 39.0 ppf
   KRP: 950 psi at 45 spm
   Pump: 5½ × 12 triplex
   SIDPP: 320 psi
   SICP: 500 psi
   Pit increase: 40 bbl
   Mud weight: 11.7 ppg

5. Hole: TVD 11,905 ft; MD 13,450 ft; size 9⅞ in.
   DP: 5 in., 19.5 ppf, XH
   Casing: 10¾ in; 55.5 ppf; TVD 9,200 ft
   KRP: 1,100 psi at 45 spm
   Pump: 6 × 12 triplex
   SIDPP: 450 psi
   SICP: 600 psi
   Pit increase: 30 bbl
   Mud weight: 12.8 ppg

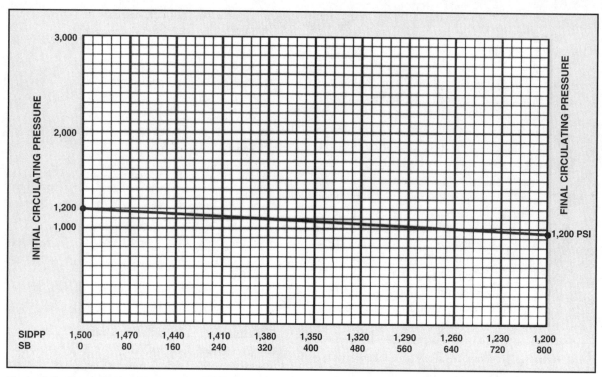

Figure 6.6   Total *SBS* is 800; thus the first division after 0 is 80 strokes, the second, 160, the third, 240, and so on, until the total of 800 is reached, wait-and-weight method.

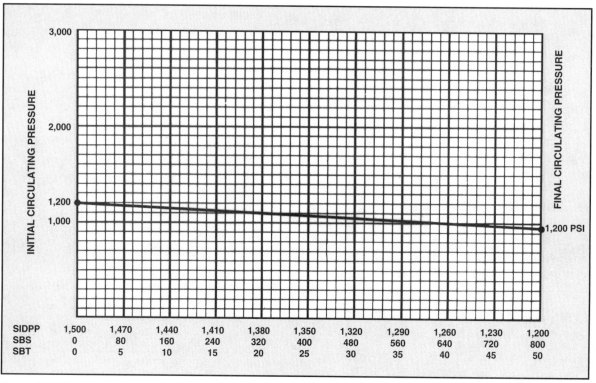

Figure 6.7   Total *SBT* is 50 min; thus the first division after 0 is 5, the second, 10, the third, 15, and so on, until the total of 50 is reached, wait-and-weight method.

Weight-and-Wait Method
Well-Control Work Sheet

Date _____

Time of shut-in _____

Name _____

1. PRERECORDED INFORMATION:

Bit, TVD_____ ft   Bit, MD_____ ft   Shoe, TVD _____ ft   Shoe, MD _____ ft

Casing: _____ Size, in.   Weight _____ ppf   Grade _____   70% Burst _____ psi   MMS _____ psi

Hole: _____ Size, in.   Choke-Line Friction Pressure (Floating Rigs Only) = _____ psi

Drill Pipe _____ Size, in.   Weight _____ ppf

Drill Pipe Capacity _____ bbl/ft   Drill Collar Capacity _____ bbl/ft

Bit-to-Shoe Stks = (_____ − _____) × _____ + _____ = _____ stks
$\quad\quad\quad\quad\quad$ Hole Depth $\quad$ Shoe Depth $\quad$ Ann. Cap., bbl/ft Pump Output, bbl/stk

$\quad\quad\quad\quad\quad\quad\quad\quad\quad\quad\quad\quad\quad\quad\quad\quad\quad\quad\quad\quad\quad\quad\quad$ Bottoms Up _____ stks

Maximum Allowable Mud Weight

(_____ + _____ + 0.052) + _____ .... = _____ ppg
$\quad$ Leak-Off Pres. $\quad$ Shoe TVD $\quad\quad\quad$ Test Mud Wt

Maximum Allowable Surface Pressure Based on Leak-Off Test

(_____ − _____) × _____ × 0.052 .... = _____ psi
$\quad$ Max. Allow. Mud Wt. $\quad$ OMW $\quad\quad$ Shoe TVD

KILL RATE PRESSURE:

Pump No. 1 _____ psi at _____ stks/min   Pump No. 2 _____ psi at _____ stks/min

Pump Output _____ bbl/stk   Pump Output _____ bbl/stk

---

2. RECORD:

Shut In Drill Pipe Pressure (SIDPP) .... = _____ psi

Shut In Casing Pressure (SICP) .... = _____ psi

Pit Volume Increase .... = _____ bbls

3. DETERMINE INITIAL CIRCULATING PRESSURE (ICP):

ICP = _____ + _____ .... = _____ psi **(ICP)**
$\quad\quad\quad$ KRP $\quad\quad$ SIDPP

4. CALCULATE MUD WEIGHT INCREASE (MWI):

MWI = _____ + _____ + _____ .... = _____ ppg
$\quad\quad\quad$ SIDPP $\quad$ 0.052 $\quad\quad$ TVD

$\quad\quad\quad\quad\quad\quad\quad\quad\quad\quad\quad\quad\quad\quad\quad\quad$ Add Old Mud Weight _____ ppg

5. KILL WEIGHT MUD (KWM): .... = _____ ppg **(KWM)**

6. DETERMINE FINAL CIRCULATING PRESSURE (FCP):

FCP = _____ × _____ + _____ .... = _____ psi **(FCP)**
$\quad\quad\quad$ KRP $\quad\quad$ KWM $\quad\quad$ OMW

7. CALCULATE SURFACE-TO-BIT STROKES (SBS):

A. DP = _____ × _____ + _____ .... = _____ stks
$\quad\quad\quad\quad$ DP Capacity $\quad$ DP Length $\quad$ Pump Output

B. DC = _____ × _____ + _____ .... = _____ stks
$\quad\quad\quad\quad$ DC Capacity $\quad$ DC Length $\quad$ Pump Output

$\quad\quad\quad\quad\quad\quad\quad\quad\quad$ Total Strokes .... = _____ **(SBS)**

C. Circulating Time to Bit = _____ + _____ .... = _____ min
$\quad\quad\quad\quad\quad\quad\quad\quad\quad$ SBS $\quad\quad$ stks/min

8. DRILL PIPE PRESSURE REDUCTION SCHEDULE:

Pressure = (ICP − FCP) ÷ 10 = _____ psi/division

Time	0										Time	
Pressure	ICP									FCP	Pressure	
Strokes	0										Strokes	

Maintain final circulating pressure (FCP) to bottoms up after kill-weight mud reaches the bit.

Figure 6.8   Wait-and-weight method work sheet

CONCURRENT METHOD

The concurrent method of well control is often thought to be the most complicated, chiefly because it requires more recordkeeping than the driller's and wait-and-weight methods. Another potential disadvantage, when compared to the wait-and-weight method, is that higher casing pressure can develop with the concurrent method if a gas kick occurs. Further, since mud is weighted up in a series of steps, mud mixing can be a problem. Not only must the mud be weighted up as the well is being circulated, but once a new weight is achieved, that weight must be maintained throughout the time it is circulated. Perhaps the concurrent method's strongest advantage is that, on rigs where the mud cannot be weighted up to kill weight all at once, the casing pressure that develops is lower than the casing pressure that develops with the driller's method when a gas kick must be handled.

Procedure for the Concurrent Method

The following is one procedure for the concurrent method:

1. After the well is closed in and the information recorded, calculate *ICP*, *FCP*, and the mud-weight increase. Fill out the kill sheet, using increments of mud weight across the bottom of the graph. Also calculate surface-to-bit strokes and time and the drill pipe pressure decrease with each of the mud weights (fig. 6.9).
2. Start the pump and bring it up to kill-rate speed while holding casing pressure constant. When the pump is up to the kill rate, adjust drill pipe pressure to the calculated value. Circulation can be started as soon as *ICP* has been determined.
3. When circulating, have mud pit personnel call up the mud weight each time mud weight in the pits is increased. Each time that the mud weight increases to one of the values at the bottom of the chart, have the choke operator adjust circulating pressure to the drill pipe pressure shown on the graph.
4. Continue circulating until mud of the required kill weight comes back to surface and the well is dead.

Completing a Graph for the Concurrent Method

Follow these steps to complete the graph for the concurrent method:

1. On the concurrent method kill sheet, record *ICP* on the left vertical line of the graph.

2. Across the bottom of the graph, indicate the mud-weight increase. Start with the present mud weight in the left-hand division and increase the mud weight by 0.1 ppg in each division until the final mud weight is reached (fig. 6.10).
3. Record *FCP* on the right vertical line of the graph.
4. Connect *ICP* and *FCP* with a straight line. The line shows the circulating pressure decrease with the mud-weight increase.

EXAMPLE PROBLEM FOR THE CONCURRENT METHOD

Given the following information,

$$TVD = 11,000 \text{ ft}$$
$$OMW = 11.0 \text{ ppg}$$
$$KWM = 11.5 \text{ ppg}$$
$$ICP = 1,600 \text{ psi}$$
$$FCP = 1,100 \text{ psi}$$
$$SBS = 1,588,$$

use the graph on a concurrent method kill sheet and show the circulating pressure decrease that occurs with each mud-weight increase.

WEIGHT-UP CONSIDERATIONS

Whether using the driller's, wait-and-weight, or concurrent method to kill a well, the mud must be weighted up to the required density. Usually, barite is added to the mud system to achieve the required weight. It is therefore important to know how many sacks of barite are required to get the desired weight increase. The following equation can be used to determine the number of sacks of barite to add:

$$sx = 1,470 \, (W_2 - W_1) \div (35 - W_2) \quad \text{(Eq. 55)}$$

where

sx = sacks of barite to add per 100 bbl of mud

W_2 = desired mud weight, ppg

W_1 = initial mud weight, ppg.

As an example, assume that the initial mud weight is 12.2 ppg and that the desired weight is 12.7 ppg. According to the equation, the number of sacks of barite required to achieve this weight is—

$$sx = 1,470 \, (12.7 - 12.2) \div (35 - 12.7)$$
$$= 1,470 \times 0.5 \div 22.3$$
$$= 735 \div 22.3$$
$$sx = 32.95, \text{ or about } 33.$$

WELL-KILL WORKSHEET
CONCURRENT METHOD

1. PRERECORDED INFORMATION

   kill-rate pressure at _____ strokes per minute .... _____ psi

   time of shut-in _____ am/pm

2. RECORD AT TIME OF SHUT-IN

   circulating time, surface to bit _____ min., _____ pump strokes

   shut-in drill pipe pressure (SIDPP) .... _____ psi

   shut-in casing pressure (SICP) .... _____ psi

3. DETERMINE INITIAL CIRCULATING PRESSURE

   kill-rate pressure + SIDPP .... _____ psi

4. CALCULATING MUD WEIGHT INCREASE

   MWI = SIDPP + TVD + 0.052 .... _____ ppg

   original mud weight .... _____ ppg

5. NEW MUD WEIGHT REQUIRED .... _____ ppg

6. DETERMINE FINAL CIRCULATING PRESSURE

   kill-rate pressure × new mud weight + old mud weight .... _____ psi

GRAPHICAL ANALYSIS

1. Plot initial circulating pressure at left edge of graph.
2. Plot final circulating pressure at right edge of graph.
3. Connect the points with a straight line.

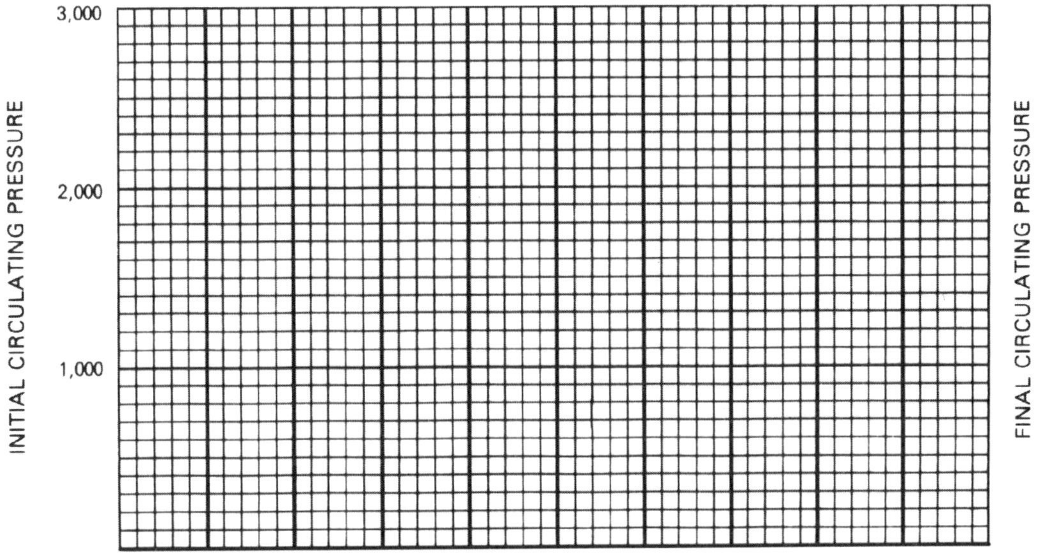

MUD WEIGHT

ICP

SBS

SBT

Figure 6.9   Well-killing work sheet, concurrent method

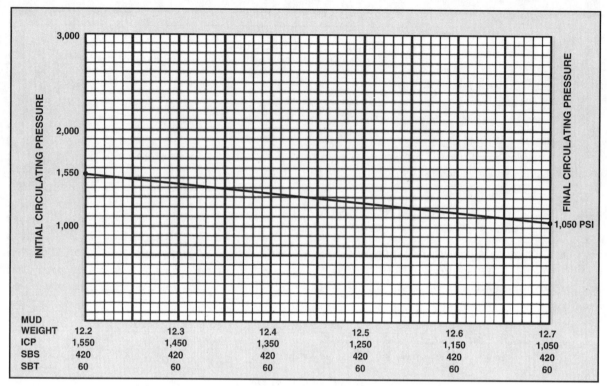

MUD WEIGHT	12.2	12.3	12.4	12.5	12.6	12.7
ICP	1,550	1,450	1,350	1,250	1,150	1,050
SBS	420	420	420	420	420	420
SBT	60	60	60	60	60	60

Figure 6.10 Well-killing work sheet, concurrent method

The answer shows that about 33 sacks of barite must be added to every 100 bbl of mud in the system to achieve an increase from 12.2 ppg to 12.7 ppg. Thus, in this example, if the mud system contains 800 bbl, 8×33, or 264, sacks of barite must be added to achieve the increase.

Just as the level of mud in a hole rises when drill pipe is run into it, the volume occupied by the mud increases when barite is added to the mud tanks, or pits. It is therefore important to know the volume of the rig's mud tanks and the volume increase that occurs when barite is added to the system to increase the mud's weight. To determine the volume of a rectanglar tank in bbl, one equation that can be used follows:

$$V_{mt} = L \times W \times H \div 5.614 \quad \text{(Eq. 56)}$$

where

V_{mt} = volume of mud tank, bbl
L = length of mud tank, ft
W = width of mud tank, ft
H = height of mud tank, ft.

Thus, a mud tank that is 16 ft long by 10 ft wide by 7 ft high, has a volume of—

$$V_{mt} = 16 \times 10 \times 7 \div 5.614$$
$$= 1,120 \div 5.614$$
$$V_{mt} = 199.5, \text{ or about 200 bbl.}$$

Bear in mind that, since several mud tanks are usually employed, the volume of each tank must be determined and the volume of each added to determine the total volume of the tanks in the system.

To determine the volume increase of the mud when barite is added to it, the following formula can be used:

$$V_i = N \div 14.9 \quad \text{(Eq. 57)}$$

where

V_i = volume of increase, bbl
N = number of sacks of barite added to the system.

For example, suppose that 264 sacks of barite were added to the mud system to achieve the desired increase in mud weight. To determine the amount these sacks of barite increase the volume of the mud—

$$V_i = 264 \div 14.9$$
$$V_i = 17.7, \text{ or about 18 bbl.}$$

The answer shows that the volume increase in the mud system is about 18 bbl.

Mud Properties Following Weight-up and Dilution

After the mud has been weighted up (or diluted to decrease its density), other mud properties are also affected. Three important ones are gel strength, plastic viscosity, and yield point. Gel strength is a measure of the mud's ability to suspend cuttings and keep them from falling in the hole when circulation is stopped. Gel strength that is too high, however, can be detrimental because it can increase surge and swab pressures in the circulating system. It is important, therefore, for crew members to measure and adjust, if necessary, the mud's gel strength to its proper value after weighting up or diluting the mud. Similarly, crew members should also ensure that the mud's viscosity and yield point are at optimum values after weighting up or diluting the mud. A mud whose viscosity and yield point are not correct can create circulating system problems detrimental to well control. For example, mud that is very thick (viscous) and has a high yield point requires a great deal pump pressure to move. Consequently, equivalent circulating density increases, which can lead to lost circulation problems.

VOLUMETRIC METHOD

The volumetric method is sometimes used for controlling bottomhole pressure when no drill pipe is in the hole or when the well cannot be circulated. It can also be used when the well is closed in and the crew is waiting on orders or equipment. Further, it can be employed when stripping in or out of the hole. The technique is not a well-killing method in the sense of being a method of circulating, since it simply coordinates the increase or decrease of annular pressure with the amount of mud displaced from the hole.

The basic premise of the volumetric method is that each barrel of mud contributes a certain pressure to the bottom of the hole, which may be measured in psi per barrel (psi/bbl). In other words, if 1 bbl of fluid is bled off, bottomhole pressure is reduced by the amount of pressure exerted by that 1 bbl; if 1 bbl of fluid is pumped in, bottomhole pressure is increased by the pressure exerted by that 1 bbl. To obtain a value for psi/bbl, the mud gradient in psi/ft is divided by the annular volume in bbl/ft,

or the hole volume in bbl/ft if no pipe is in the hole. The formula follows:

$$P_{bbl} = (MW \times 0.052) \div AV \qquad \text{(Eq. 58)}$$

where

P_{bbl} = psi/bbl of mud

MW = mud weight, ppg

AV = annular volume, bbl/ft.

The volumetric method is needed only if gas is in the well. The annular volume used for volumetric calculations should be that part of the annulus in which the gas is located. Because it is not often known exactly where the gas is, some operators recommend using the annular volume just below the BOP stack.

To use the volumetric method, the well is closed in under pressure and $SICP$ is recorded. If $SICP$ does not rise, bottomhole pressure remains constant and the volumetric method is not usable. If $SICP$ rises, however, mud can be bled from the well corresponding to casing pressure increases according to the calculated psi/bbl value. Bleeding mud from the well while allowing $SICP$ to increase maintains constant bottomhole pressure, because bottomhole pressure is the sum of casing pressure and pressure exerted by the mud column.

In summary, the volumetric method may be used when (1) no pipe is in the hole, (2) the bit is on or off the bottom, (3) the bit or the drill stem is plugged, (4) stripping in or out of the hole, or (5) awaiting the arrival of a snubbing unit, perforating truck, and so on.

Procedure for the Volumetric Method

When a well is shut in under pressure, the following step-by-step procedure for the volumetric method may be used:

1. Record $SICP$.
2. Rig up the choke-line output to discharge fluid into a tank. The tank must be graduated so that fluid can be measured accurately to the nearest gallon.
3. Calculate the volume of fluid to be bled or added to the well that provides 100 psi hydrostatic pressure (smaller increments can be used if desired).
4. Monitor $SICP$ and allow it to increase by the same increment (100 psi, in this case).
5. When pressure has increased by 100 psi (or whatever increment is selected), note the new

SICP. Hold this new casing pressure constant by opening and closing the choke to bleed off small volumes of mud as necessary. (Use of a hand-adjustable choke is recommended.) Measure the amount of accumulated mud bled off and when its volume equals the volume calculated in step 3, close the well in and repeat steps 4 and 5.

As the gas rises, it may be necessary to bleed off more frequently. When gas reaches the surface, a similar procedure may be used to lubricate mud into the hole to compensate for reductions in casing pressure as the gas is bled off.

The volumetric method holds bottomhole pressure almost constant; its accuracy depends on knowing the actual capacity of the hole and the location of the influx. Hole washouts, tapered drill stems, or different sizes of casing can lead to inaccuracies. Thus, the volumetric method is only used if other methods cannot be.

The volumetric method simply provides a way to control bottomhole pressure as gas expands and migrates up the hole. If the gas reaches the surface, it may become desirable to reduce surface casing pressure. To reduce surface casing pressure, a measured volume of fluid is pumped, or lubricated, into the annulus at a pressure below that which would cause formation fracture. The injected fluid pressure should show an increase and then a decrease as fluid is injected. Gas (not fluid) should be bled until the pressure is reduced by the amount of the hydrostatic pressure of the injected fluid, plus any initial increase in casing pressure during injection.

EXAMPLE PROBLEM FOR THE VOLUMETRIC METHOD

Calculate psi/bbl of mud (P_{bbl}) under the following conditions:

Mud weight	=	12.0 ppg
Casing	=	9⅝ in., 53.5 ppf, grade P-110
Casing annular capacity	=	0.0459 bbl/ft
Kick size	=	10 bbl
SICP	=	800 psi (allowed to increase to 900 psi).

The solution is—

$$P_{bbl} = (MW \times 0.052) \div AV$$

$$= (12.0 \times 0.052) \div 0.0459$$
$$= 0.624 \div 0.0459$$
$$P_{bbl} = 13.59 \text{ psi/bbl.}$$

As per step 5, the volume of mud to bleed is about 7.36 bbl because 100 psi divided by 13.59 psi/bbl equals 7.36 bbl.

As an example, suppose SICP starts at 800 psi and is allowed to rise to 900 psi. As it increases, say to 920 psi, the choke is opened and a small amount of fluid is bled to reduce SICP back to 900 psi. When SICP goes up again, the choke is cracked open again to drop the pressure back to 900 psi. These actions are repeated until pit watchers advise that the calculated 7.36 bbl has accumulated from the small bleed offs. The well is then shut in until SICP reaches 1,000 psi, and the choke is cycled as necessary to 1,000 psi until another 7.36 bbl has been bled off. This action continues until SICP stops increasing, which should happen when gas accumulates below the BOPs.

An equation is available that can be used to calculate the amount of drop in bottomhole pressure as a result of bleeding fluid from the well:

$$BHP = SICP + (F_b \times P_{bbl}) \quad \text{(Eq. 59)}$$

where

BHP = amount bottomhole pressure is reduced after fluid is bled, psi

$SICP$ = amount of pressure drop in SICP after fluid is bled, psi

F_b = amount of fluid bled, bbl

P_{bbl} = pressure exerted by mud, psi/bbl.

Using the example in which ¼ bbl of fluid is bled from the well to make SICP 50 psi, and in which P_{bbl} is 13.59 psi/bbl, a value for BHP can be determined as follows:

$$BHP = 50 + (¼ \times 13.59)$$
$$= 50 + 3.4$$
$$BHP = 53.4 \text{ psi.}$$

LOW CHOKE-PRESSURE METHOD

When SICP begins to rise while circulating a kick out and while holding bottomhole pressure constant, inexperienced personnel sometimes believe that allowing SICP to decrease is safer. Unfortunately, opening the choke to reduce SICP also reduces bottomhole pressure and allows formation fluids to flow once again into the well. In some cases, however, deliberately opening the choke to reduce back-pressure on the well can be used successfully to control the flow of fluids from a well. Any procedure to kill a kick in which the choke is adjusted so that *SIDPP* falls below the value required to maintain bottomhole pressure at or above formation pressure is a low choke-pressure method. In effect, the choke is opened, casing pressure is lowered, and additional formation fluids enter the hole. Formation fluids continue to enter the well as long as choke pressure is maintained at a low value.

Low choke-pressure methods are sometimes employed in areas where it is possible to drill underbalanced. Usually, drilling underbalanced is applied in tight, low-permeability formations with which the operator is very familiar. By maintaining low choke pressure, the crew can continue to drill underbalanced and keep the drilling rate high. Low choke-pressure methods have also been used to avoid damaging tight but fractured formations. For the method to be successful, the kicking formation must be tight—of low permeability so that influxes are of relatively low volume—and the operator must have drilled a sufficient number of wells in the area to know the characteristics of the kicking formation.

In cases where the hole has penetrated a high-permeability formation or when formation permeability is not known, attempting to kill a kick with lowered choke pressure is an uncertain procedure that can rapidly cause an unmanageable situation. Therefore, most operators do not recommend that low choke-pressure methods be used except in carefully specified circumstances or when *SICP* begins to exceed equipment limitations. Such methods should not be used simply because pressures during initial closure threaten to exceed formation fracture gradients below surface casing that has been set too

shallow to avoid broaching. Such cases are often the result of poor planning and, should they occur, might be better handled by using a diverter procedure.

REVERSE CIRCULATION METHOD

When a well is being tested or completed, it is often advantageous to reverse circulate the well—that is, circulate down the annulus and up the drill string (often called the work string during completion operations). From a well control point of view, advantages of reverse circulation include reduction in the time needed to circulate the kick fluids from the well and better control of high surface pressures because drill pipe and tubing generally have higher burst strengths than casing. Disadvantages of reverse circulation include possible plugging of bit jets or circulating ports in a work string. Also, because circulation rates are slow during a well-control procedure, gas may migrate up the hole faster than the downward flow rate of the drilling or completion fluid. As a result, the gas kick will not be circulated out of the hole.

The reverse circulation method uses *SICP*, instead of *SIDPP*, as the primary method of monitoring the well-control operation. Further, reverse circulation requires placing a choke in the standpipe or arranging the choke manifold in such a way that the kick fluids are conducted through the standpipe to the choke and choke manifold. Gas migration rates must also be kept in mind when reverse circulating a kick. The rate of gas migration can range from 30 ft/min when it migrates through fresh water to as little as 2 ft/min or less when migrating through a heavy viscous mud. The gas migration rate can be controlled by increasing the pumping rate or by adding a viscosifier to the kill fluid.

Chapter 7

Unusual Well-Control Operations

During a well-control operation, the characteristics of the well or kick may call for procedures that deviate from normal. Rig supervisors and crews should be aware of such procedures and be prepared to initiate them if necessary. While it is impossible to cover every unusual situation that could occur, this chapter covers several.

A HOLE IN THE DRILL STRING

When a kick is being circulated out of the well and a hole is washed through the drill string, a decrease in *SIDPP* occurs without a corresponding decrease in *SICP*. Because *SIDPP* decreases, the choke operator may close the choke in an attempt to bring *SIDPP* back to the previous value. Closing the choke, however, causes casing pressure to increase to a value higher than that required to prevent the entry of additional kick fluids into the well. If the hole in the pipe is large, the choke may be closed to the point that the additional back-pressure causes formation fracture and lost circulation. Therefore, it is important to be alert to the possibility of a hole's being washed through the pipe and to be able to react properly to the problem.

Once it is certain that a hole has appeared in the string, the next step is to determine whether the hole is above or below the kick fluids, because the location of the hole bears on how the situation should be handled. For example, if the hole in the pipe is above the kick, it becomes difficult or impossible to maintain constant bottomhole pressure while circulating the kick in the conventional manner.

Since the hole opens the pipe to annular pressure, drill pipe pressure gauge readings may help to locate the hole. For example, if *SIDPP* is much higher than expected and does not decrease when a small amount of mud is bled from the well, it is likely that the hole

in the pipe is above the kick. In fact, if no kill-weight mud is in the drill pipe, *SIDPP* may be the same as *SICP*. If, on the other hand, the hole in the pipe is below the kick, it is likely that *SIDPP* will be near the previous shut-in value.

Since a slower-than-normal pump speed is usually employed when circulating a kick out, and since the hole in the drill stem may be quite small, detection of the problem may be difficult. If the location of the hole is determined to be below the kick, however, many operators recommend that circulation be continued until the kick has been circulated out. A change to the slower kill rate reduces the flow rate of mud through the hole in the pipe and reduces the likelihood of the hole's being washed larger. Keeping the hole in the pipe from getting larger may make it possible to continue circulating the well without excessive back-pressure.

A PLUGGED BIT

When a well kicks and a large amount of barite and chemicals are added to the mud in the pits, and when mud in the pits is stirred during weight-up, it is possible for relatively large lumps of solid material to form. When circulated down the drill stem, these solid masses may totally or partially plug the jets of the bit. Fortunately, plugging is not common and, when it does occur, the jets usually are only partially plugged.

When the bit plugs, either partially or totally, while circulating a kick out of the well, pump pressure suddenly goes up; however, casing pressure remains relatively constant. It is important for the choke operator to recognize bit plugging and not to open the choke in an attempt to reduce SIDPP to its proper value. Opening the choke could reduce bottomhole pressure and allow the entry of additional formation fluids. If the bit is completely plugged, the following suggested methods could help to overcome the problem:

1. Rapidly increase, then decrease, the pump rate; the pressure surge may clear the jets.
2. Perforate the drill stem above the plugged bit.
3. Use a string shot near the bit; the shot may clear the jets.
4. Use a shaped charge near the bit; the explosion may clear the jets.

If the bit is partially plugged, some operators suggest that the best course of action may be the following:

1. Stop the pump and close the choke to shut the well in completely.

2. Record SIDPP and SICP.
3. Open the choke, start the pump, and bring its rate up to the original kill rate while keeping SICP constant.
4. When the pump is up to kill-rate speed, note SIDPP; it now becomes the new circulating pressure. If it appears that this pressure is too high for the pumps to handle, pick a new kill rate at a slower pump rate and repeat the steps.

OTHER PROBLEMS

In addition to a hole in the drill string and a plugged bit, other unusual situations can occur. An alert crew can often identify the problem by noting its effect on SIDPP, SICP, drill string weight, pit level, and pump rate. Table 1 shows several problems and their effects, as well as what occurs when a gas kick reaches the surface. As an example, note that a washed out bit nozzle causes SIDPP to decrease and the pump rate to increase. It has no effect on SICP, drill stem weight, or pit level.

Table 1
Effect of Various Problems on Pressures, Weight, Pit Level, and Pump Rate

Problem	Drill Pipe Pressure	Casing Pressure	Drill Stem Weight	Pit Level	Pump Rate SPM
Choke washes out	decreases	decreases	no change	increases	increases
Choke plugs	increases	increases	no change	no change	decreases
Gas reaches surface	decreases	decreases	decreases	decreases	no change
Loss of circulation	decreases	decreases	increases	decreases	increases
Hole in drill stem	decreases	no change	no change	no change	increases
Pipe parted	decreases	no change	decreases	no change	increases
Bit nozzle plugs	increases	no change	no change	no change	decreases
Bit nozzle washes out	decreases	no change	no change	no change	increases
Pump damage or gas-cut mud	decreases	no change	no change	no change	decreases
Gas feeding in	no change	increases	increases	increases	no change
Hole caved in	increases	no change	stuck	decreases	decreases

PRESSURE BETWEEN CASING STRINGS

When two or more casing strings exist in a well, and formation gas enters the wellbore, it is possible for the gas to become trapped in the annular space between the strings. Suppose, for example, that an intermediate (protection) liner is hung inside surface casing. Further, suppose that the liner hanger packing does not make a good seal between the liner and the casing in which the liner is hung. In such a case, formation gas can leak past the hanger's packing and enter the annulus of the surface casing.

A poor cement job can also allow pressure to become trapped between two casing strings. Cement voids around the casing shoe of the second string that extend upward, can allow formation gas to enter the annulus between the two casing strings. Also, a hole worn into a string of casing that lies inside another casing string can allow formation gas to become trapped inside the annulus between the two strings.

In whatever way kick fluids enter the annulus between casing strings, they become trapped there as long as the annulus valve at the top of the casing is closed. The danger comes when it is necessary to open the annulus valve to nipple down the BOP, set a new casing string, and the like. If the annulus valve is opened and gas is trapped behind it, the gas escapes suddenly with great force, which can harm or kill personnel and damage the rig. Crew members should therefore open the annulus valve slowly and carefully to bleed off any gas trapped in the casing annulus.

PUMP FAILURE

If a pump fails during the time a kick is being circulated out of the well, it is usually simply a matter of changing to a backup pump. Most operators require crew members to obtain kill rate pressures for all pumps on the rig, in case a pump fails. During the time required for changing to another pump, any gas influx will continue to rise up the hole, so it is important to maintain close surveillance of the casing pressure. If it approaches the maximum allowable shut-in pressure, it may be necessary to bleed pressure, using the volumetric method described in chapter 6. Recall that the volumetric method maintains constant bottomhole pressure by bleeding small measured amounts of mud from the annulus to control *SICP*. If all pumps fail, crew

members can use the volumetric method to maintain control of the well while pumps repairs are carried out. Also, keep in mind that anytime a pump or other equipment malfunctions or fails, the foremost concern of the rig crew is to do everything possible to keep the well under control at all times.

BOP FAILURE
Flange Failure

Failure of the blowout preventer system can occur for many reasons. For example, a flange seal (the pressure-tight seal at the bottom of the annular BOP and on top and bottom of the ram BOPs) can fail, resulting in a high-pressure stream of fluid exiting with great force at the failure point. At the same time, back pressure on the well is reduced, and additional kick fluids can enter the well. If the failure occurs at the annular BOP's flange, one solution is to close the pipe ram BOP (assuming that drill pipe is in the hole) after ensuring that the rams will close on the body of the pipe and not on a tool joint. Another possible solution is to pump a graded sealant into the wellhead and then bullhead down the annulus. Should the bottommost BOP be closed and flange failure occur there, then one solution may be to drop the pipe into the hole and close the blind rams. If the blind rams fail to hold, one possible last resort is to pump cement to plug the well. Most operators and contractors prepare emergency response plans (ERPs) for such possibilities; rig crews should follow their ERP.

Weephole Leakage

Most ram BOPs in a surface BOP stack have weepholes. When the main seal of the ram shaft fails, hydraulic fluid leaks from the weephole. Because a failed shaft seal can lead to failure of a positive closure of the rams around the pipe or on open hole, manufacturers provide weepholes to alert crew members to the problem. Manufacturers also provide a temporary way in which to repair the leak, because the leak could occur when the preventer is shut in on a kick. Usually, the BOP has a hex screw located at the weephole, which, when tightened, injects sealant to stop the shaft seals from leaking. Therefore, during a well-control situation, if leakage from a BOP's weephole is noted, a crew member should use the proper hex wrench and screw in the weephole packing to stop the leak.

Tightening the weephole packing is only a temporary measure. After the well is killed and routine operations resume, crew members should repair the preventer shaft seals.

Failure of BOPs to Close

When the BOPs fail to close, the chances are good that the hydraulic BOP operating unit (accumulator) has malfunctioned. Virtually all operating units have nitrogen precharged accumulator bottles, electric pumps, and pneumatic pumps that move hydraulic closing fluid through lines to the stack. Crew members should therefore be certain the charging system is operating as it should. If it is not, manual means of closing may be required. The ram preventers in some surface stacks can be closed by turning large wheels; a pipe ram can thus be closed around the pipe in spite of applying maximum closing pressure. It takes longer to close the BOP manually than with a hydraulic operating unit, so the kick influx will be large. If the BOPs cannot be shut manually, it may be necessary to manifold a high-pressure test pump or a cement pump to the stack's closing lines. By connecting a test or other type of pump to the closing unit's hydraulic system, it may be possible to close the preventers.

Failure of BOP Seals

If, upon closing a preventer, the packing in the preventer fails to make a good seal around the drill pipe or, in rare cases, on open hole, immediate steps should be taken to close the well in completely. For example, suppose a kick is detected and crew members close the annular preventer, only to discover that the annular packing element is not properly sealing around the pipe in spite of applying maximum closing pressure. In this case, after ensuring proper space out, the pipe ram preventers should be closed. Once the ram preventers are closed, control of the well is maintained and, if required, the failed packing element in the annular preventer of a surface stack can be replaced.

FLOW PROBLEMS DOWNSTREAM FROM CHOKE

In the event that the flow line downstream from the choke being used to control the well plugs or otherwise becomes unusable, it will be necessary to switch to a backup choke and choke line. Because of the possibility of choke malfunction or difficulties in the line downstream from a choke, most operators and contractors install backup chokes in the manifold. Because of the difficulty of changing failed seals in subsea BOP stacks, more BOP elements (often including a second annular preventer) are often available. Ideally, another remote-adjustable choke will be installed because they are very convenient to use in maintaining correct backpressure on the well while the kick is being circulated out of the hole. If a remote adjustable choke is not installed, it will be necessary to use a manual choke.

Also, if the well is being circulated through a mud-gas separator, it is possible for the line to the separator to become plugged. If this line becomes plugged, crew members may have to redirect the flow into the flare line and completely bypass the line to the mud-gas separator.

PRESSURE GAUGE FAILURE

Although rare, surface pressure gauges can malfunction or fail. For this reason, most rigs have several gauges that personnel can use to read shut-in drill pipe pressure, shut-in casing pressure, and pump circulating pressure. It is important to remember, however, that when changing from one gauge to another, it is necessary to take new readings because of gauge variation. For example, when changing from one pump pressure gauge to another, personnel should determine and record the pump pressure at the reduced circulating rate using the new gauge. In the same way, new readings should be recorded for drill pipe pressure and casing pressure gauges if they are involved.

ANNULUS BLOCKED

If the annulus becomes completely blocked (packed off) while a kick is being circulated, it is impossible for the mud and kick fluids to exit the well from the annulus. Should annular pack off occur, one possible action is to perforate the drill pipe at a depth above the pack off. After determining the mud weight required to kill the well at the depth of the perforations, the mud is circulated through the perforations and back to the surface. With the well under control, it may be possible to wash over the plug in the annulus and reestablish full circulation.

PIPE OFF BOTTOM

When a well kicks during a trip, an error in procedure has occurred. When tripping out, mud must be put into the well to replace the drill stem, and the hole must take the proper amount of fill-up mud. If formation fluids enter the hole during a trip out, the hole will not take enough mud. Similarly, during a trip in, the proper amount of mud must be displaced from the well. Moreover, when no pipe is in the hole and formation fluids enter it, mud will flow from the well. An alert crew, therefore, will recognize that an influx has occurred and take steps to prevent further intrusions.

If a well kicks while a trip is being made, pressure-control measures can be more difficult, but the quicker the reaction to the problem, the less difficult the solution will likely be. A full-opening drill pipe safety valve, inside BOP, operating wrenches, and the proper crossover subs should be available immediately on the rig floor. Further, all equipment should be in good working condition and be placed on the floor where it can be installed quickly and correctly. Before the safety valve is installed, the annulus should be open. If the annulus is partially or completely closed, fluids from inside the drill stem will likely flow at such a high rate that it will be difficult or impossible to stab a safety valve.

If the well kicks with the pipe off bottom, and if pipe can be stripped back to bottom, then the kick can be controlled with the mud weight in use when drilling before the trip. If the pipe cannot be stripped back to bottom, then a higher mud weight will be needed to kill well pressure with the shortened drill string. It is important that *TVD* to the bit and not *TVD* to the bottom of the hole be used when calculating the mud weight needed to kill the well. In many cases, the extra-heavy mud needed to kill the well with the shortened string may be enough to cause lost circulation. In any event, care must be taken to anticipate the consequences of moving volumes of original- and kill-weight muds when pipe is later run back into the hole.

PIPE OUT OF THE HOLE

If a kick occurs with the drill stem out of the hole, most operators recommend that the well be shut in immediately and preparations be made for stripping or snubbing the pipe back into the hole. During these preparations, it is also usually recommended that *SICP* be noted and recorded every 15 min. If *SICP* rises, which is likely if gas is in the kick fluids and migrates up the hole, the problem is aggravated.

Migration of gas to the surface can cause an excessive increase in bottomhole pressure unless the gas is allowed to expand. The volumetric method of well control can be used to control casing pressure by bleeding fluid from the well to exactly compensate for increasing casing pressure.

A FLOAT IN THE DRILL STEM

A float, or back-pressure, valve can be installed in the drill stem to prevent kick fluids from entering it. It is usually installed between the bit and the drill collar (fig. 7.1). One problem with drill pipe float valves is that if a kick is experienced and the well is shut in, *SIDPP* may read zero; or *SIDPP* may actually indicate some pressure. Regardless of the reading, however, it is not reliable, because accurate pressure indications from below cannot get through the closed valve. (Available from several manufacturers are float valves with special ports that may allow *SIDPP* to be read without opening the float valve.) Since *SIDPP* is essential to most well-control procedures, it is necessary to determine its value. Several methods have been used to overcome the problem; the following is one:

1. Rig up a cement pump to pump mud into the drill stem.

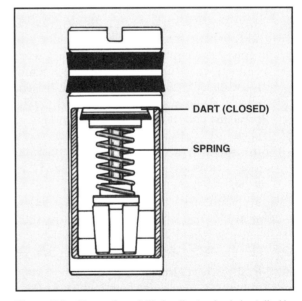

Figure 7.1   Normally, a drill pipe float valve is installed in a special sub, or float body, just above the bit. Circulating pressure overcomes spring pressure to keep the dart open. When circulation stops, the dart springs closed.

2. While holding *SICP* constant with the choke, pump as slowly as possible and keep a close watch on *SIDPP*. It will rise to a certain value and then stop rising. When it stops rising, stop the pump. The pressure noted after the pump is stopped should be *SIDPP*.

Other methods of determining *SIDPP* use the mud pump. One such method is to pump as slowly as possible until *SICP* starts to increase, then stop pumping. The pressure indicated on the drill pipe pressure gauge should be *SIDPP*.

STRIPPING AND SNUBBING OPERATIONS

To kill a well properly, the drill stem must be at or near the bottom. With pressure at the surface, it may become necessary to run the drill stem into or remove it from the well under pressure. This action is called *stripping*. When well pressure exerts so much upward force that the weight of the string is not sufficient to allow it to be stripped into the wellbore, then snubbing becomes necessary. Snubbing requires the use of special equipment to force the pipe through the preventer or preventers used in the stripping operation.

Preparing for Stripping

Before any stripping operation begins, thorough preparations should be made to reduce the chances of error. The following procedure has been used successfully to prepare for a stripping job:

1. Reduce closing pressure on the annular preventer to minimum sealing pressure. Except with subsea stacks, minimum sealing pressure is usually determined by allowing the preventer to weep fluid between the drill stem and the preventer packer. With subsea stacks, a table of operating characteristics for the preventer in use must be employed to determine minimum sealing pressure. (See tables B.15, B.16 in Appendix B for examples.)

2. Record *SICP*.

3. Make sure that an inside BOP or an inside BOP and a drill pipe safety valve are available in good working order and in full open position.

4. If pipe is to be stripped out of the hole, install a back-pressure valve, or float, in the lower section of the string.

5. Remove all drill pipe casing protectors (rubbers) from the string before attempting to casing protectors strip in.

6. Rig up to use a hand-adjustable choke whether stripping in or out.

7. Use a trip tank for accurate measurement of mud volumes bled from or added to the well.

8. If stripping in, calculate the amount of mud to bleed from the well as the volume of drill pipe replaces the volume of mud in the hole. Remember to use the closed-end displacement of the pipe being run.

9. Be prepared to fill the drill string with mud periodically.

10. When the stripping operation involves the use of two preventers, the distance between them must provide sufficient clearance for tool joints. Moreover, when stripping in, bear in mind that the first joint stripped in will have an inside BOP and tool joint, or an inside BOP, drill pipe safety valve, and tool joint that must fit between the preventers.

11. Be aware of company policies in reference to stripping operations. Senior personnel may have to make decisions based on such policies and they must be prepared to adhere to them.

12. If the drill pipe has rubbers installed, consider carefully whether to strip out. Problems could arise if the rubbers strip off or accumulate under the preventers.

13. Since the life of the packing element in an annular preventer can be extended by limiting the maximum well pressure imposed on it, many companies set such limits during stripping operations. Some operators use a limit of 2,000 psi maximum well pressure for stripping; however, recent tests reveal that adequate performance is obtained from annular BOPs exposed to 3,000 psi during stripping operations. In any case, the crew should be aware of policy limits and adhere to them.

Stripping into the Hole

If the rig crew does not fully understand the stripping process and its limitations, then stripping into the hole can be a hazardous operation. Yet workover rigs with a minimum of equipment and small crews routinely strip in and out of a well with no difficulty. Therefore, senior personnel on a drilling rig have the responsibility of explaining to the crew exactly what they are, and are not, to do during a stripping job.

Stripping in with Annular Preventer

Stripping into the hole using the annular preventer is not difficult, but several recommendations should be kept in mind:

1. The pressure-regulating valve in the annular BOP system is designed so that hydraulic fluid can pass through it in two directions. Fluid flows through the valve and to the preventer to operate the preventer. Then, to allow the preventer to open slightly when a tool joint passes through it, fluid is reversed and flows back through the valve (fig. 7.2). Therefore, the pressure-regulating valve must be in good operating condition. Also, the lines from the valve to the annular BOP should be large enough to allow fluid to flow with a minimum of restriction. On subsea stacks, an accumulator bottle

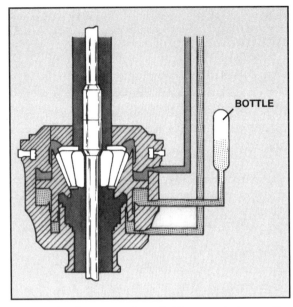

Figure 7.3   An accumulator bottle mounted near the annular preventer allows operating fluid to pass back into the bottle to facilitate stripping tool joints.

can be installed near the preventer to allow fluid to pass back into the bottle freely (fig. 7.3). If stripping with the annular BOP is part of the company's policy, a stack-mounted accumulator bottle should be considered.

2. As stated before, use the lowest possible closing, or operating, pressure on the annular preventer. Low closing pressure helps prevent wear on the packer. The operating pressure should be reduced until the annular BOP weeps when the pipe is being stripped through it. On subsea stacks, a table of operating characteristics (available from the manufacturer or in Appendix B) must be used to obtain the closing pressure.

3. Keep water or oil on top of the packer as a lubricant.

4. Well pressure can be so high that it pushes pipe out of the hole or prevents it from being stripped in without a pull-down, or snubbing, device. To strip into the hole with an annular preventer, the weight of the drill stem must be greater than the pressure exerted upward against the tool joints by annular pressure. An equation is available that can be used to estimate whether the drill stem weighs enough to be stripped into the hole:

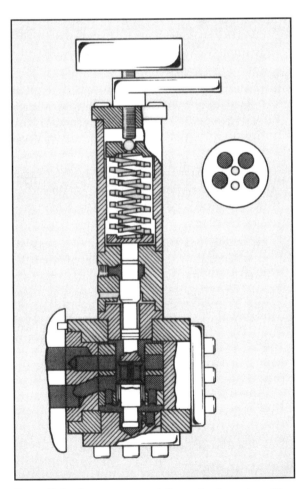

Figure 7.2   A pressure-regulating valve in vent position allows a tool joint to pass through the preventer easily.

$$WBF = (OD_{dp})^2 \times 0.7854 \qquad \text{(Eq. 60)}$$
$$\times SICP + F$$

where

WBF = wellbore force, lb

OD_{dp} = outside diameter of drill pipe, in.

$SICP$ = shut-in casing pressure, psi

F = friction factor, 1,000 lb.

As an example, assume that—

OD_{dp} = 5 in.

$SICP$ = 750 psi.

Therefore:

$$WBF = 5^2 \times 0.7854 \times 750 + 1,000$$
$$= 25 \times 0.7854 \times 750 + 1,000$$
$$WBF = 15,726 \text{ lb.}$$

The example problem shows that the buoyed weight of the string must be greater than 24,008 lb if it is to be stripped into the hole successfully through the annular preventer.

The length of pipe required to achieve sufficient weight to overcome wellbore force can also be calculated. However, since the hole is full of mud, and since the drill stem tends to float in a fluid-filled hole, the buoyancy of the pipe in the hole must also be considered, because it adds an additional upward force against which the pipe must be stripped. Following is a formula for determining the buoyancy factor:

$$BF = (65.5 - MW) \div 65.5 \qquad \text{(Eq. 61)}$$

where

BF = buoyancy factor, dimensionless

MW = mud weight, ppg.

As an example, if 12-ppg mud is in the hole, then,

$$BF = (65.5 - 12) \div 65.5$$
$$BF = 0.82.$$

With the buoyancy factor and wellbore force known, the length that the drill pipe string must be to be stripped in can be calculated with the following formula:

$$L_{dp} = WBF \div (W_{dp} \times BF) \qquad \text{(Eq. 62)}$$

where

L_{dp} = length of drill pipe string, ft

WBF = wellbore force, lb

W_{dp} = weight of drill pipe, pounds per foot (ppf)

BF = buoyancy factor, dimensionless.

As an example, if the drill pipe weighs 20.9 ppf, then,

$$L_{dp} = 24,008 \div (20.9 \times 0.82)$$
$$= 24,008 \div 17.14$$
$$L_{dp} = 1,401 \text{ ft.}$$

Therefore, in the example, the drill pipe string must be at least 1,401 ft long to achieve a sufficient weight for stripping through the annular preventer.

If it appears that pipe can be stripped, the following set of procedures can be used to strip pipe in with the annular preventer:

1. Once again, be sure that closing pressure on the annular BOP is adjusted so that it weeps fluid when pipe is going into the hole. On rigs with subsea stacks, use the value from the preventer's operating characteristics table.

2. Run the pipe no faster than about 1 ft/sec. Run it more slowly when a tool joint passes through the preventer. With subsea stacks, measurements are required to determine when a tool joint is passing through the BOP. If the vessel is heaving a great deal, some operators recommend that running speed not be slowed to avoid reversing the direction of the tool joint in the preventer.

3. Maintain constant $SICP$ with the choke as pipe goes into the hole. Mud displaced from the hole can be measured and corrections made to get exact annular pressure changes as stripping continues. Equations 58 and 63 can be used for calculating the volumetric displacement correction to maintain constant $SICP$:

$$P_{bbl} = (MW \times 0.052) \div AV \qquad \text{(Eq. 58)}$$

where

P_{bbl} = pressure per bbl of mud, psi

MW = mud weight, ppg

AV = annular volume, bbl/ft.

As an example, if—

MW = 15 ppg

AV = 0.045 bbl/ft,

then,

$$P_{bbl} = (15 \times 0.052) \div 0.045$$

$$= 17.33 \text{ psi/bbl.}$$

$$\Delta SICP = PV_c \times P_{bbl} + SICP_i \quad \text{(Eq. 63)}$$

where

$\Delta SICP$ = new casing pressure, psi

PV_c = pit-volume change, bbl

P_{bbl} = pressure per bbl of mud, psi

$SICP_i$ = initial shut-in casing pressure, psi.

As an example, if—

PV_c = 20 bbl

P_{bbl} = 17.33 psi

$SICP_i$ = 1,000 psi,

then,

$$\Delta SICP = 20 \times 17.33 + 1,000$$

$$= 1,346.6 \text{ psi.}$$

For many stripping jobs, simply holding *SICP* constant with the choke should be adequate. Because gas migrates up the hole, however, a correction may be needed; if so, these equations or similar ones can be used to calculate the correction.

4. Every stand of pipe stripped into the hole should displace mud; if not, circulation has probably been lost. As pipe is stripped into the hole, the fluid in the hole gains in height because the pipe displaces the fluid. Since the volume of the hole and the pipe can be determined, as well as the displacement of the pipe, it is not difficult to calculate the gain in the fluid's height. One equation that can be used follows:

$$h = L \times (C_{dp} + D_{dp}) \div AV \quad \text{(Eq. 64)}$$

where

h = height gain, ft

L = length of pipe stripped, ft

C_{dp} = drill pipe or drill collar capacity, bbl/ft

D_{dp} = drill pipe or drill collar displacement, bbl/ft

AV = annular volume, bbl/ft.

As an example, assume that 2,500 ft of 5-in. 19.5 ppf drill pipe with 6⅜-in. tool joints is stripped into an influx in a 12¼-in. hole. How much will the influx gain in height? To solve the problem,

first use tables B1 and B2 in Appendix B to find the drill pipe's capacity and displacement. In this example, the pipe's capacity is 0.01776 bbl/ft and its displacement is 0.00750 bbl/ft. Next, determine the annular volume with 5-in. drill pipe in it. Annular volume equals hole diameter squared, minus pipe diameter squared, divided by 1,029.4. In this case, it is $12.25^2 - 5^2 \div 1,029.4 = 150.06 - 25 \div 1.029.4 = 125.06 \div 1,029.4 = 0.1215$ bbl/ft. With annular volume known, use equation 64 to find the solution. Thus —

$$h = 2,500 \times (0.01776 + 0.00750) \div 0.1215$$

$$= 2,500 \times 0.0253 \div 0.1215$$

$$= 63.25 \div 0.1215$$

$$h = 521 \text{ ft.}$$

5. If *SICP* does not stop rising even though mud displacement stops between stands, use the volumetric correction equations (equations 58 and 63).

Stripping in with Ram Preventers

Stripping into the hole using ram preventers requires good judgment and careful measurements. Ram BOPs can be used for stripping if the pressure in the annulus is too high to strip tool joints through the annular BOP, if rubbers on the drill pipe cannot be removed, or if the annular preventer is inoperable or unavailable. An estimate of how much the string has to weigh to be stripped into the hole successfully with ram BOPs can be made by using equations 60, 61, and 62. When determining the length of pipe required to make the proper stripping weight, however, a modification to the formula is needed. Since rams close on the body of the pipe instead of the tool joints, and since the tool joints cannot be stripped through closed rams, drill pipe OD should be used rather than tool joint OD. For example, with 5-in. 20.9-ppf drill pipe, an *SICP* of 750 psi, and a mud weight of 12 ppg, the lightest the string could weigh and still be stripped in with the ram preventers can be calculated using the following equation:

$$WBF = (OD_{dp})^2 \times 0.7854 \times SICP + F \quad \text{(Eq. 65)}$$

where

WBF = wellbore force

OD_{dp} = OD of drill pipe, in.

$SICP$ = shut-in casing pressure, psi

F = friction factor, 1,000 lb.

As an example,

$$WBF = 5^2 \times 0.7854 \times 750 + 1,000$$
$$= 25 \times 0.7854 \times 750 + 1,000$$
$$WBF = 24,873 \text{ lb.}$$

With 12-ppg mud in the hole, the buoyancy factor (using equation 61) is—

$$BF = (65.5 - 12) \div 65.5$$
$$BF = 0.82.$$

With drill pipe that weighs 20.9 ppf, the minimum length required (use equation 62) is—

$$L_{dp} = 24,873 \div (20.9 \times .82)$$
$$= 24,873 \div 17.14$$
$$L_{dp} = 1,451 \text{ ft.}$$

Note that the length and weight required for stripping with ram preventers are less than required to strip through the annular preventer, because the upward wellbore pressure is acting against the cross-sectional area of the pipe rather than the tool joints; since the cross-sectional area of the pipe is smaller than the cross-sectional area of the tool joints, wellbore pressure is acting on a smaller area and thus is lessened.

If it is determined that pipe can be stripped using the ram preventers, the following procedure can be used:

1. Select the two rams to be used and measure from the rotary table to the top of the upper ram and to the top of the lower ram. (An annular preventer can be used in place of the top set of rams.)
2. Reduce the closing pressure on the rams to 500 psi or less.
3. With the upper ram closed, lower a joint of pipe slowly while measuring it until the tool joint is 2 ft above the upper ram. (On floating rigs, the distance must also be great enough to allow for vessel heave.)
4. Stop lowering and close the lower pipe ram.
5. Bleed off the pressure between the upper and lower rams and open the upper ram.
6. Carefully measuring the joint, lower it until the tool joint is between the two rams.
7. Stop lowering and close the top ram.
8. Using a test pump, pressure up the space between the two rams to the same value as well pressure. Open the bottom ram.
9. Continue the stripping process by going back to step 3 and repeating the steps.

During the stripping operation, maintain constant *SICP* by bleeding mud though the choke. The mud displaced from the hole by the pipe can be measured and corrections made to get the exact annular pressure changes as pipe is stripped in. For most stripping jobs, holding casing pressure constant should be adequate; however, migrating gas may require corrections.

During strip in, every stand of pipe should displace mud; mud displacement and the rise in pressure should stop when no stand is being stripped. If a stand does not displace mud, circulation has been lost. If the pressure does not stop rising and mud displacement stops between stands, use a volumetric correction equation (such as equation 58 or 63).

Stripping Out of the Hole

Stripping out of the hole follows the same general procedures as stripping into the hole; however, a drill pipe float or a pump-down inside BOP is necessary to seal the pipe before coming out of the hole. Stripping out of the hole with a gas kick should be carefully reviewed before a decision is made to proceed; indeed, most operators do not recommend stripping out with gas in the wellbore.

Snubbing

When the upward force that is generated by wellbore pressure acting on the cross-sectional area of the tool joints or drill string is greater than the weight of the drill string, snubbing equipment should be rigged up to force the pipe into the well through the preventers. Equations 60, 61, and 62 can be used to confirm how much pipe will have to be snubbed before stripping operations can commence. Before snubbing operations begin, thorough preparations should be made to ensure that all of those involved in the operation know their duties and positions. A review of the operator's and contractor's procedures is essential and all equipment must be in good working order. In general, the same preparations that are made for stripping operations should be made for snubbing.

Two general types of snubbing, or pull-down, units are available: mechanical and hydraulic. Whether a mechanical or a hydraulic unit is used, the usual procedure is to snub pipe into the hole until the pipe's weight is sufficient to allow stripping operations to begin. Usually, the pipe will start to fall through the snubbers by means of its own weight.

Once the pipe begins to fall of its own weight, stripping procedures can be followed. Keep in mind that comparatively high casing pressure will continue at the surface.

Mechanical Snubbing Units

Available in several sizes, mechanical units are designed to use the hoisting equipment on the rig. The smallest units are capable of exerting about 50,000 lb of force. Larger sizes range upward to units capable of exerting 350,000 lb of force. One type relies on rig power to snub pipe in or out of the hole through a system of pulleys and cables controlled by the rig's drawworks. Basic components of a typical mechanical unit are the blowout preventers, or control heads; stationary and traveling snubbers; operating manifold; power package; snub line; and balance weights (fig. 7.4).

Downward thrust to force pipe into the hole against well pressure is achieved by means of the pulley system. Raising the traveling block causes the traveling snubbers that grip the pipe to move down and pull pipe into the hole. After each downward stroke, the stationary snubbers attached to the top control head grip the pipe until the traveling snubbers are raised or lowered to grab another portion of the pipe. Flow around the pipe is shut off by three hydraulically controlled control heads. The two upper heads are opened and closed to lubricate pipe in or out of the hole. The bottom head is closed to change packing in the upper heads. Drill pipe must be plugged by use of a landing nipple and plug assembly, slip-type plug, or bridge plug, depending on whether pipe is being run in or pulled out of the hole.

Hydraulic Snubbing Units

Most hydraulic units are self-contained and thus are operated with or without the rig's being in place on the well. One such unit features a blowout preventer, or control-head, stack similar to that used on a mechanical unit. A multicylinder hydraulic jack raises or lowers traveling slips that grip the pipe and snub it into or out of the hole. Stationary slips below the traveling slips hold the pipe in place while the traveling slips are being repositioned on the pipe prior to another pull. An integral stripper controls pressures up to 3,000 psi. The control heads are used if pressures higher than 3,000 psi are involved and as a backup to the stripper. All operations are carried out at a control console in the work basket (fig. 7.5).

PIPE RECIPROCATION DURING A WELL KILL

Most operators and contractors stress that the first concern during a well killing operation is to gain control of the well first and worry about other problems, such as sticking the drill stem, later. If it becomes necessary, however, to reciprocate (move up and down) the drill stem during a well killing operation, it is important to pay attention to detail. If the annular preventer is closed on the well, the lowest possible closing pressure should be used to prevent as much wear as possible to the packing element. The same holds true for ram preventers. Also, the weight of the drill stem must be heavy enough to overcome the upward force of well pressure; otherwise, it is not possible to strip the drill stem downward into the well—it will have to be snubbed (see the earlier discussion of stripping and snubbing in this chapter). Further, moving the drill stem when the well is closed in on a gas kick is not without risk, because gas continues to migrate upward and increase *SICP* during the period of pipe reciprocation.

LOST CIRCULATION

Lost circulation is a condition in which whole mud is lost to a formation. Well kicks cause additional pressure in the hole, so special care should be taken to avoid or to minimize lost circulation during a well kick. Most well-control procedures are designed for the purpose of circulating heavy mud to kill the kick. If circulation is lost, it can be difficult or impossible to circulate the annulus full of heavy mud. When a well is shut in after a kick, *SIDPP* is used to calculate the mud-weight increase needed to kill the kick. Shut-in pressures can also indicate the likelihood of lost circulation. The way to find out if circulation has been lost is to attempt to circulate mud; if returns are reduced or fail to come back to the surface at all, it is safe to assume that circulation is lost.

Conditions for Lost Circulation

In general, three conditions are responsible for lost circulation: bad cement jobs, induced fractures, and vuggy or fractured formations.

Bad Cement Jobs

One of the most common causes of lost circulation during a well kick is a bad cement job at the base of the last string of casing run into the well. Because a bad cement job can cause lost circulation, most operators

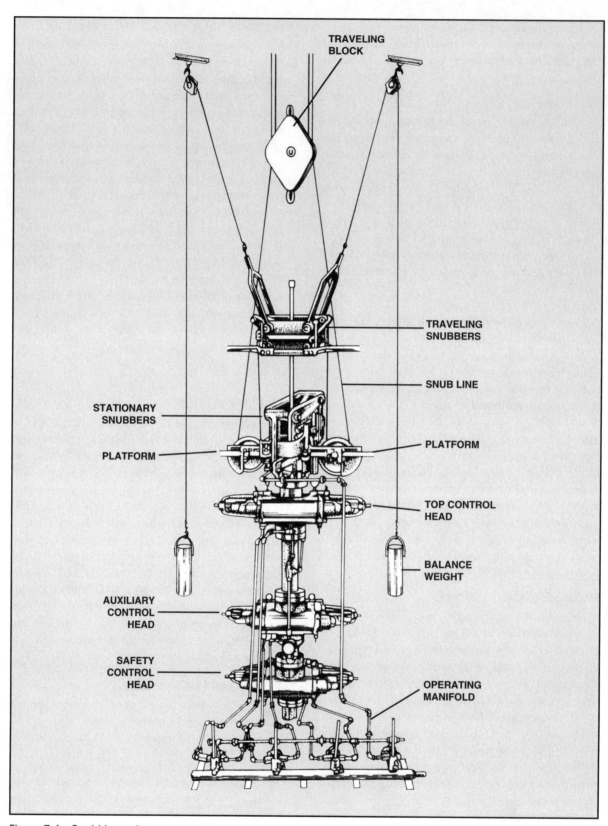

Figure 7.4   Snubbing unit

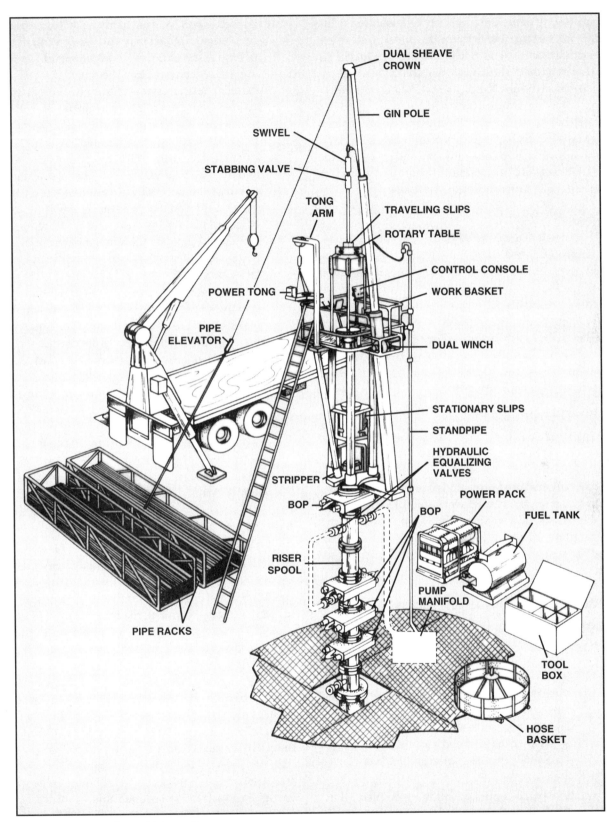

Figure 7.5   Hydraulic snubbing unit

run leak-off or pressure-integrity tests just after drilling out the shoe to determine the pressure at which fracture occurs. The test is usually conducted to determine the highest mud-weight equivalent that is expected to be used before the next string of casing is set; sometimes the test is carried out to pressures set by statutory requirement. In any case, a poor cement job is especially hazardous, because it may allow kick fluids to broach around the casing and under the rig. Major losses both on land and offshore have occurred because gas surfaced around the casing.

Induced Fractures

Fractures that are induced by drilling and well-control procedures can also cause lost circulation. Such fractures can be caused by pressure surges, mud weight that is too high, or other well-control procedures. In most cases, induced fractures close, or heal, on their own in a short period of time if pressure is relieved. Induced fractures occur at the weakest point in the hole—usually at the casing shoe; therefore, an induced fracture can cause the same problems associated with a poor cement job.

Vuggy or Fractured Formations

When drilling in hard-rock country, some formations are vuggy; that is, they have large natural openings into which vast amounts of whole mud can flow. Naturally fractured formations can also take mud in high quantities. Such formations are difficult or impossible to seal and a waiting period often does not help. Many times, the entire formation is vuggy or fractured, so that the pressure required to kill the well is very close to the pressure that causes lost returns.

Well Control with Partial Lost Returns

Lost returns during a well kick can first be detected when mud level in the pits drops. If some returns continue to come back to the surface, several techniques are recommended by well-control personnel:

1. After notifying the supervisor, crew members can try to keep the mud volume up by mixing. The pressure on the zone of lost returns will go down after the intruded kick fluids are circulated above the zone, so the problem may solve itself.

2. Some well-control operators recommend that the pump be stopped and the well shut in if the lost circulation continues to worsen. If the hole is given from 30 min to 4 hr to sit quietly, the lost circulation problem may cure itself. With this technique, most operators recommend keeping *SIDPP* constant by relieving choke pressure. If choke pressure goes up by more than 100 psi, some operators recommend that the next technique be tried.

3. Pick a slower circulating rate and a new initial circulating pressure. With the pump stopped and the well shut in, open the choke, start the pump at the new, slower rate, and close the choke until annular pressure is the same as when shut in. (With subsea wellheads, reduce annular pressure by the amount of choke-line friction loss.) Then shift to the new initial circulating pressure on the drill pipe.

4. Mix a slug of lost circulation material that is effective in the area. In general, lost circulation material is more effective in hard-rock country than in areas where the rocks are plastic.

5. Mix a slug of heavy mud to try to kill the kick. A heavy mud slug may work with a small kick if the zone of loss is well above the zone that is kicking. After the kick is killed, solve the lost circulation problem.

6. If severe partial returns (60% to 90% loss) cannot be stopped, use a barite or gunk plug to seal off the kick zone, then work on the lost circulation problem.

Barite Plugs

Sometimes, after a kick has occurred and is being circulated out of the well, lost circulation occurs. Lost circulation can present a complex and dangerous condition in that gas could be replacing drilling mud (drilling mud is, of course, the first line of defense against blowouts). If a plug can be placed in the wellbore to seal off the zone that is giving up gas, the lost circulation problem can be worked on.

Barite plugs with weights of 18 ppg to 22 ppg have been found to be effective in controlling active zones. A barite plug usually consists of barite, fresh water, and a thinner slurry. The plug is spotted as close to the active zone as possible. A barite plug should have high density—up to 22 ppg—a rapid settling rate, and a high filter loss. Finally, it should be large enough to fill about 500 ft of open hole. By increasing the density of the mud up to 22 ppg using barite, the increase in hydrostatic pressure may control the pressure in the formation. The plug of barite, sodium acid pyrophosphate (SAPP), and caustic soda should be mixed in fresh water. Barite is the weighting material, SAPP is a thinner, and caustic raises the pH of the water. The components are mixed to form the 18 ppg to 22 ppg slurry. Hydrostatic pressure must be sufficiently high to stop the flow; otherwise, the barite cannot settle out to form a solid plug. Materials with densities higher than barite, such as hematite, ilmenite, and galena, can also be used to mix plugs. Even though cement can be used, it may be difficult to set a plug that will hold, because gas influxes tend to channel cement.

To spot a barite plug successfully, special mixing and pumping equipment, such as a cement hopper and a high-capacity cementing pump, are required. Pilot testing should also be conducted to ensure that the quantities of SAPP and caustic added to the slurry are sufficient to allow the barite to settle out properly. If the barite settling rate is too slow, more SAPP can be added to speed the settling rate. The following is one recommended procedure for mixing a barite plug:

1. Prepare to mix the barite slurry through the hopper of the cementing unit and to pump directly into the drill pipe.
2. Calculate the volume required to yield a settled plug of barite 500 ft long in open hole. If necessary, increase the volume to allow for severe hole washouts (see table 2).
3. Mix about 0.7 lb of SAPP per bbl of fresh water.
4. Adjust the pH of the fresh water to 9 with caustic soda; use about 0.25 lb of caustic for every bbl of water.

5. Mix the barite plug to achieve a slurry weight of 18 ppg to 22 ppg, preferably 22 ppg.

Once the slurry is mixed, the following suggestions for pumping it may be employed:

1. Pump the barite slurry at a rate of 5 bbl/min to 10 bbl/min.
2. Underdisplace the slurry mixture by about 2 bbl to 4 bbl to avoid contamination by drilling fluid.
3. Displace the barite slurry out of the drill stem with a high-density slug to reduce the possibility of backflow and bit plugging.
4. Rapidly pull one stand of pipe; continue tripping pipe out of the hole until the bit is above the top of the barite plug. It may be necessary to strip the pipe out.
5. Hold back-pressure on the annulus.
6. When the bit is above the barite plug, begin circulating while continuing to hold back-pressure on the well.
7. Wait for the barite to settle and form a solid plug. The length of the wait depends on the additives in the slurry.
8. To determine whether the plug is holding, circulate the well using the first circulation of the driller's method and check the returns. If returns are free of gas, the high-pressure zone has very likely been properly sealed off by the barite plug. Once the underground flow problem is solved, attention can be turned to solving the lost circulation problem.

Table 2
Barite Plug Yields

| Slurry Density (ppg) | Water (gal/sk) | Sk Barite/ Bbl Slurry | Bbl/Sk | Slurry Yields | | | Ft³/Sk |
				Bbl/200 Sk	Bbl/300 Sk	Bbl/400 Sk	
18.0	5.10	5.30	0.189	37.8	56.2	75.5	1.060
20.0	3.70	6.43	0.156	31.1	46.6	62.1	0.873
21.0	3.20	6.95	0.144	27.8	43.2	57.5	0.807
22.0	2.75	7.50	0.133	26.6	40.0	53.3	0.748

Gunk Plugs

For underground water flows, gunk plugs have been successful in providing a seal in the annulus. A *gunk plug* is a mixture of diesel oil and bentonite. When dry bentonite is added to diesel oil, the bentonite does not yield, and the slurry remains very fluid; thus it can be pumped to the bit with relatively low pressure. When the slurry leaves the bit and is exposed to the water in the annulus, the bentonite hydrates, or swells, rapidly and causes the slurry to become extremely viscous. Its extreme viscosity slows formation flow and, as more slurry enters the annulus, seals completely.

One recommendation is that the bentonite-diesel oil slurry be jet mixed with a cementing unit to 11.0 ppg. To make this weight of slurry, use about three sacks, or 300 lb, of bentonite per bbl of diesel oil. Some operators also prefer to add about 15 lb of mica per bbl to increase the strength of the plug. The volume of slurry to be pumped usually ranges from 20 bbl to 150 bbl, depending on the rate of underground flow and the amount of open hole.

The biggest problem in using gunk plugs is the danger of the slurry's contacting water inside the drill stem. If this happens, the bentonite hydrates, causes excessive pump pressure, and usually plugs the drill stem. It is important, therefore, that not only the drill stem but also the pumping and mixing equipment be free of water. To avoid plugging problems in the drill stem, it is recommended that diesel oil spacers be pumped ahead of and behind the slurry.

Gunk plugs tend to lose strength with time under downhole conditions; therefore, many operators recommend that cement slurry be squeezed through the bit to provide a permanent seal as soon as it has been determined that underground flow has been shut off.

The following procedure has been used successfully to spot a gunk plug across a lost circulation zone:

1. Run survey tools, such as flow and temperature logs, to locate the flowing or lost circulation zone accurately.
2. Rig up both the cementing unit and the rig pumps so that either can be used to displace the slurry. Some operators also recommend that a third pump be connected to the annulus to pump down the annulus to keep casing pressures low.
3. Using a cementing unit, jet mix the slurry to 11.0 ppg. The slurry can be batch mixed or mixed on the run.
4. Pump 5 bbl to 10 bbl of diesel oil into the drill stem to serve as a spearhead, or spacer, between the drilling mud and slurry.
5. Displace the slurry down the drill stem at a rate of 3 bbl/min to 5 bbl/min. Follow the slurry with a 10 bbl to 20 bbl diesel oil spacer.
6. When the slurry reaches the bit, begin pumping water-base mud down the annulus at a rate of ½ bbl/min. Pumping water-base mud down the annulus lowers surface pressure and could provide water for slurry hydration.
7. Wait from 6 hr to 8 hr, or run a temperature survey to determine whether the plug is effective.
8. Release pressure on the annulus and pull the drill stem slowly.
9. Squeeze cement through the bit to provide a permanent seal.

EXCESSIVE CASING PRESSURE

In the drilling industry, maximum allowable surface pressure (MASP) can range from zero to as high as 100 percent of the casing burst rating. Regardless of how MASP is determined, when a large gas kick is circulated out, the fracture pressure at the casing seat, or MASP, may be exceeded. Decisions on whether to shut the well in when pressures exceed stated limits should be based on casing design and the depth at which it is set, knowledge of the characteristics and contents of the formation, company policies, and prior analysis of the possible consequences of different courses of action.

Accurate calculation of pressures in a well at various depths is difficult, because such calculations are based on the assumption that gas migrates or is circulated up the hole in slug or bubble form. Experience has shown that gas does not move upward as a slug or bubble; rather, it tends to disperse into the drilling fluid, the extent depending on the type of gas and type of mud in the hole. Because gas disperses unevenly over the length of the annulus, pressure calculations cannot be accurate; therefore, the decision on whether to shut in when pressure limits are reached must be based on factors that often cannot be calculated accurately.

BULLHEADING

Bullheading into a well is forcing gas or other wellbore fluids back into a formation by pumping into the annulus from the surface. The well remains closed in so that mud and kick fluids are displaced into the weakest exposed formation. Bullheading is not a routine procedure, but it may be useful when anticipated surface pressures are expected to exceed the pressure limitations of the surface equipment, when kick fluids are hazardous if circulated to the surface, when the drill pipe is plugged or parted so that kill mud cannot be circulated to bottom, or when a weak zone below the kick takes mud too fast for the well to be killed. Bullheading is perhaps most appropriate for wells with very short open-hole sections, since in such cases the influx is most likely to be squeezed back into the formation from which it came.

When bullheading, the pumping rate must exceed the rate at which the gas migrates up the hole to clean the annulus. One indication of too low a pumping rate is an increase, rather than a decrease, in pump pressure. To determine the required pump rate, the rate of gas migration must be calculated. One way in which to calculate the gas migration rate is to use equation 31:

$$R_{gm} = \Delta SICP \div MG \qquad \text{(Eq. 31)}$$

where

R_{gm} = rate of gas migration, ft/h
$\Delta SICP$ = change in *SICP* after 1 h, psi
MG = mud gradient, psi/ft.

Ideally, bullheading will fracture a formation, and continued pump pressure will force gas back into the formation. Anytime high pressure is applied at the surface, however, formation breakdown at the casing shoe, rather than at a formation lower in the hole, is possible. Should fracture at the shoe occur, an underground blowout may develop, and broaching around the casing is a possibility. Therefore, bullheading is not without risk, and caution should be exercised whenever the procedure is used.

SNUBBING INTO THE DRILL STEM

Under certain circumstances, small-diameter tubing may have to be snubbed into the drill stem. For example, assume that a well is shut in and is being circulated, and that a large hole or a washout occurs in the drill stem. Further assume that the washout is very large and is at a depth in the drill stem that makes it impossible to continue to circulate the well. One solution is to snub small-diameter tubing into the drill stem to reestablish circulation.

One way to snub tubing is with a coiled tubing unit. A coiled tubing unit is a portable machine that eliminates the need for making up and breaking out individual tubing joints, since the unit's tubing is a continuous length of small-diameter pipe coiled onto a reel. Coiled tubing ODs generally range from ½ in. to 2 in. The primary advantage of coiled tubing is that it affords a great savings in time, since individual joints of tubing do not have to be made up during the snubbing operation. Further, coiled tubing units are light in weight and are equipped with blowout preventers and other pressure-control devices that make them ideal for snubbing. A disadvantage is that coiled tubing has relatively low collapse and yield strength. In high-pressure situations, therefore, the tubing may collapse or burst. Also, a coiled tubing unit does not allow the tubing to be rotated.

If a coiled tubing unit cannot be employed, then it may be possible to snub small-diameter, individual tubing joints into the drill stem. A special snubbing unit that maintains pressure control as the tubing is snubbed into the drill pipe is required. Although snubbing tubing a joint or a stand at a time into the drill stem takes longer than snubbing with coiled tubing, jointed tubing can be rotated and it is stronger.

UNDERBALANCED DRILLING

Underbalanced drilling is drilling ahead while a formation in the well is kicking (producing). In cases where the influx is of relatively low volume, it is sometimes possible to seal the annulus of the well with a rotating head (fig. 7.6) or a rotating annular BOP (fig. 7.7) and drill ahead. They provide a way to seal around the kelly, or in the case of top drives, around a joint of drill pipe, while the kelly or pipe is rotating. Usually, high penetration rates are achieved because the influx of fluids from the formation moves cuttings rapidly away from the bit, allowing its cutters to remain in constant contact with uncut formation. In other words, the bit cutters do not have to redrill old cuttings that have not had time to be moved away from the bit. What is more, underbalanced drilling into the zone to be produced reduces formation damage allowing increased production when the well is completed.

Another benefit of underbalanced drilling is the re-duced risk of differential sticking of the drill stem.

Rotating Heads

A typical rotating head consists of a bowl, stripper rubber, bearing assembly, and a kelly driver (see fig. 7.6). The rubber and bearing assembly form a pressure-tight seal in the bowl. On rigs using kellies, the kelly fits inside the kelly driver, which, in turn, seats in the

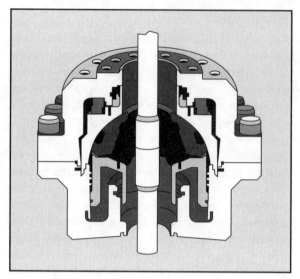

Figure 7.7   Rotating annular BOP

bearing assembly. Annular fluids exit through the outlet on the side of the bowl. Normally, drilling fluid and gas go to a mud-gas separator, where the gas is flared and the mud returned to the active system.

Rotating Annular Preventers

A rotating annular preventer, like a rotating head, seals around the kelly or drill pipe and allows the kelly or pipe to rotate. One type of rotating annular BOP uses a packing element to seal around the drill stem member when hydraulic pressure is applied below the packer, much as in a conventional annular BOP (see fig. 7.7). However, in the rotating annular, an interior assembly rotates within the preventer body as the drill stem rotates. Returning fluids exit the well through a return line mounted in the stack below the rotating annular (not shown in the figure).

OTHER SPECIAL WELL-CONTROL CONSIDERATIONS

Slim Holes

To reduce costs, many wells are being drilled with coiled tubing units and with conventional drill stem and bits that are relatively small in diameter (slim). Indeed, some well diameters are less than 3-in. From the standpoint of well control, the main concerns with small-diameter holes is that intruded gas and other fluids extend long distances up the hole and circulating times are reduced; in short, fluid move-ment occurs rapidly.

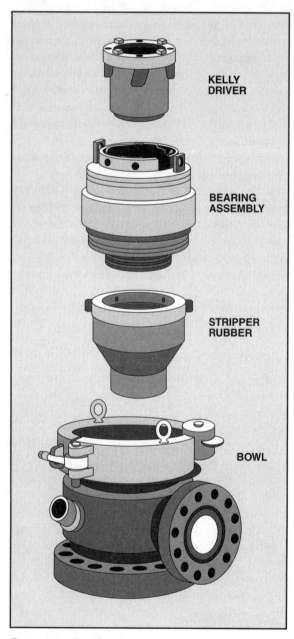

KELLY DRIVER

BEARING ASSEMBLY

STRIPPER RUBBER

BOWL

Figure 7.6   Rotating head

Tapered Holes and Tapered Strings

A tapered hole is a wellbore that has sections that are drilled with different diameters; in fact, most wells are tapered. The top part of the hole is large in diameter, the middle parts are of medium diameter, and the last, deepest part is small in diameter. Tapered strings are drill strings that vary in diameter. Sometimes, drill pipe and drill collar IDs can vary widely over the depth of a well. For example, a drill pipe string might be made up of 3½-in., 4-in., 4½-in., and 5-in. OD elements. Drill collar ODs and IDs can also vary. The main concern to keep in mind with tapered strings and tapered holes is that fluids extend over longer distances in small diameters and over shorter distances in large diameters. As a result, fluid flow happens quicker in small diameters than in large diameters. Another concern in tapered hole is to remember that when rising gas goes from a narrow diameter to a larger diameter section of annulus, *SICP* will drop because the gas may occupy less vertical space in the large diameter section. This drop may lead well-control personnel to believe that lost circulation has occurred.

HORIZONTAL WELL-CONTROL CONSIDERATIONS

Horizontal wells present some unique well-control conditions. For one thing, crew members handling a kick in a horizontal well must remember that measured depth *(MD)* is quite different from true vertical depth *(TVD)*. For another, drilling through a formation horizontally can expose long sections of hydrocarbon formations, making inflow potential during a kick much greater. Also, drilling through a long section of exposed formation increases the flow potential of the well. Because so much formation is exposed to the wellbore, when a kick occurs its volume can be very large, even if the crew detects it quickly and promptly shuts in the well.

Kicks may be hard to detect in the horizontal portion of the well. Gas separates and migrates from other formation fluids when drilling stops. As a result, gas can accumulate in pockets in the top part of the horizontal borehole (fig. 7.8). This accumulation of gas in the upper parts of the hole, as well as in washouts and vertical fractures in the formation, may disguise the size of the kick. Crew members may not recognize a kick until enough gas has filled the upper sections of the horizontal hole and then begins to flow into the vertical part of the hole. In short, a large volume of gas may enter the well before the well is shut in. As a result, *SICP* may reach or exceed *MASP*.

After gas separates and migrates to the top of the horizontal hole, it stops migrating. So, with a gas kick, *SIDPP* and *SICP* should stabilize after the well is shut in, if the gas portion of the kick does not extend to the vertical section of the hole. If gas does extend into the vertical and near-vertical hole portions, gas migration occurs and *SIDPP* and *SICP* increase accordingly. Further, if the influx is entirely in the horizontal part of the well, then hydrostatic pressure in the annulus and drill stem are the same, so *SIDPP* and *SICP* are the same or very close to the same. On the other hand, if *SICP* is higher than *SIDPP*, then the gas is also in the curved and vertical section of the hole.

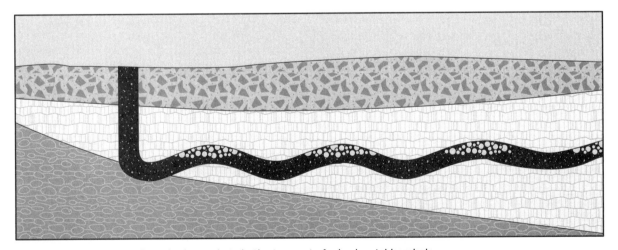

Figure 7.8   Gas accumulates in the pockets in the top part of a horizontal borehole.

As mentioned earlier, in horizontal wells, kicks can occur anywhere along the long horizontal path of the wellbore, which means that a kick can occur at a shallower *MD* than normally expected. As a result, gas can reach the surface much sooner than expected. What is more, a kick can enter weak zones, such as faults, vertical fractures, and formation vugs (openings) along the horizontal wellbore, which means that accurate pressure readings cannot be obtained at the surface. After shutting in a horizontal well, crew members should look for fluctuations in *SIDPP* and *SICP*. Lost circulation immediately after shutting in a horizontal well is possible, which could lead to an underground blowout and stuck pipe in the horizontal section.

When drilling a horizontal well underbalanced, a rotating head or a rotating annular preventer is used to control the formation flow. The head or annular BOP behaves as a choke, in that it exerts back-pressure against the well, and it may therefore cause lost circulation. If the well is kicking, kick pressure may also cause the kick fluids to enter a weak zone. When the pumps are shut down, circulating pressure loss in the annulus and back-pressure created by the rotating head are lost. Under these circumstances of reduced bottomhole pressure, it may be possible for the fluids lost to the weak formation to reenter the well.

When circulating a kick out of a horizontal well and maintaining constant bottomhole pressure, a gas influx in the horizontal section of the hole will expand little if at all. Thus, very little choke adjustment is required. Once the gas influx is circulated in the vertical section, however, it expands and choke adjustments become necessary to maintain constant bottomhole pressure. Also, in a well with a long horizontal section, gas may be more strung out than it is in a vertical well. Gas will therefore vent from a horizontal well for a longer period than from a vertical well.

In horizontal wells, the placement of drill stem elements is often the reverse of the placement in vertical wells. That is, drill collars, if used, are near the surface, heavy-walled drill pipe is below the collars, and drill pipe and the bottomhole assembly tools are below the heavy-walled drill pipe. Therefore annular velocities and volumes are reversed as well—that is, annular velocities and volumes are lower near the bottom than they are near the surface, which is the opposite of the situation in a vertical well. Specifically, when the influx reaches the heavy-walled drill pipe and collars in the vertical section, the influx elongates

and its velocity increases because the clearance (space) between the pipe and wall of the hole is reduced. As the gas elongates, hydrostatic pressure in the annulus is reduced, which, if not offset by quickly adjusting the choke, leads to another influx. The choke operator must be alert and quickly adjust the choke to maintain proper bottomhole pressure and to minimize pressure on the casing shoe.

Many horizontal wells are drilled underbalanced, or, as engineers say, they are produced while drilled. Drilling underbalanced is also termed pressurized drilling. In any case, allowing a well to produce while it is being drilled minimizes formation damage. Formation damage can occur when the solids in a drilling mud, which is being circulated at a pressure higher than the formation with which it is in contact, reduce the formation's porosity and permeability. The higher hydrostatic pressure of the mud forces the solids into the formation, which is lower in pressure. Formation damage can therefore often be minimized by keeping hydrostatic pressure lower than formation pressure.

To drill underbalanced, a rotating head or a rotating blowout preventer seals the annulus of the well against pressure while at the same time allows the kelly or drill stem to rotate and put weight on the bit. Usually, the rotating head is mounted on top of the BOP stack. Normally, a well can be drilled underbalanced as long as the volume of production from the formation is not too high. The produced fluids exit the well through a line to a mud-gas separator and the physical size limitations of this flow equipment limits the volume they can handle. A well can be killed while drilling underbalanced, using conventional means. If the well has vertical fractures, however, the kill fluid may go into empty or depleted fractures, and make it difficult to kill the well.

Bear in mind that when a kick occurs in a horizontal well, it is important to know the well's *TVD*. A horizontal well always has an *MD* that is longer than *TVD*. *TVD* is, however, used to calculate the density of the kill-weight mud. On the other hand, *MD* is used to calculate hole volumes.

Killing Horizontal Wells

The wait-and-weight method is widely used to control wells (see chapter 6). In theory, the wait-and-weight method regains control of the well in one circulation. In reality, however, and especially in horizontal wells, two or more circulations are needed. More circulations are usually needed because of inefficient hole displacement, gas pockets, and uneven ascent of gas to the surface. The wait-and-weight method often uses a chart or graph of calculated values to predetermine the drop in *SIDPP* as kill-weight mud is pumped down the drill stem. Calculating the values for the graph assumes (1) that the length of the column of kill-weight mud increases the same amount for each incremental increase in pump strokes and (2) that the true vertical height of the kill-weight mud column increases the same amount for each incremental increase in pump strokes. The first assumption is true as long as the ID of the drill pipe is all the same and that the ID of heavy-walled drill pipe, and drill collars are the same as drill pipe ID. The second assumption is true as long as the hole is vertical and the first assumption is correct. Usually, any differences in ID often are not significant enough to make a big difference in plotting the pressure decrease in the drill stem as heavy mud is pumped down it.

While killing in directional or horizontal wells, *SIDPP* decreases just as it does in a vertical well until the mud reaches the bottom of the kick-off point, KOP, (fig. 7.9). At the bottom of the KOP, *SIDPP* should be the same as final circulating pressure, since *TVD* has been achieved. However, kill-weight mud still has to fill the drill string. Thus, at the bottom of a horizontal well's radius, where the well becomes horizontal, *SIDPP* may be slightly below *FCP*. *SIDPP* may be lower than *FCP* because the kill-weight mud's hydrostatic pressure has reached *TVD* but the pressure caused by friction losses through the string are not realized until the kill mud reaches and exits the bit.

Because it is more difficult to plot the *SIDPP*'s pressure drop when using the wait-and-weight method in horizontal drilling, many contractors and operators prefer using the driller's method. As described in chapter 6, in the driller's method, the pumps are brought up to kill-rate speed, *SICP* is adjusted to its stable, pre-circulation value, and the well-control operator holds *SIDPP* constant until the kick is circulated out. Depending on the reach of the horizontal well, it may require a considerable amount of circulating to displace the influx from the horizontal portion.

Procedure for Off-Bottom Kill

As mentioned earlier, when pipe is off the bottom and cannot be stripped back to bottom, killing the well to the depth of the pipe may be necessary. This

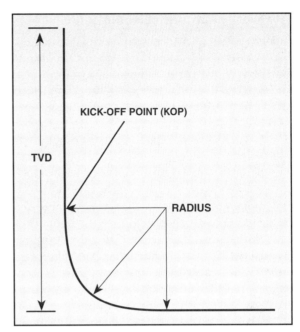

Figure 7.9   Kick-off point (KOP)

procedure is often called an off-bottom kill and consists of using the depth to the bottom of the pipe (not the hole) to determine how much to weight up the mud to kill the well. Similarly, when a kick occurs in a horizontal well, it is particularly important to keep in mind that *TVD* and not measured depth is used to determine the weight of kill-weight mud.

Gas Behavior in Horizontal Section

A horizontal well is virtually never perfectly horizontal; instead, it tends to vary up and down from horizontal, which results in portions of hole that arch upward and downward. What is more, some horizontal holes are intentionally deflected upward into a formation at a depth higher than the horizontal section. Gas influxes can become trapped in upward arching portions of the hole, as well as in portions of the hole deflected upward, because gas is lighter than drilling fluid and rises to the top of the arched hole. This gas tends to stay there—drilling fluid bypasses it and does not move it up to the surface, particularly if circulating at reduced pump rates. It is important to anticipate that an increase in pump rate when returning to drilling may sweep gas into the vertical hole, essentially behaving as another gas kick that may require careful circulating to remove.

THIEF AND KICK ZONE COMBINATIONS

In some cases, a formation is exposed to the wellbore that, when the drilling mud is weighted up to kill a kicking formation or zone, breaks down (fractures) and circulation is lost, either partially or completely. The formation into which wellbore fluids are being lost is called a thief zone. A thief zone can occur above or below a kick zone. Either situation complicates well control operations.

Where (1) the thief zone is above the kick zone, (2) the losses are partial, and (3) crew members can mix enough mud to replace that being taken by the thief zone, one procedure is to continue normal kill procedures—that is, maintain the kill rate pump speed and *SIDPP* as determined earlier. This procedure maintains a constant bottomhole pressure above that of the kicking formation. Once the kick fluids go above the thief zone, the zone may heal.

If, however, the loss rate is so high that not enough mud can be mixed to replace that going into the thief zone, another procedure is to add lost circulation material (LCM) to the mud and circulate it into the well. LCM includes such substances as mica flakes, walnut hulls, and other materials that can plug the permeability of the thief zone. Keep in mind that it may not be possible to pump LCM through small bit nozzles or other restrictions in the drill stem. In fact, many well-control specialists recommend installing a special circulating sub in the drill stem in areas where a kick and lost circulation may be expected.

If it is not possible to pump LCM, another way to handle lost circulation is to stop pumping and shut in the well completely. Sometimes, thief zones can heal themselves if given enough time in a static condition. Solids in the mud opposite the thief zone may plug the zone's permeability. One indication that the zone has cured itself is that *SIDPP* and *SICP* will stop decreasing. (Recall that one indication of lost circulation is that *SIDPP* and *SICP* drop and do not respond as they should to choke adjustments.) If the problem continues and circulation losses become severe or complete, it may be necessary to set a barite or gunk (cement) plug to heal the zone, as covered earlier in this chapter.

Sometimes, a thief zone can exist below a kick zone. While drilling into the thief zone, lost circulation occurs, which may cause the upper kick zone to flow. If the mud weight is lowered to avoid losses into the thief zone, the upper zone would kick, but control would be easy since the thief zone has not been penetrated.

Most contractors and operators agree that the primary concern is to control the well, so upon seeing the signs of the kick, the driller should shut in the well. If the pressure of the shut-in well is high enough to cause mud to flow into the lost circulation zone, *SIDPP* may be zero because the drill stem is full of mud, the kick zone is not at the bottom of the hole, and mud is being lost at the bottom of the hole. *SICP* may, however, be well above zero; the value for *SICP* depends on the size and fluid content of the kick.

Chapter 8

Well Control for Completion and Workover

Just as in drilling operations, the fluid that is circulated in a well being completed or worked over has many applications. For example, fluids are employed in perforating, cementing, fracturing, and acidizing. They are also used in well killing, recompletion, drilling, deepening, plugging back, and cleaning out. Further, fluids serve as packer fluids, completion fluids, and circulating fluids. Completion and workover fluids may be gases, oils, brines, muds, or other chemical solutions.

Packer and completion fluids are different from the fluids used in drilling and working over a well. Packer fluids are placed in the well between the tubing and the casing above the packer to offset formation pressure below the packer. Usually, a packer fluid remains in the annulus over the life of the well. Since packer fluids stay in the well for a long time, they are specially formulated to remain liquid. They must remain liquid so they can be circulated months or years after they are placed in the completed well. Packer fluids must also be noncorrosive to prevent them from harming the casing and the tubing with which they are in contact. Completion fluids are similar to packer fluids, but they are used opposite productive formations to prevent permanent damage to the zone.

CHARACTERISTICS

A drilling or a completion and workover fluid has several important characteristics. For example, it should be dense enough to control well pressures but not so heavy that it fractures the formation and flows into it. It should balance formation pressures but not fracture a formation. Further, the fluid should be cost effective. Sometimes, expensive fluids are necessary to prevent damage to especially sensitive formations; less expensive fluids, however, may also be

available that cause little or no formation damage. Experience in the area is very valuable in determining which fluid to use. Also, a fluid should be as free of solid particles as possible. Solids can plug perforations as well as reduce production after fracturing or gravel packing. Moreover, it should be noncorrosive to prevent failure of tubular goods and subsequent fishing jobs. It should also be stable if it is to be left in the hole for an extended period. Fishing for packers and tubing that are stuck because of fluid breakdown can be expensive and may even lead to abandonment of the well before production is fully depleted. Completion and workover fluids should also be filtered or cleaned and have few or no solids. Some fluids have large amounts of suspended solid particles, which can be harmful to the producing formation, as well as being abrasive to equipment. Even though a fluid has a low solids content, it can still cause plugging if it reacts adversely with the formation.

Some fluids that are excellent for normal operations can be incompatible with cement slurries or acids. In such cases, it may be necessary to use a fluid spacer to separate them. The spacer, which is usually salt water or a special mud, is placed behind the cement or acid to keep the completion or workover fluid from contacting, and thus contaminating, the cement or acid.

In addition, some chemical additives, as well as the fluid to which they are added, can cause health and environmental problems. Environmental and health considerations may require using a less effective or more expensive product that presents less of a hazard to personnel and to the environment.

FUNCTIONS

The fluids used in drilling, workover, and completion are, for the most part, not unusual or exotic; they are, however, extremely important to the success of most jobs. They must not damage the producing formation, the equipment, the personnel, or the environment. It is essential that the fluids be applied properly, controlled, and monitored.

Drilling, workover, and completion fluids range in weight, or density, from very light, or low in density, to very heavy, or high in density. The lightest fluids are gases; the heaviest are liquids to which a weighting material, such as barite, has been added. Regardless of its density and whether it is a gas or a liquid, a workover and completion fluid has several functions. Some important ones are—

1. transportation of wanted and unwanted materials into and out of the well;
2. suspension of wanted and unwanted materials when circulation is stopped;
3. pressure control to prevent kicks;
4. absorption of heat and lubrication of pipe, bits, and mills;
5. delivery of hydraulic energy;
6. provision of a suitable medium for wireline tools, and for logging and perforating tools;
7. allowing downhole equipment to be run safely in a reasonable amount of time;
8. avoiding damage to producible formations;
9. avoiding damage to downhole equipment;
10. avoiding damage to surface equipment; and
11. avoiding damage to personnel and the environment.

Transportation of Materials into and out of the Well

To perform many operations, materials must be circulated both into and out of the well. Materials such as acid, cement, gelled pills, plastic, gravel, fracture sand, sealers, and other fluids, are injected and circulated.

Other materials, which may be damaging, must be removed to keep the wellbore clean. Potentially damaging materials include dry cement, corrosive fluids, cuttings, debris, gravels, gas, metals, contaminated mud, plastics, sand, and unused wet cement. What is more, an accumulation of material in the wellbore can cause sticking or failure of the work string, pipe plugging or bridging, increased torque or drag, lost circulation, fill, perforation or formation plugging, and excessive equipment wear.

Suspension of Materials When Circulation Is Stopped

If a fluid has high gel strength, it has good suspension capabilities when circulation is stopped. Its gel-like structure resists the settling of solids and cuttings until circulation is resumed. Suspension of solids and cuttings reduces the amount of fill and thus cuts down on the likelihood of tools, tubulars, and wireline getting stuck. In workover and completion operations, however, high gel strength can be a hindrance because a high gel-strength fluid develops high swab and surge pressures. Excessive swab and surge pressures can cause the formation to kick or to break down. In either case, the producing zone can be damaged by the procedures used to control the kick or seal the fracture.

Pressure Control

At any time during drilling, completion, and workover, the wellbore may expose producible formations, which are characterized by high pressure and high permeability and which contain hydrocarbons. In fact, work is sometimes performed on a live well under pressure. The vast majority of work, however, requires that the well be killed—that is, that formation pressure be balanced by a column of fluid in the well to prevent the formation from kicking. Weighting materials can be added to obtain a balanced condition; however, the fluid's weight must not be so high that it leaks or flows into the formation and damages it.

Heat Removal and Lubrication

As a bit or mill and work string turn in a well, extreme heat develops from friction as the assembly contacts the walls or the bottom of the hole. The completion or workover fluid must absorb this heat to cool the workover assembly and to prolong the life of the bit or mill, since heat can weaken and damage the metals used in bits and other downhole tools.

Delivery of Hydraulic Energy

Many routine and special activities during completion and workover require that pressure be applied at the wellhead and transmitted through the fluid to a downhole location. The only way in which pressure can be transmitted from the rig's pumps to the wellhead or downhole is through the workover or completion fluid.

Provision of a Suitable Medium for Tools

Much of the activity associated with well completion and workover is done by wireline—that is, tools such as perforating guns, logging tools, and packers, are often placed in the well by running them on wireline. In any wireline operation, the fluid must allow ready access to the wireline equipment. To provide access to equipment, the fluid must be kept in good condition; that is, it must meet the specifications of the job. For example, it must contain no more than the recommended solids, it must be of the proper density, it must have the proper gel strength, and it must be of the proper viscosity.

Allowing Downhole Equipment to Be Run

If the fluid is not properly conditioned—if it is too thick and viscous—surging, swabbing, formation damage, or circulation problems can occur when attempting to run downhole tools and equipment. Time is money, and nowhere is this more evident than when completion or workover activities go wrong. An excessive number of rig hours spent on a particular operation is often caused by poor fluid application.

Avoiding Damage to Producible Formations

A completion or workover fluid must not cause permanent damage to the productive zone by leaving silts, or fines, sludge, gum, or resins in the formation. Further, it should not change the wettability of the reservoir sand or rock. (*Wettability* is a rock grain's tendency to be coated by one of the liquids that occurs within the reservoir. Most rock grains are water wet, which means that they are coated with water, usually salt water. Some are oil wet, which means they are coated with oil.) If the wettability is changed by the completion or workover fluid, production may be restricted.

Moreover, if fresh water is used as a completion or workover fluid, it can cause a flow-blocking emulsion in some gas- and oil-producing formations. Further, some completion and workover fluids can cause sensitive formations to swell, which can lead to a decrease in productivity. High fluid flow rates can cause hole erosion.

Avoiding Damage to Downhole Equipment

Much consideration is given to packer fluids that are left in the hole. Such fluids must be nonsettling and noncorrosive. The expected life of the well usually dictates the type of fluid and additives left in the well. During completion and workover activities, however, the packer fluid is sometimes altered, diluted, or replaced. If it is not properly treated, it can become corrosive, which can shorten the expected life of seals and equipment.

Avoiding Damage to Surface Equipment

Corrosive fluids can lead to failure of sealing elements in many types of surface equipment. Additionally, sand-laden fluids can be very abrasive and can erode and cut valves, swabs, and other equipment in a short time; sand should therefore be removed from the fluid at the surface. Completion and workover fluids should also be conditioned to remove silt and other erosive and corrosive elements.

Avoiding Damage to Personnel and the Environment

Fluids used in remedial activities can be very hazardous to personnel. Acids, caustic, bromides, some chlorides, and other chemicals can cause serious burns. Such reagents can also be toxic and can cause vision and respiratory problems. Care and safety clothing should be used when handling and mixing these chemicals and when handling tubulars pulled out of well fluids.

The environment is a very precious resource. It can be damaged easily by fluids used in and produced from the well. Spill prevention and reporting, safe hauling, and proper disposal of fluids used on the rig are regulated by law. Every person on the rig must be aware of the regulations concerning the use and disposal of completion and workover fluids. Safety cannot be stressed enough, as many of the fluids are very dangerous. Protection of personnel and the environment is critical.

OIL FLUIDS

In most producing areas, oil is plentiful and economical to use. It is usually noncorrosive and does not cause clay swelling in the producing zone. It weighs about 7 ppg, which is excellent for low-pressure oilwells. Oil does, however, have some disadvantages. For example, it can contain wax or fine particles of sand, solids, or asphalts that can damage formations. Moreover, it may be corrosive if H_2S or CO_2 are present. Further, oil may be too light to hold well pressure in some areas and too heavy in others. In addition, it is a fire hazard and it is very slippery, which can be a problem when pipe is pulled wet. Also, oil pollutes if it is spilled. Bear in mind, too, that oil may not be compatible with the reservoir oil if it is obtained elsewhere in the field.

Diesel fuel and kerosene are sometimes used as workover fluids. They are inexpensive, very clean, and noncorrosive. They can be hazardous, however. Proper fire extinguishing equipment should therefore be readily accessible and crews should be well trained in its use.

OIL-EMULSION FLUIDS

The most common oil-emulsion fluid used in workover and completion is an oil-in-water emulsion. In an oil-in-water emulsion, oil is dispersed as small droplets in water. The water phase may be fresh- or salt water. To keep the emulsion stable (to keep it from separating) emulsifying agents, such as starch, soap, or organic colloids, are used. Although other oils are available, diesel oil is the most commonly used oil for the dispersed phase.

Water-in-oil emulsion fluids are sometimes used as completion and workover fluids. A water-in-oil emulsion is the opposite, or inverse, of an oil-in-water emulsion and is therefore called an inverse-emulsion fluid. In an inverse-emulsion fluid, water is dispersed as droplets in oil. This type of emulsion is very unstable above 200°F, and if it contains a lot of solids, the solids can cause formation plugging.

SYNTHETIC FLUIDS

Operators, rig owners, and rig personnel need to be aware of the effects of drilling, completion, and workover fluids on the environment. Almost all such fluids contain components that are toxic to human, animal, and plant life. Oil fluids are especially damaging.

Some of the harmful effects are invisible and some are observable. Unseen effects, for example, include pollution of groundwater, which may be a drinking water supply. Offshore, release of fluids can damage marine life. Visible effects include pollution of surface water and land, affecting soil productivity. The U.S. and other governments have enacted legislation to protect the environment, and rig personnel must be familiar with the laws in order to comply.

At the same time, the search for petroleum has resulted in drilling deeper wells in increasingly difficult environments, such as in deep water and arctic locations. These locations present special problems—for example, high downhole pressures and temperatures. Traditional oil fluids that can handle these adverse conditions are also the most environmentally hazardous.

So engineers developed synthetic fluids that have a lower impact on the environment and are easier to dispose of safely. For example, some new invert-oil fluids use vegetable oil or ester oil, both of which are biodegradable. Others use mineral oil, which is less toxic than diesel oil.

Mud companies have also developed environmentally safe additives for water-base fluids. Such additives include PAL (polyanionic lignin), polymers such as VA/VS (vinyl amide/vinyl sulfonate copolymer) and MPT (modified polyacrylate terpolymer), and an oxygen scavenger to control corrosion. (Corrosion can be a big problem at high temperatures.) Using environmentally safe fluids lowers the normally high cost of disposing of cuttings, which are coated with the fluid. Also, programmers have developed computer software that helps operators understand and monitor compliance with environmental regulations.

GAS FLUIDS

Gas can be used as a completion and workover fluid in some low-pressure reservoirs. During operations, flow from the well is controlled only by surface backpressure. In some fields, natural gas is used because it is readily available and cheap; it is, however, extremely flammable and thus quite hazardous. Often, it is better to use nitrogen instead of natural gas. Nitrogen is inert, or nonreactive, and it usually does not harm the formation, metal goods, or rubber seals.

Cleaning trash from the well can be a problem with gas; therefore, foam mixed by the service company supplying the nitrogen is available. It has good-to-excellent hole cleaning characteristics and carrying capacity.

WATER-BASE FLUIDS

Water-base fluids include (1) fresh water, (2) salt water, or brine, and (3) water-base muds. Fresh water is not often used in completion and workover activities because it can cause clays to hydrate and severely damage formations. (*Clay hydration* is the swelling that occurs when clays in the formation take on water.) Low-salinity water is usually plentiful and inexpensive, however, and it normally requires little treatment. In cases in which the water has a high solids content, it can be filtered to remove the solids.

Brines

Brines are widely used because they are readily available and can be mixed easily. Their cost is usually low and no explosion or fire danger is present. Brines can, however, be an environmental hazard in some areas.

When salt is added to water, the density increases and creates a greater hydrostatic head in the well. Adding salt does not add solids to water, as long as all of the salt dissolves into solution. An increase in the salt concentration of water usually inhibits clay hydration within a formation. In some areas, however, salt water that contains sodium chloride (NaCl) causes shales and clays to swell. In such cases, a calcium or potassium salt, such as calcium chloride or potassium chloride ($CaCl_2$ or KCl), instead of NaCl, can be used to prevent the hydration of formation clays.

Single-salt brines such as those that contain only NaCl, KCl, or $CaCl_2$ are generally low in density. Single-salt brines such as those containing calcium bromide ($CaBr_2$) and zinc bromide ($ZnBr_2$) can, however, be mixed to form a high-density solution. The most commonly used single-salt brine is one that contains water and NaCl. The density of a single-salt brine can be increased by adding more salt until a saturation point at a given temperature is reached. Beyond the saturation point, the salt fails to go into solution and either crystallizes or settles out as a solid. Multisalt brines, in which two or more salts are added to water, can be used where higher densities are needed. Table 3 lists the density ranges of several fluids.

As stated earlier, the most commonly used brine is NaCl and water. The maximum density of NaCl brines is 10 ppg at 60°F. Preparing NaCl brines with densities up to 9.7 ppg is relatively easy, because the amount of NaCl required to obtain 9.7 ppg readily dissolves in water. From 9.7 ppg to 10 ppg, however, additional NaCl dissolves very slowly in the solution.

Table 3
Fluid Density Ranges

Fluid	Approx. Minimum Density (ppg)	Approx. Maximum Density (ppg)	Practical Maximum Density (ppg)
Oil		8.5	8.0*
Diesel oil		7.0	7.0
Fresh water			8.3
Seawater	8.4	8.6	8.5
Brine–sodium chloride (NaCl)	8.3	10.0	9.8
Brine–potassium chloride (KCl)	8.3	9.8	9.7
Brine–calcium chloride ($CaCl_2$)	11.0	11.7	11.5
Brine–calcium bromide ($CaBr_2$)	11.5	15.1	15.0
Brine–zinc bromide ($ZnBr_2$)	14.0	19.2	18.1

* Some oils sink in water.

KCl brines are often used where a water-sensitive reservoir is exposed to the wellbore and where densities over 9.7 ppg are not needed. Corrosion rates of KCl brines are reasonably low but can be further reduced by keeping the pH between 7 and 10 and by adding corrosion inhibitors.

CaCl$_2$ brines are easily mixed at densities up to 11.6 ppg. At greater densities, CaCl$_2$ brines tend to freeze or crystallize at relatively high temperatures—for example, an 11.6 ppg CaCl$_2$ brine freezes at 44°F. In winter, such high freezing temperatures may cause operating problems. In general, CaCl$_2$ is available in two grades of purity: 94–97% pure and 77–80% pure. Most operators prefer to use 94–97% pure because it is easier to mix into solutions. From a safety standpoint, it is important to keep in mind that a considerable amount of heat is generated when dry CaCl$_2$ is mixed with water. CaCl$_2$ brines are not particularly corrosive if their pH range is maintained between 7 and 10. Also, any good-quality corrosion inhibitor can be used to retard corrosion further.

Crystallization

The formation of crystals in a completion and workover fluid can be a real hazard. When mixing a fluid, many different salt and mineral combinations may be used to get the desired fluid weight at the most economical and safest condition. The mixture often contains all of the material water can hold at a given temperature—the *saturation point*. No further weight is gained by adding more material.

Should more material be added and the temperature held constant, one of two things happen: either the material falls to the bottom of the tank, or crystallization occurs. Crystallization in a completion or workover fluid looks like ice forming and is sometimes called freezing. Should the temperature of the fluid in the tanks be reduced by a change in weather or other conditions, crystallization can occur, reducing not only the fluid density, but also its ability to be pumped.

Variations in temperature and variations in brine solutions themselves affect the crystallization point; it is vital, therefore, to get the crystallization point for a particular solution from the fluid supplier. It is possible, however, to give general crystallization or freeze points for some brines (table 4). Nevertheless, for specific brine solutions, be sure to consult charts from the supplier.

Water-Base Muds

Water-base muds are a mixture of water, clays, and chemicals that are sometimes used in completion and workover operations. Some water-base muds are, however, laden with solids and, as a result, can cause extensive formation damage by causing water loss and blocking pore spaces. Water-base muds are used, however, because their cost is low and they are often easy to work with. Moreover, water-base muds make controlling high-pressure, high-permeability gas wells much simpler. Further, it is sometimes necessary to use mud if very expensive clear fluid is lost to a formation.

CEMENT

Strictly speaking, cement is not a completion or workover fluid; rather, it is a liquid slurry, which, when set, bonds casing to the wall of the wellbore and prevents fluids from migrating behind the cemented casing. Nevertheless, during cementing operations when cement displaces the drilling or other fluid in the well, cement becomes the primary fluid available to control formation pressures. It is important therefore for all personnel involved in the cementing operation to be aware that the density of the cement must be adequate to control formation pressures. Among cement's many other characteristics that require confirming prior to the well's being cemented, it is important that cement density not be overlooked. What is more, it is important to ensure that any weighting materials added to the cement be kept in suspension during the cementing operation. If weighting material is allowed to settle out of the slurry, it is possible that the cement circulated into the well will not have sufficient density to control formation pressures.

PACKER FLUIDS

One of the most important steps in a workover is often the last one before putting the well back on production: the placing of a packer fluid in the annular space between the casing and the tubing. The packer fluid usually remains in the well until the well is reworked or abandoned. A packer fluid provides formation pressure control and prevents the casing from collapsing. To be effective, a packer fluid should be (1) noncorrosive, (2) stable with time and temperature, (3) reasonably economical, (4) pumpable at the time it is placed in the well and remain pumpable for

Table 4
Crystallization Point of Brines

Weight (ppg)	Crystallization or Freeze Point (°F)
Sodium Chloride (NaCl)	
8.5	29
9.0	19
9.5	6
10.0	25
Calcium Chloride (CaCl$_2$)	
8.5	30
9.0	31
9.5	9
10.0	−8
10.5	−36
11.0	−22
11.5	35
Calcium Chloride/Bromide (CaCl$_2$/CaBr$_2$)	
12.0	54
12.5	57
13.0	59
13.5	61
14.0	64
14.5	65
15.0	67

a long period, (5) of sufficient density to control well pressures, (6) unharmful to packer seals, and (7) capable of keeping solids suspended in it so that they cannot settle on top of the packer.

In older wells, drilling mud was often left in the well as the packer fluid. Drilling muds used as packer fluids often caused expensive fishing jobs when it came time to work over the well, because the mud's solids separated from the liquid as time passed and collected and solidified on top of the packer. Perhaps the worst cases occurred with lime-based muds when they were exposed to high temperatures; the lime reacted with the clays in the mud and became a cementlike substance. Such problems led to the development of the many good packer fluids that are now available.

PLUGS

Plugs, or pills, are used much as mechanical plugs are used to solve or control many downhole problems; they are also used for downhole treatment. For example, plugs seal casing leaks, correct the injection profile in water-injection or disposal wells, and stop lost circulation in highly permeable sands. Plugs, or pills, can also divert acid during well cleanup or stimulation and shut off saltwater flows. Further, a plug can be formed that blocks flow inside of the tubing or work string but can be removed readily or worked through with concentric tubing or coil tubing. What is more, plugs can stabilize unconsolidated gravel zones, seal fractures, and improve cement jobs by sealing thief zones into which low-viscosity cement would be lost. Plugs can also be used to kill underground blowouts.

Many types of soft, or pumpable, plugs are available: neat cement, thickened oil-base mud, diesel oil–cement, diesel oil–bentonite, bentonite-cement, silica-clay, polymers, plastics, acids, and various lost circulation, plugging, and treatment chemicals are examples. Weighting materials and viscosifiers are often added to plugs to make them dense and highly viscous. Further, a retarder or accelerator may be added to slow down or to speed up setting time, depending on temperatures and pumping times.

Sometimes, a time-delayed self-complexing plug may be required; if necessary, a breaker can be added to provide a predictable plug breakdown time, usually from 1 to 10 days. To get a predictable breakdown time with polymer pills, an enzyme is used as a breaker, which reduces the large polysaccharide, or sugar, molecules to low-molecular-weight polymers and simple sugar. Most operators recommend that a breaker always be added to a polymer plug that is in contact with a producing zone.

As an example of using a plug, consider a dual-completion well in which one of the two producing zones requires fluid of relatively high density to kill it. Unfortunately, the high-density fluid causes lost circulation in the other zone. To solve the problem, a small plug can be spotted in the weak zone. The plug protects the zone and prevents it from being fractured. A breaker is added to the plug so that it dissolves when the weak zone is ready to be produced. For typical operations, a plug of about 5 bbl is usually sufficient. Frequently, 1- or 2-bbl plugs are adequate.

Polymers that flash set are sometimes used to create a pack-off in the tubing or work string. The tubing or work string is filled from the surface with a special plastic polymer, which is weighted to the required density. The polymer forms a very tough and rubbery pack-off. A macaroni string, which is tubing that is smaller in diameter than the work string, is pushed through the pack-off in the work string and withdrawn, rotated, or reciprocated as much as desired. Once the macaroni string is withdrawn from the pack-off, the hole in the pack-off that was created by the macaroni string closes back up.

GENERAL FLUID SAFETY

During the mixing of any completion and workover fluid, all personnel should be informed of the hazards involved in handling and mixing the chemical solutions. As mentioned previously, some of these chemicals can cause serious burns, can be toxic to people and to the environment, and can cause visual and respiratory problems. Protective clothing—goggles, vinyl or rubber gloves, aprons, boots, and so on—should be used when handling and mixing chemicals. When chemicals are to be mixed with water or other fluids, mix the chemicals into the water or fluid to reduce the possibility of a violent reaction.

Always have a method of washing eyes and skin near the mixing point. In case of contact with eyes and skin, immediately flush with water and contact a supervisor for further instructions. Pit jetting, or mixing, guns should be secured in one position while unattended. Material should be stacked to a reasonable height to minimize handling and danger.

REMEDIAL OPERATIONS

Introduction

Remedial operations cover a wide range of activities: workover, wireline work, stripping, snubbing, completion, perforating, and so on. A remedial operation is any work done on a well that occurs after the well is drilled until it is plugged and abandoned. The most common remedial operations are squeeze cementing, perforating, drill stem testing, acidizing, sand control, fracturing, plugging back, plugging and abandoning, deepening, and sidetracking.

Squeeze Cementing

Squeeze cementing excludes water or gas from the well, improves the primary cementing job, allows a new zone to be recompleted, or repairs casing that has been corroded or damaged. Squeeze cementing is also referred to as remedial or secondary cementing.

Squeeze cementing is accomplished by running a squeeze tool on the work string to a point above an area into which cement is to be placed. Once at the desired depth, cement is circulated down to the squeeze tool and the tool is set to isolate and protect the casing from high pressure. Cement is then pumped to the area to be sealed off. Hydraulic pressure is applied to force or squeeze the cement slurry into contact with the formation, either in open hole or through perforations in the casing or liner. Excess cement is then reverse-circulated out of the well. The cement should be left in the casing opposite the perforations or damaged area and not be drilled out after the squeeze operation.

In a squeeze job, only the water in the cement is forced, or squeezed, into the formation under pressure. The cement itself remains across the face of the formation. The loss of water sets up, or hardens, the cement. If enough pressure is applied, however, whole cement may fracture the formation and cement may enter the fractures.

The most important criterion for a good squeeze cement job is clean perforations or channels. Minimal blockage and clean surfaces assure a better bond.

A wide selection of oilwell cements for squeeze operations are available. They can be formulated to be extremely heavy or very light in weight. Additives can adjust water-cement ratios, viscosity, set strength, pumpability time, temperature tolerance, and other factors.

Many methods of applying cement under pressure can be used:

1. Bradenhead squeeze: no packer is in the hole. Casinghead valves are closed and the well is pressured up on the casing and the work string during the operation.

2. Bullhead squeeze: a packer is set when the job starts and all fluid in the work string is pumped into the formation ahead of the cement. The casing may be pressured if necessary to reduce the differential pressure across the packer.

3. Hesitation squeeze: cement is pumped in and the pumps are stopped for a few minutes. Pumping is then started and stopped until the desired pressure is reached.

4. Set-through squeeze: after the squeeze, the work string is lowered past the perforations where the cement was squeezed. Excess cement in the casing is washed out to allow reperforating without having to drill out the cement. Set-through squeezing requires a special low-water-loss cement.

5. High-pressure or low-pressure squeeze: jobs done with a high or low final squeeze pressure; low pressure never breaks the well down.

6. Circulation squeeze: perforations are made below and above the zone of interest. A retainer is set between the perforations and circulation is established across the zone. The cement is circulated in place, and the work string is pulled out of the retainer about 10 stands above the cement. The work string is slugged (heavy fluid is placed in the string) and tripped out.

Because of the high pressures encountered on many squeeze jobs, rig pumps are usually not adequate. As a result, a service company's high-pressure, low-volume pumping units are required.

All fluids should be in excellent condition prior to a squeeze job and should be compatible with the cement or other materials being used. If fluids are not compatible, a buffer solution should be run ahead of and behind the squeeze medium.

As with any operation, planning and safety should be a primary consideration. Care should be taken to follow instructions exactly when mixing cement additives such as accelerators and retarders. Further, no one should ever ride through the V-door with the cement manifold when it is being lifted to the floor. In addition, only those who are necessary to the operation should be in the immediate area during the pumping operation. Finally, no one should ever hammer on any union or other part of the surface equipment while it is under pressure.

Perforating

Perforating is putting holes through the casing, cement, and into the formation. Perforations create a way for fluids to flow from the formation to the wellbore. Sometimes, after a well is initially perforated, additional perforations are made to increase the flow area.

Originally, perforations were made by lowering a special gun that fired bullets through the casing, cement, and formation. Today, most perforating is

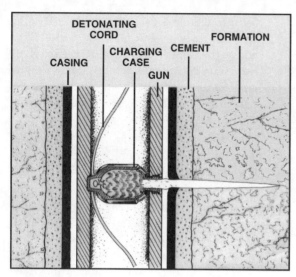

Figure 8.1   Jet perforating

done with jet charges (fig. 8.1); bullet guns are, however, still used in soft formations. Perforating guns may be run on wireline, tubing, or coiled tubing. In highly deviated directional wells or in horizontal wells, the extreme angle necessitates that the perforating guns be run on tubing.

Many types of perforating guns are available. Some are expendable, some are retrievable, and some are expendable-retrievable. Expendable guns disintegrate when its charges are fired. Retrievable guns are constructed so that after the charges are fired, the gun is pulled back to the surface and reused. An expendable-retrievable gun is constructed so that some of the gun disintegrates when the charges are fired, but a large portion remains intact for reuse.

Any type of gun may be fired with pressure higher in the casing than in the formation or with pressure lower in the casing than in the formation. If the pressure is lower in the casing than it is in the formation, any debris left in the perforation is washed out as the formation flows into the casing. Since clean perforations are essential to a good perforating job, perforating underbalanced is the preferred method. Be aware, however, that some formations contain small particles, or fines, that can flow with the formation fluids. These fines may pack into and close off the perforations.

The fluid in the casing opposite the zone to be perforated should be as free of solids as possible to prevent plugging of the perforations. Nitrogen is

sometimes used because it has low density and it is clean.

Perforating equipment should be handled only by experienced personnel. No one should stand near or handle the perforating guns unless it is necessary. All radios should be turned off and welding discontinued during perforating. Special warnings concerning radio restrictions should be given to tugs, boats, and helicopters in the area.

Drill Stem Testing

Drill stem testing is a temporary completion to determine the productivity of a zone and to estimate its size. Open-hole drill stem testing is often done when the well is first drilled to determine the economic feasibility of setting casing to produce a zone. A drill stem test (DST) helps select a completion method because it yields information about the reservoir's fluid, pressure, and ability to produce.

To run a DST in a cased well, a packer is run to a point above the zone to be tested after the zone has been perforated. Once the zone has been isolated with the packer, a surface-operated valve is opened to allow flow into the work string. Flow may or may not be allowed to come to the surface. The valve is then closed and time is allowed for the bottomhole pressure to build up and be recorded. From this information, it is possible to estimate reservoir size. After the test, the packer is released. Fluids are either circulated out of the hole or brought to the surface as the string is pulled. Regulations require that the DST string be reverse-circulated before the DST assembly is pulled from the well.

Often, a water cushion is run to reduce the amount of hydrostatic head in the work string. Reducing hydrostatic pressure in the work string allows the well to flow easily. Note that if the reservoir pressure is very low, it may be necessary to swab formation fluids into the string even when a water cushion is used.

The length of the required water cushion can be calculated using the estimated bottomhole pressure (BHP) and the following formula:

$$WC_{ft} = P_d \div F_g \qquad \text{(Eq. 66)}$$

where

WC_{ft} = water cushion, ft

P_d = pressure differential, psi

F_g = fluid gradient, psi/ft.

As an example of determining the length of a water cushion, consider the following. A well to be tested has an estimated BHP of 2,852 psi at a formation depth of 10,000 feet. The operator specifies that a 200-psi differential into the work string be provided to allow the well to start flowing. Thus, the well requires 2,652 psi of water cushion. If an 8.5-ppg salt water is available for the cushion, then its length can be calculated as follows:

$$WC_{ft} = 2,652 \div (8.5 \times 0.052)$$
$$= 2,652 \div 0.442$$
$$WC_{ft} = 6,000 \text{ ft.}$$

To determine the quantity of salt water that is required to provide a water cushion of the proper length, the length of the cushion is simply multiplied by the capacity of the tubing. As an example, determine how many bbl of salt water are required to fill 6,000 ft of 2⅞-in. tubing with a capacity of 0.00579 bbl/ft. The answer is 6,000 ft × 0.00579 = 34.74 bbl.

When a formation has very low pressure, the tubing may be run dry. If no cushion is used, be sure the hydrostatic pressure in the annulus is not high enough to collapse the tubing. Running the work string dry may result in too great a surge when the DST tool is opened. A string choke of some type should always be run on bottom. The swivel and kelly hose should not be used as part of the test line.

Drill stem testing should only be done during daylight hours. Two pressure recorders should be used; if both work well, their results can be compared. If one fails, the other serves as a backup. Always record all data possible during the test period and take samples of any recovered fluids. When running DSTs, all available safety devices and procedures should be implemented.

Acidizing

Productivity can be lost because of damage in the reservoir around the wellbore. The mud used to drill the well often has a high solids content and a density greater than that needed to prevent kicks. Such characteristics can damage the pay zone. Usually, acids can be pumped into a formation to remove some or all of the damage caused by drilling mud. What is more, some formations, such as those comprising limestone or dolomite, can be treated with acid to improve their permeability.

For a successful acid job, the nature of the problem must first be determined and the character of the producing formation understood. This information is obtained from core analysis data and from well logs. If the problem is mud solids that have damaged the formation and not low permeability, damage may also have resulted from the drilling mud's high water loss. High water loss may have caused swelling of the formation's own natural bentonite clays. Acid treatment can reduce this swelling. By reducing or eliminating formation damage, the drainage area is increased and the producing zone may clear itself of solids that create blockage.

When formation damage is present, the acid injection pressure should be kept at pressures below the fracture, or breakdown, pressure. If the formation fractures, the acid rapidly flows through the damaged area and into the fracture without cleaning the damaged zone, which occurs only in the first few inches around the wellbore. Acidizing without fracturing the formation is usually referred to as *matrix acidizing*.

When a formation lacks permeability, matrix acidizing is usually not adequate. Instead, the formation's fracture pressure is exceeded to force acid into the formation to create a larger flow area by dissolving some of the formation's bedding material. This procedure is referred to as an *acid frac job*.

Care must be taken when acidizing oil-producing sands in which movable water also exists, because more vertical permeability than radial permeability can be created. If more vertical than radial permeability is created, then water from zones below the oil zone can also be produced.

The time of acid exposure depends on the nature of the material being dissolved and on the acid in use. Hydrochloric acid (HCl) is the most commonly used, but others include hydrofluoric acid (HF), acetic acid (CH_3COOH), and formic acid (HCOOH). Information on their use and hazards associated with them should be obtained from the supplier.

Surfactants and solvents are other additives used in cleaning up the formation. They act as a soap, or cleaning solution, and aid in preventing gels and emulsions that form by fines or silts mixing with the spent acid.

Pumps and equipment for acidizing, the product to be used, and the time requirements are normally handled by the acidizing service company. Planning and safety precautions are essential. General safety considerations should include the following:

1. Use only steel hoses and have only essential personnel in the area. A readily available supply of water should be nearby to wash anyone who is contaminated by acid or other harmful chemicals.

2. Test all lines to pressures in excess of those to be used during the job. Tie down all lines. Check for any leaks. Do not overfill tanks.

3. Be sure pressure gauges are installed and working. Avoid acid spills and clean up at once if they do occur.

4. A check valve should always be installed at the wellhead. Whenever a check valve is placed in a line, be sure that a tee and valve or some other means of releasing the trapped pressure between the wellhead and the check valve is available; otherwise, the valve cannot be removed when it is no longer needed.

5. Almost all materials used in acidizing are hazardous. Always pour acid into water, not water into acid. Never breathe acid fumes.

6. Safety clothing and equipment should be used and in good working order.

7. Respirators should be available and wind direction monitored.

8. Have a safety meeting prior to the acid job and know what to do in case of burns, eye injury, or fume poisoning. Do not swallow any of the materials used. It must be remembered that accidental mixing of some acidizing materials may cause an explosion. Some corrosion inhibitors can be fatal even if absorbed through the skin. In some instances, hydrogen sulfide (H_2S) and other toxic gases may be formed.

Sand Control

The production of sand with reservoir fluids is a major problem in some areas. It can cut or plug the choke and flow lines, cause excessive equipment failure, complicate well clean out, and cause malfunctions of downhole equipment. Sand disposal can also be a problem in some locations. Methods used to control the production of sand include running screens or slotted liners, packing with gravel, and sand consolidation using a plastic resin.

Screens are the simplest to install in most cases. The job consists of hanging a slotted liner or wire-wrapped screen opposite the sand-producing interval. The screen is sized to allow formation fluids to flow but not sand.

Gravel packing is probably the most common sand control method (fig. 8.2). The well is cleaned out, perforated with large holes, and gravel is pumped in to hold formation sand in place. The gravel should be sieved on location and tested for silts, clays, and fines. It should be round and contain only a small percentage of flat or angular grains. Only silica should be used because of its grain strength. Gravel size is determined by the service company and based on core or produced-sand analysis.

Plastics of various types are used to consolidate sands, some with materials such as ground walnut hulls blended into them. Plastic leaves a permeability small enough to prevent sand inflow but large enough to allow fluid flow. Reservoirs that produce sand usually have poor consolidating material. The plastic glues the sand around the wellbore area to hold the sand in place. Plastic consolidation projects are usually designed by the service company and the operating company's engineers. Keep in mind that sand consolidation plastics can contain chemicals that are highly irritating to the eyes, lungs, and skin.

Fracturing

Fracturing consists of pumping a propping material into a hydraulic pressure-created crack in the formation to improve the performance of the well. Fractures are usually vertical and extend out in two directions opposite each other from the wellbore. In tight formations, fractures increase the flow area to the well.

Various fluids are used during the pumping operation as carriers for the proppant. The proppant is usually sand, but metal or glass beads may be used. The proppant must be as round as possible and contain no fines or clay. Since it must hold the fracture open, compressive strength is important.

A typical frac job starts out by filling the tubing with salt water. Pump pressure is increased until formation fracture pressure is reached. A steady injection rate is established and maintained, and a proppant is added downstream of the pump in low concentrations. The amount of proppant is increased until the required amount of fluid and proppant is in the well. When all the frac material has been injected, fluid pushes it into the formation. The final amount

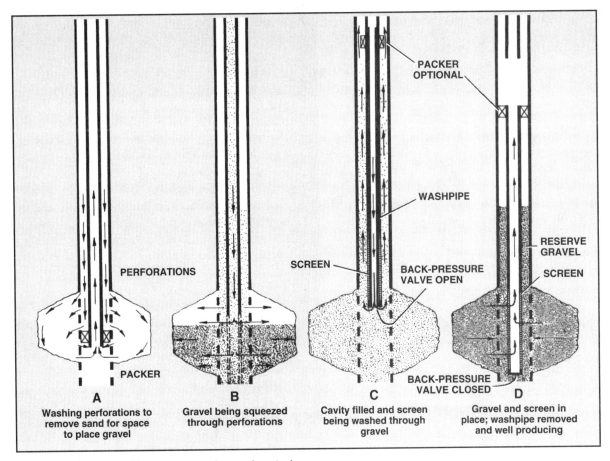

Figure 8.2   Gravel-packing procedures for sand control

should be carefully measured to ensure that the well is not overflushed, since the fracture should not close near the wellbore.

Radioactive sand is sometimes used as a proppant so that the well can be logged to determine the height of the fracture. Care must be taken to prevent exposure to this material while it is on location waiting to be used. Also, do not handle any sand that may be produced after radioactive sand is pumped.

Thorough planning is essential to any frac treatment. Ensure that adequate pump pressure and hydraulic horsepower (HHP) are available to initiate and propagate the fracture. Meetings should be held to plan the activity, and the following items should be discussed:

1. Hazards and safety precautions;
2. Each person's job and location;
3. Proper testing methods and pressures;
4. Contingency plans;

5. Proper clothing and personal safety equipment, such as ear protection, safety glasses, and rubber gloves;
6. Lines of communication;
7. The placement of No Smoking signs and warnings;
8. Precautions if radioactive material is used;
9. Emergency handling of personnel in the event of an accident; and
10. Possible evacuation procedures.

Plugging Back

Moving the completion interval from a lower zone to a formation higher in the well is known as *plugging back*. Plugging back is a routine operation when handled properly. Once the rig is on location and the well has been killed, the lower, or old, producing formation is squeezed off. Squeezing is often done through an old permanent packer already in the well.

The seals and stinger are pulled; the seals are repaired; the stinger and seals are run back into the packer; and cement is pumped down the tubing, through the packer, and out the perforations. The packer is left in the hole as a plug on top of the cement. Federal and state regulations often require that a cement plug be left on top of the old packer for added safety.

Retrievable packers are often squeezed through or a special cementing tool can be used. Regulations require that a cement plug be run for safety. After cement in the lower section of the hole has had time to set up properly, the upper new reservoir is perforated and put on production.

Plugging and Abandonment

A time comes in the life of every well when it will never be produced again or when it is uneconomical to continue production. The operator could simply close the master valves on the Christmas tree and declare the well abandoned. Many reasons exist, however, to explain why simply closing valves to abandon a well is not good practice:

1. If the well is left as is, the casing eventually deteriorates and fluid migrates from one zone to another.
2. Pressured formations eventually contaminate freshwater zones.
3. Blowouts could occur, with pollution and human hazard.
4. Water locations could become a navigation hazard.

In good plugging and abandoning practice, the producing perforations are squeeze cemented. The required cement plugs are placed in the casing as the tubing, or work string, is pulled out of the hole. Often, the upper, uncemented portion of the casing is cut off and recovered; then cement plugs are set in the upper area of the hole. The wellhead is removed as required by most regulations.

Deepening

Many reasons exist for deepening wells. In some cases, the technology of a relatively few years ago was not sufficient to allow wells to be drilled to depths that today are routinely achieved. In other cases, economics dictated the total depth of a well—that is, a well that, say, five years ago could not economically be drilled to a given depth because of the low price of oil, could, in today's economic climate, be drilled deeper.

In still other cases, production from shallow wells can be adversely affected by offset production from nearby, deeper wells. The shallow wells therefore must be deepened to prevent offset drainage.

Deepening requires that all well-control information for drilling operations be understood and applied. If the well has been on production before deepening, the perforations are squeezed off with cement. The well is then deepened, logged, and tested.

When the new depth is reached, a liner is usually run from a point above the bottom of the casing to total depth. This liner is cemented in place and the well is perforated in the new interval. The new formation is then put on production after running a packer and tubing.

Sidetracking

Sidetracking is a way in which the lower part of an existing well can be abandoned or bypassed. Many reasons for sidetracking exist (fig. 8.3): damaged or collapsed casing; irretrievable junk in the hole; a damaged production zone in the old well; or a less-depleted drainage area lying adjacent to a depleted drainage area. To sidetrack an old well, the first step is to cut a window in the casing of the old well. When cutting a window in an old well, it is easier to cut the casing if it is backed up with a good sheath of cement. If no cement is opposite the window, cement should be squeezed or circulated into place.

A window is cut in the casing after a kickoff tool or a whipstock packer is set at the proper depth. Drilling is then directed out of the hole and to the desired location by setting the tapered whipstock at different points to change the route of the new hole. When the desired depth and target are reached, the new hole is logged and a liner is run and cemented in place. Completion is then carried out in the normal manner with a packer and tubing.

MULTIPLE COMPLETIONS

A multiple completion is one in which two or more pay intervals are produced simultaneously through the same wellbore without commingling the fluids.

Zone separation not only is desirable for reservoir control, it is also compulsory under most state regulations. Multiple completions have the advantage of allowing additional production from each zone. In addition, (1) two or more marginal zones that may not warrant separate wells can sometimes

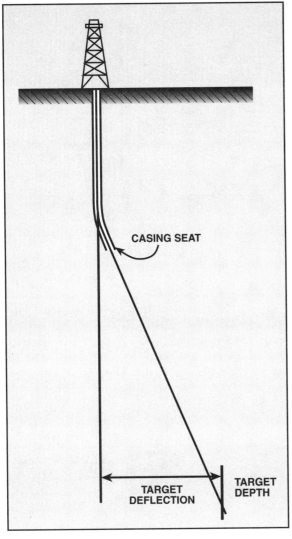

Figure 8.3. A sidetracked well

be produced economically; (2) more production for a given amount of pipe can be achieved because the equivalent of several producers can be obtained from one well; and (3) fields can be developed more quickly because fewer wells need be drilled. Multiple completions are particularly effective offshore, where platforms are very expensive and drilling costs are high.

In the past, many operators were reluctant to apply multizone completion techniques because the procedures were complicated. Moreover, considerable doubt existed that the desired separation between zones was actually obtained. With the advent of modern packers and other downhole equipment, most doubts have been eliminated. In fact, equipment is now available for running five strings of tubing inside casing.

In any multiple completion, advanced planning is vital to ensure that the drilling and casing program is adequate to meet the requirements of the completion. In short, the well should be carefully designed if it is to be successful.

The first multiple completions were duals, in which the lower zone was produced through the tubing and the upper zone was produced through the annulus between the casing and the tubing. This method was generally satisfactory, but producing through tubing is more dependable and permits swabbing, the running of bottomhole pressure tests, and other operations. Thus, an individual tubing string for each zone is usually preferred. A parallel-string multiple completion may use many different sizes of casing and tubing, but several popular combinations are shown in table 5.

Table 5

Multizone Completion Casing and Tubing Sizes (in.)

API Casing Size	API Tubing Size		
	Dual Completion	Triple Completion	Quadruple Completion
5½	1½*		
7	2⅜	2¹⁄₁₆*	
7	2⅞	2⅜	
9⅝		2⅞	2⅜

*Non-API tubing

A triple- or a quadruple-zone completion with parallel tubing strings is basically the same as a dual, except for the additional strings of tubing and packers. Because of the similarities, only dual completions are described in this section, but remember that where duals are discussed, they could be made triples and quadruples by simply adding additional tubing strings and related equipment.

In a typical dual completion, casing is set and cemented through the two productive intervals, the wellhead is nippled up, and blowout preventers are installed. The zones proposed for completion are logged and often drill stem tested to ensure satisfactory production from each zone before the two strings of tubing and related equipment are run into the well. Drill pipe suitable for a work string is picked up, the well is cleaned out to the desired depth, and the drilling mud inside the casing is displaced by salt water or other suitable completion fluid. Salt water is often ideal for perforating because it holds most formation pressures, is usually readily available, and generally causes only minimal formation damage.

The completion plan usually includes running a perforating depth-control log, such as a gamma ray and collar locator log, then perforating the upper zone and running a formation test. The formation test should include shut-in pressures for reservoir evaluation. Flow from the well may be routed to a separator so that fluid and gas samples can be drawn, the gas-oil ratio can be determined, and the oil and gas volumes produced can be gauged.

The lower zone is then perforated and a DST taken in the same way as the first one. If either zone produces water, squeezing or plugging back to minimize or eliminate water production should be considered. If both zones prove acceptable for completion, before running dual completion equipment, a bit and casing scraper should be run to smooth the rough spots in the casing, particularly through the perforated intervals, so damage to the packer is minimized.

Although various types of mechanical-set packer may be employed for multiple completions, equipment and procedures described here concern a nonretrievable packer for the lower zone, and a hydraulic-set, retrievable dual packer for the upper position (fig. 8.4).

The first step of the completion procedure is to run and set the lower packer. In this case, a nonretrievable,

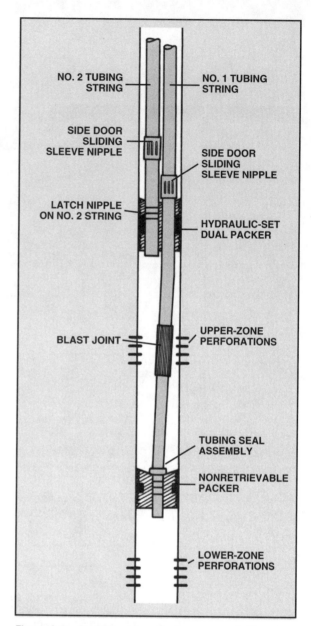

Figure 8.4   Equipment in a dual-zone completion

or permanent, type is set by wireline or run on drill pipe. This packer is set about 30 ft above the lower perforations. The No. 1 string—the long string—with the tubing seal assembly seats in the lower packer. Other well equipment assembled in order includes sufficient tubing to place the upper packer about 30 ft above the upper perforations, with a specially prepared blast joint opposite the perforations, a dual packer (a sliding sleeve device), and tubing to extend to the surface. The sliding sleeve tool is used to open ports to permit circulation between the tubing and the casing. Figure 8.5 illustrates one type of hydraulic-set, retrievable dual packer.

When the first string has been completely run, the tubing seal assembly is seated in the lower packer. Tubing nipples of various lengths are used as spacers so the string can be landed in the tubing hanger with the proper tension. Some of the details of a dual tubing head are shown in figure 8.6. Sometimes, both tubing strings are run simultaneously with the dual packer, using special slips and elevators for handling the two strings of pipe at one time.

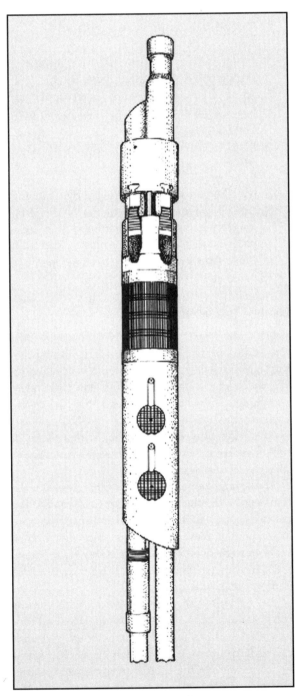

Figure 8.5   Hydraulic-set dual packer

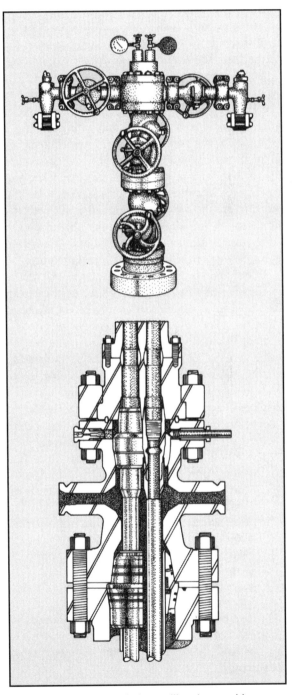

Figure 8.6   Dual-completion wellhead assembly

The No. 2 string is then assembled, starting with a latch nipple to seat and seal in the flow tube of the dual packer. Tubing is picked up to extend to the surface and spaced out and suspended at the tubing head, as was done with the first string. Plugs are then placed in each tubing string at the hanger, the blowout preventers are removed, and the top of the Christmas tree is installed.

The lowermost valve in this assembly controls flow through the No. 1 string, which produces the lower zone. The next valve controls the No. 2 string, which handles the flow from the upper zone. The wellhead fittings are pressure tested, all seals are checked for pressure tightness, and the tubing plugs are removed. Fresh water is pumped into the No. 2 string, following a setting ball, which seats in a catcher in the upper, hydraulic-set packer. Fluid circulation up to this point is around the upper packer and through the annulus to the surface. When the ball seats, hydraulic pressure is applied to set both the upper and lower slips of the upper packer, and to compress the resilient sealing element. The setting ball is flowed out of the string by well pressure.

If formation pressure is sufficient, the well can be made to flow through the short, No. 2 string to clean up the upper zone; otherwise, this zone can be swabbed into production. The tubing can also be unloaded by opening the sliding sleeve and displacing the fluid by using gas. The long, No. 1 string can be swabbed to permit the lower zone to flow, or the sliding sleeve ports in the dual packer can be opened and load fluid displaced, as described for the No. 2 string. The sliding sleeves are then closed using wire-line tools before cleaning up the zones by forcing flow to the pit.

The following summarizes the steps in a dual completion. To run the No. 1 string:

1. Run and set lower packer about 30 ft above lower perforations.
2. Assemble enough tubing to place upper packer about 30 ft above upper perforations.
3. Place specially prepared blast joints opposite perforations.
4. Place dual packer.
5. Place tubing that extends to surface.
6. Seat tubing seal assembly in lower packer.
7. Use tubing nipples as spacers to allow string to be landed with proper tension in tubing hanger.

To run the No. 2 string:

1. Place latch nipple to seat and seal in flow tube of dual packer.

2. Pick up tubing to extend to surface.
3. Space out and suspend string at tubing head.
4. Place plugs in each string at hanger.

After both strings are run:

1. Remove blowout preventers.
2. Install top of Christmas tree.
3. Pressure test wellhead fittings.
4. Test all seals for pressure tightness.
5. Remove tubing plugs.
6. Pump fresh water into No. 2 string after a setting ball that seats in upper packer.
7. Apply hydraulic pressure to set upper and lower slips of upper packer and to compress sealing element.
8. Force setting ball out of string with well pressure.
9. Bring upper zone into production by making well flow through short string or by swabbing.
10. Bring lower zone into production by swabbing long string (No. 1) or by opening sliding sleeve ports in dual packer and displacing load fluid.
11. Close sliding sleeves with wireline tools.
12. Force flow to pits to clean up both zones.

Multiple Tubingless Completions

A *tubingless completion* is a completion in which the flow path for the reservoir fluids is established through the casing alone. No tubing is installed inside the casing. Usually, the casing set in a tubingless completion is smaller in diameter than the casing used in a conventional completion, in which tubing is used as the flow string. A major advantage of tubingless completions is the cost savings made possible by not running tubing.

The savings gained from single tubingless completions also holds when multiple strings of small casing are run in open hole. Procedures and equipment are available that allow numerous strings of pipe to be set in the same borehole. Hole sizes are usually 7⅞-in. to 9⅞-in., but may be larger. Table 6 gives some hole and pipe combinations.

Casing with a diameter of 4½ in., with various combinations of small pipe sizes, has been set, but inner strings of tubing are usually employed with casing larger than 3½ in. In one Texas Gulf Coast field, for example, the hole for an eight-zone completion was drilled, logged, and tested in the usual manner. Then eight strings of casing were suspended

Table 6

Multiple Tubingless Completion

Hole and Pipe Combinations

Hole Size (in.)	No. of Strings	Size (in.)
7⅞	2	2⅞
8¾	3	2⅞
9⅞	4	2⅞
9⅞	3	3½

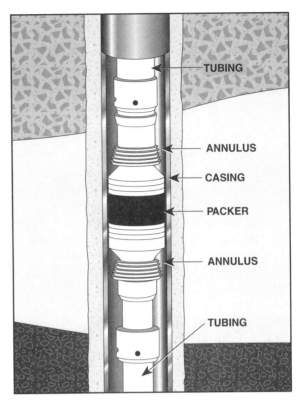

Figure 8.7   A packer goes around tubing and seals the casing-tubing annulus.

to produce each zone separately. A joint of larger casing at the top of the well provided room for running and hanging the multiple strings.

Either conventional float shoes and float collars, or float shoes with latch-down plug devices may be used, but reciprocating scratchers instead of the rotating type must be employed. Various types of centralizers provide standoff from the wall of the hole and between strings. Each string is run independently, but the longest is run first. Most operators set the first string in the hanger at the wellhead and circulate while running the second string. They usually run two strings to the deepest zone and switch strings if necessary.

TESTING AND COMPLETING PRESSURE-CONTROL EQUIPMENT

Packers

A packer seals the annular space between the casing and tubing (fig. 8.7). It provides a secure seal between everything above and below where it is set, and keeps well fluids and pressure away from the casing above it. The sealing element of a packer, the packing element, is a dense synthetic rubber ring that expands against the side of the casing. A packer may have one packing element or several, separated by metal rings.

If the packer moves up or down while the packing elements are expanded, they will rub off against the casing, much like an automobile tire rubs against a curb. To prevent packer movement, packers have slips to hold them in place. Slips are serrated pieces of metal that grip the side of the casing (fig. 8.8).

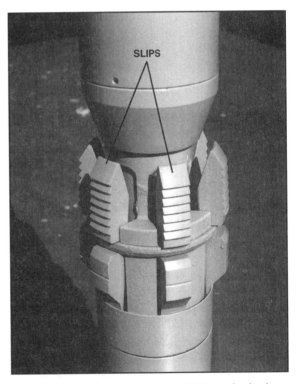

Figure 8.8   The slips grip casing to hold the packer in place.

Upward-pointing slips prevent upward movement and downward-pointing slips prevent downward movement. While the packer is being run into the hole, the slips are held against the packer. A setting mechanism then moves them outward to grip the casing when the packer is ready to be set at the desired depth.

Most packers have a circulation valve that allows fluids to pass through them. Although a packer usually seals fluids off from the annulus, sometimes crew members open the circulation valve and use a circulating fluid to set or retrieve the packer.

When packer failure occurs, or when a well is to be worked over, crew members remove the packer. How they remove it depends on whether it is a retrievable or a permanent packer. They unseat and pull retrievable packers, but they must remove permanent packers with a milling tool. A service and supply representative can redress most retrievable packers with seals and slips at the job site.

Specially designed packers are available for specific jobs or circumstances. For example, whipstock packers, which are used for sidetracking out of cased hole, can be oriented directionally from the rig floor. Inflatable test packers can be run into the hole and inflated or deflated from the surface through external ¼-in. tubing. Dual packer assemblies are often run for drill stem testing.

Packers can be set in several ways. The most common setting mechanisms are—

1. *hydraulic.* A pump-out ball seat near the bottom of the tubing provides a means of applying setting pressure. After the packer is set and the ball and seat are pumped out, a seal assembly is run as part of the tubing string.
2. *mechanical.* The packer, plus a seal assembly, is run on tubing to the setting depth. The upper slips are released by right-hand rotation of the tubing. An upward pull on the tubing sets the packing element and lower slips.
3. *electric wireline.* A small charge of electrical current, transmitted through wireline, ignites a charge in the setting assembly, gradually building up gas pressure. The pressure sets the packer. When it is necessary to retrieve the packer, a prescribed setting force is applied, which causes a release stud to part and free the setting equipment from the packer, allowing it to be pulled from the hole.

4. *sandline* or *slickline.* A pressure-setting assembly is installed in the packer and run to the desired setting depth on the line. A go-devil is installed on the slickline or sandline and dropped into the well. The go-devil firing head mechanically activates the setting assembly by firing a blank cartridge, thus detonating a secondary igniter and powder charge in the setting assembly.

Lubricators

A lubricator uses grease to form a seal between the wireline and tubing or casing. A lubricator consists of several components: a stuffing box, top catch tool, risers (lubricators), valves, unions, a tool trap, and, usually, a BOP (fig. 8.9). The stuffing box provides a seal against well pressure. Riser joints are lengths of tubing whose rated working pressures are higher than wellhead pressure. Riser joints are also selected to be of sufficient diameter and length to accommodate the workover tools anticipated for a particular job. A wireline BOP that can be closed quickly to form a seal around the wireline is also part of a lubricator assembly.

Lubricators are normally installed above the master valve on the Christmas tree; or, if a blowout preventer is in use, it is installed on top of the preventer. To use a lubricator, a string of tools is made up on wireline and inserted into the lubricator. The wireline is then threaded through the stuffing box with the tools positioned below the box. The well's master valve is closed, and the lubricator assembly is installed on the wellhead. The lubricator is then pressured up to anticipated well pressure. The master valve on the tree is opened, and the tools are lowered into the hole to perform the work. A lubricator may also have a tool trap, which catches tools if the wire parts or if it is pulled from the rope socket.

Shop Tests

Some operating companies require that a lubricator be pressure-tested by a service company in its shop every 6 months to 1½ times its working pressure. A dated report with the service company supervisor's signature that certifies the test accompanies the lubricator assembly when it is carried to the job. Most service companies inspect and test each lubricator when it is returned to the shop. The shop test is usually performed by filling the lubricator with

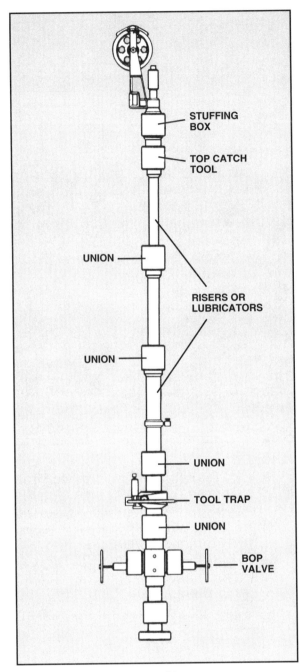

STUFFING
BOX

TOP CATCH
TOOL

UNION

RISERS OR
LUBRICATORS

UNION

UNION

TOOL TRAP

UNION

BOP
VALVE

Figure 8.9 Lubricator assembly

water and pressuring it up with a hand pump. Test pressures up to 22,500 psi may be applied, depending on the particular lubricator.

Field Tests

Most operators also require that a lubricator be field-tested before it is used on a well. A service company normally furnishes a pump to make the test. The test

comprises the following steps:

1. Record the tubing pressure.
2. Close all valves on the well.
3. Place the wireline BOP with its adapter on the Christmas tree.
4. Pressure below the BOP to the working pressure of the lubricator or Christmas tree. Do not exceed the lower working pressure. Hold for 15 min. If test pressure holds, bleed it off.
5. Install the stuffing box and the lubricator, with the line and wireline tools above the BOP. Fill the lubricator with fluid and bleed off air. Test the assembly to working pressure, as in step 4. If the test is good, bleed off the pressure.
6. Proceed with wireline operations.

In addition to the test procedures, most operators also recommend that—

1. the lubricator be secured at all times.
2. the lubricator cover the entire string of tools and fish when possible.
3. the lubricator support be in position before wireline tools are pulled up.
4. where possible, a platform be provided so that personnel do not stand on wellhead connections.
5. the lubricator have a pump in-line below it and be equipped with a high-pressure, low-torque valve.
6. when bleeding down the lubricator, the valve be opened and closed several times to prevent valve freeze.

Christmas Trees

A Christmas tree is an assembly of valves, spools, flanges, and connections that control the flow of fluids from the well (fig. 8.10). Because a Christmas tree controls the flow of fluids from a well, crew members must be careful not to damage it when they move in and rig up. Carelessness could prove fatal to personnel and could destroy the rig.

Many types of Christmas tree are available. Some, like those on pumping wells, may be simple and consist mainly of a stuffing box. On the other hand, complex trees with numerous master and wing valves may be required on deep, high-pressure gas wells. Each well is unique and requires a specific type of tree. In spite of the wide variety of trees available, they share certain basic components:

1. *Pressure gauges.* Pressure gauges monitor tubing pressure and casing, or annular, pressure.

Figure 8.10 The Christmas tree is mounted on top of the tubing head.

2. *Gauge flange, or cap.* The gauge flange seals the top of the tree and has a fitting for a pressure gauge. When the gauge flange is removed, the tubing becomes accessible and bottomhole test or lubricator equipment can be installed.

3. *Crown, or swab, valve.* The crown valve shuts off pressure and allows access to the well for wireline, coil tubing, or other workover units to be rigged up.

4. *Flow, or cross, tee.* The flow tee allows tools to be run into the well while the well is producing.

5. *Wing valve.* A wing valve shuts in the well for most routine operations. Wing valves are the easiest valve to replace on the tree.

6. *Choke.* The choke controls the amount of flow from the well.

7. *Master valves.* Master valves are main shut-off valves. They are open most of a well's life and are used as little as possible, especially the lower master valve, to avoid wear or damage to them.

8. *Tubing hanger.* The tubing hanger supports the tubing string, seals off the casing annulus, and allows flow to the Christmas tree.

9. *Casing valve.* The casing valve gives access to the area between the tubing and the casing.

10. *Casing hanger.* The casing hanger is a slip-and-seal assembly from which the casing string is suspended.

11. *Casing.* Casing is a string of pipe that keeps the wellbore from caving in and prevents communication from one zone into another.

12. *Tubing.* Tubing is a string of pipe through which produced fluids flow.

Christmas Tree Removal

Removing a Christmas tree requires careful planning. All procedures should be defined, reviewed, and understood by company, rig, and service personnel involved in the job. When planning the removal of a tree, questions that should be considered include—

1. Is the tree to be removed before or after the rig is moved in?
2. If service on the tree is required, will it be sent to a shop, or will it be serviced at the location?
3. Is the tree manufacturer's service representative present, and are replacement parts available?
4. Is the rig's BOP equipment ready for immediate installation?
5. Is the well to be killed or is it to be worked on under pressure?

Other points to consider are—

1. All exposed tree flanges should be protected and all BOP flanges should be inspected and cleaned.
2. New seal rings should be available, for once a metal seal ring has been used, it is permanently distorted and must be replaced.
3. Tubing and casing pressures should be checked with gauges known to be working properly.
4. If the well is to be killed, the casing should be installed and cemented properly.
5. No communication between tubing and casing should exist; if communication exists, all points of failure should be repaired.

Once the job has been planned and preliminary steps taken, work can begin. Assuming that the well is to be killed prior to the tree's removal, the first step is to pump kill fluid into the tubing and bullhead the fluid into the formation. *Bullheading* is pumping fluid down the tubing and displacing the fluid into the formation. One way to ensure that the fluid has been bullheaded into the formation is to calculate the tubing's volume and then note when that volume of kill fluid has been pumped. Bear in mind, however, that clear fluids may fall faster than they are pumped. Also remember that gas migrates faster than it is bullheaded. Finally, be aware that if too much fluid is pumped, formation damage may occur. An increase in pump pressure should be a sign that the kill fluid has reached bottom.

Record the volumes pumped and the pressures. When the well is dead, set a wireline plug in the tubing. Shut in the well and check for pressure build up for about 1 hour (hr). Have a full opening valve available with proper threads and size to fit the tree. If, after 1 hr, no pressure build up occurs, remove the tree and install a BOP stack.

Test Trees

Test trees are used in well completion operations to determine a well's producing potential. In effect, a test tree allows the well to be temporarily completed before final completion procedures are initiated. Since the test tree will undergo well pressures just as a Christmas tree and other wellhead equipment, the same precautions in setting, pulling, and using test trees should be observed as those used when installing, maintaining, or removing a Christmas tree.

Subsea test trees are designed to be used in conjunction with the subsea BOP stack. A subsea test tree is used as a temporary master valve during well testing from a floating drilling vessel. The device is run into the well through the marine riser and seated in the well's subsea wellhead assembly. Special latches and hangers suspend the tree in the well, and special valves allow the operator to open and close the well and control flow from it. Most subsea test trees are designed to allow the shear rams in the rig's BOP stack to be closed, which means that, in an emergency, the rig can be moved off location, leaving the well shut in.

Chapter 9

Well Control and Floating Drilling Rigs

Drilling operations from floating vessels such as drill ships and semisubmersibles present special problems in well control. The problems occur because of hole depth, water depth, geology, and the design and operation of the subsea BOP stack and control system. While well-control procedures on floaters and land rigs are similar, several additional factors occur on floaters that must be taken into account if well control is to be successful.

SHALLOW-HOLE CONSIDERATIONS

When the first part of a well is being drilled offshore, this shallow section of hole presents a number of problems that must be dealt with. Two of the most serious are controlling a kick when only a short casing string has been set, and drilling an open-hole section prior to setting a long protective string. A number of blowouts have been caused by influxes of overpressured shallow gas into the wellbore. Because the hole is shallow, gas can quickly reach the surface with little warning. Often, because of pressure limitations at the casing shoe, it is not advisable to shut the well in on a shallow-gas kick. In such cases, the gas can be vented through some type of diverter system; or pilot holes may be drilled. Pilot holes do not involve using diverters.

Diverters are special annular BOPs that can be used on top of the marine riser or on top of the well in a subsea position. Surface diverters close the annulus around the tubular in the hole and direct (divert) gas flow into the atmosphere through vent lines. A pilot hole is a small diameter hole drilled out below drive pipe or conductor casing run into the well. No riser or subsea BOP stack is run; instead, the pilot hole is drilled and if it encounters shallow gas, the gas is allowed to flow into the sea.

Shallow Gas

The existence of gas in shallow zones can be an especially dangerous situation when drilling. Because the zone is shallow, the gas can escape to the surface within a very short period. Warning signs exist, but prompt action is necessary to prevent a blowout. Also, because of the possibility of formation fracture and broaching in shallow zones, the well often cannot be shut in safely. Since a shallow-gas kick can be so dangerous, rig crews should be especially alert for signs of a kick when surface hole is being drilled. Most well-control specialists recommend immediately shutting down the pump and making a flow check if any doubt exists about whether a shallow well has kicked. Since shallow gas reaches the surface so fast, the driller should also be especially careful to fill the hole properly when pulling the first few stands off bottom in the upper part of the hole. The hole should be filled carefully and watched between stands.

Most BOP stacks are capable of handling much more pressure than is found in underground formations. In the case of shallow gas, where the pressures are usually not excessively high, most BOPs are adequate. Further, the conductor or surface casing on which the BOP stack is mounted is usually capable of handling the pressures associated with shallow gas.

Unfortunately, while the stack and casing can withstand the pressures associated with shallow depths, shallow formations often cannot—they tend to fracture when the well is completely shut in. This tendency, combined with the fact that the casing is set at a shallow depth, can cause the fractured zone to extend to the surface, where fluids broach around the rig.

Planning

Because shallow gas can be particularly hazardous, most operators emphasize careful planning in areas where shallow-gas zones are known or suspected to exist. Planning for shallow-hole well control problems *before* the well is spudded is essential. During the planning phase, all available information about the area to be drilled should be studied closely to minimize the possibility of a blowout. Well files and the history of wells drilled in the area should be checked, daily drilling reports from nearby wells should be read, scout tickets should be located and studied, and geological data should be obtained and evaluated. Surveys that reveal the presence of shallow hazards are often available; they should be examined carefully. Area mud and casing programs should be evaluated, as well as bit records, logs, pressure and temperature measurements, and seismic data. Discussions should be held with operating and drilling personnel who have had experience in the area. And finally, crew training should be an essential part of planning. Training should be systematic and thorough so that everyone knows where to be, what to do, and what not to do if a shallow-gas kick occurs.

Broaching Gas and Buoyancy

When shallow gas broaches to the mud line and into the water, some well-control literature maintains that the water beneath the drilling vessel can be aerated enough to reduce the water's buoyancy and cause the rig to heel over and sink; however, at least one laboratory study has cast doubt on this assertion. The degree of risk depends on vessel design, load, weather conditions, and so on. In general, semisubmersibles are less affected by aerated water, or "gas boils." On the other hand, moored drilling vessels, with decks closer to the sea's surface, may be at more risk. Regardless of whether broaching gas reduces buoyancy to a critical point, shallow-gas

accumulations are potentially very dangerous because of the fire and pollution hazards, and every effort should be made to control them when they are encountered.

Diverter Systems

To overcome the problem of possible broaching, diverter systems have been used extensively. At least two types of diverter systems are available. One type is mounted on top of the marine riser above the water line (fig. 9.1A). The other type is mounted on the seafloor on top of the wellhead (fig. 9.1B). Using surface diverters, the well is not shut in; instead, flow is diverted a safe distance from the rig through a large-diameter diverter line. The diverter system is usually a part of the riser slip joint package. Since the top of the slip joint is fixed, it is a convenient place to cross over to the diverter lines. The diverter is designed to divert high-volume, low-pressure gas and prevent the buildup of pressures that would fracture the formation and cause an underground blowout.

Several brands of diverters are available. Regardless of manufacturer, however, most diverters consist of a packer insert that, when activated, seals around the drill stem; two or more diverter lines 6 in. to 12 in. in diameter with full-opening valves; and a control system. The diverter assembly is connected to the inner barrel of the slip joint and is positioned directly under the rotary table.

The diverter line valves may be of various types, such as ball, gate, diaphragm, knife, or switchable three-way target; others are an integral part of the diverter unit. Regardless of the type of valves in the diverter lines, the lines provide a means of diverting the flow of fluid overboard and downwind from the drilling vessel. When the packer is engaged, well fluids are free to flow out of one of the diverter lines. Some diverter systems are provided with a small surge chamber on the hydraulic control system to permit drill pipe and tool joints to be stripped through the closed diverter packer.

The basic design of a diverter system is described in *API RP 64, Diverter Systems Equipment and Operations. RP 64* states that the diverter lines should be sized to minimize, as much as practical, back-pressure on the wellbore while diverting well fluids. The Minerals Management Service (MMS) requires that diverter lines be at least 10 in. on jackups and 12 in. on

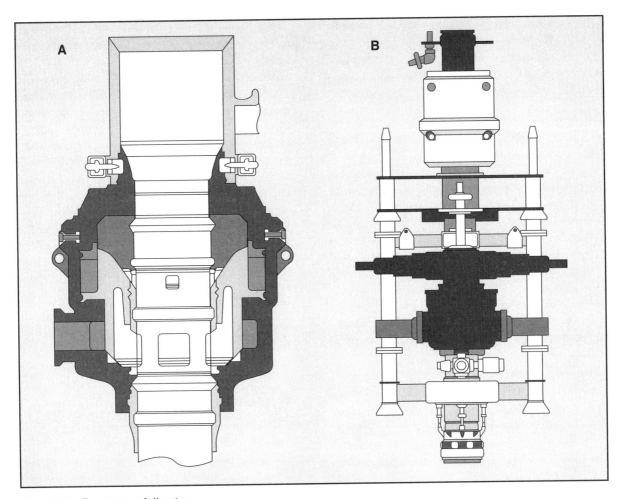

Figure 9.1   Two types of diverters

floaters that are drilling on the Outer Continental Shelf (OCS) of the United States.

RP 64 also states that the diverter system should be capable of diverting well flow overboard on the downwind side without causing excessive back-pressure on the system. Most diverter-control systems are designed to open either a port or a starboard diverter line as soon as the diverter packer is closed.

1. Response time of the diverter closing system should not exceed 45 sec on large-diameter holes and should be less on smaller wellbores.

2. The internal diameter of the vent lines usually should be from 8 in. to 12 in. (Remember that on the OCS of the United States, the MMS requires that the diameter of the diverter lines on jack-ups be at least 10 in. and that the diameter of the diverter lines on floaters be at least 12 in.)

3. A backup control panel to activate the diverter system from a remote location should be considered.

4. Flow and pressure tests of the diverter system are recommended.

5. Training and instruction of personnel in proper operation of the diverter system should be conducted to ensure that they are able to react to the short time requirements for closing the diverter.

6. The diverter should be tested each day and all valves verified to be fully open or fully closed. The following are some testing procedures:

   A. With drill pipe in the diverter, close the diverter.

   B. Record the time it takes for the diverter element to close.

C. Record the time it takes for both vent line valves to open.

D. Observe all valve openings and closings.

E. Observe whether the diverter element opens.

Maintaining and Testing Diverters

Because of the hostile saltwater environment and rugged operations that offshore diverter equipment is subjected to, most operators require a thorough maintenance and testing program to keep the equipment in good operating condition. Testing and maintenance are essential to protect the life of the crew and the rig; further, downtime can be minimized if the diverter system and, indeed, all blowout prevention equipment, is kept in excellent working condition. Maintenance data and recommendations from the manufacturer of the system should be included in the maintenance program. Moreover, *API RP 64,* should be referred to for its recommendations. Many operators follow these general guidelines:

1. Perform a function test daily.

2. Pressure test the diverter system at least once a week. Consider running more frequent tests in the shallow part of the hole.

3. Flush out the unit and visually inspect and lubricate it as required. Run a pressure test after the system has been inspected.

4. After prolonged operations, conduct an overhaul and make needed repairs. A factory representative should supervise the operation. Some operators recommend that major repairs, such as field welding, not be made on a failed unit unless supervised by the manufacturer's service representative.

5. Since kicks that require a diverter usually occur quickly, the driller and floor crew should thoroughly understand their assignments. Most operators recommend that diverter drills be a basic part of training when the shallow portion of the hole is being drilled. A diverter drill should include measuring the time it takes personnel to activate the diverter successfully.

Pilot Hole Drilling

Because of the hazards of handling shallow gas with diverter systems, some operators and contractors use pilot hole drilling to manage shallow gas. In this technique, crew members drill a relatively small-diameter borehole below the casing or conductor pipe or, when no pipe is set, directly into the seafloor. With floating rigs, normally no marine riser and BOP stack are used when drilling a pilot hole. If the pilot hole encounters shallow gas, the gas is allowed to blow from the hole and into the water. If the volume of gas is large, it may be necessary to move a floating rig off location a short distance. Sometimes, the move can be accomplished by winching off location, whereby slack is let off the anchor chains on one side of the rig and taken up on the other. This action moves the floater to a new position that is not directly over the hole.

Often, a flowing pilot hole will bridge over or the gas will blow to depletion in a relatively short period. In some cases, however, it may be necessary to stop the flow by setting a barite or cement plug in the pilot hole. Techniques for setting barite or gunk plugs are covered in chapter 7. For setting a cement plug, the crew usually runs a quick- or flash-setting cement. They pump the cement down the drill stem, out of the bit, and to the bottom of the pilot hole.

DEEPWATER CONSIDERATIONS

Deepwater well-control problems can occur as a result of difficulty in kick detection, loss of hydrostatic head because of riser disconnect, riser collapse because of gas unloading, and reduced fracture gradients because of deep water. Drilling vessel motion, especially heave, can cause additional wear to the BOP equipment and may part the drill pipe if the pipe becomes stuck.

Additional problems are associated with the fact that a subsea stack and control system are commonly used with floaters. Problem areas related to the subsea system include the diverter system, the effect of the riser choke line, the blowout preventer control system, the effect of water depth on annular preventers, drilling vessel movement, weather effects, hanging off drill pipe, and reaction time in operating the preventers.

Riser Choke-Line Effects

One of the more important factors in killing a well in deep water is the effect of the riser choke line on well-control techniques when circulating a kick out of the well. Attached to the marine riser, the choke line is usually 3-in. ID, extra-heavy-duty tubing with internal-flush stab joints. It runs from the stack to crossover lines located in the moon pool. A number of arrangements can be used with regard to line size, joints, and crossover and expansion lines.

Regardless of how the riser choke line is installed, the line and fittings between the subsea stack and the choke restrict fluid as it is circulated through the line during well-control procedures. The restriction causes a pressure loss due to the friction of the fluid on the walls of the line and internal friction within the fluid itself. In effect, the choke line is a fixed choke between the annulus of the hole and the casing pressure gauge on the rig floor.

The deeper the water, the longer the choke line is, and the greater is the choking effect of the friction losses. The choking effect of the riser choke line adds additional back-pressure on the well, which must be compensated for; otherwise, formation breakdown and loss of circulation could occur. Because choke-line friction losses increase with length, many operators recommend that, when drilling in deep water, kill-rate pressures be recorded at half the normal pump speed and at one or two still-slower speeds. By recording very slow kill-rate speeds, circulating pressure can be reduced to values low enough to reduce the riser choke line's choking effect.

If the friction pressure caused by the choke line is known (it can be determined by several methods), a formula is available that allows calculation of the equivalent mud weight this friction pressure represents at the casing shoe:

$$EMW_{cs} = (P_{cl} \div 0.052 \div D_{cs}) + MW \quad \text{(Eq. 67)}$$

where

EMW_{cs} = equivalent mud weight at casing shoe, ppg

P_{cl} = choke-line friction pressure, psi

D_{cs} = depth of casing shoe, ft.

MW = mud weight in well and choke lines

As an example, suppose that—

P_{cl} = 400 psi

D_{cs} = 4,500 ft.

MW = 12 ppg

What is the equivalent mud weight at the casing shoe that is caused by choke-line friction pressure?

$$EMW_{cs} = (400 \div 0.052 \div 4,500) + 12 \text{ ppg}$$
$$= 1.71 \text{ ppg} + 12 \text{ ppg}$$
$$EMW_{cs} = 13.71 \text{ ppg.}$$

The solution shows that choke-line friction of 400 psi adds 1.71 ppg the equivalent of mud weight at the casing shoe. Therefore, if 12-ppg mud is being circulated, the equivalent mud weight at the shoe is 13.71 ppg, a significant increase that could lead to fracture at the shoe.

Determining Choke-Line Pressure Loss

Several methods, including the following, can be used to determine choke-line pressure loss:

1. Circulate the well at the kill rate through the riser and record both pump rate and pump pressure.
2. Shut down the pump, close the annular BOP, and open the choke.
3. Open the valves on the riser choke line and circulate the well at the kill rate.
4. Pump pressure is greater when pumping through the choke line, so the difference between kill-rate pressure and the extra pressure required to circulate through the choke line is recorded as choke-line friction pressure.

This method should not be used if open hole is exposed, since it increases pressure on the formation by an amount equal to choke-line friction loss and could cause lost circulation. If open hole exists, another method may be used:

1. Close a preventer below the choke line.
2. Pump down the choke line and through the riser. The indicated pump pressure is approximately equal to choke-line friction pressure.

Gas in Choke Line

When influx fluids, particularly gas, are circulated from the relatively large annular space below the BOPs into the relatively small-diameter choke line, several things occur that makes precise control of pressures difficult. First, the replacement of drilling mud with low-density influx fluids in the choke line can cause a substantial reduction in hydrostatic pressure that occurs over a relatively short time. This reduction may affect bottomhole pressure and therefore *SIDPP* and may require compensating choke adjustments.

Second, if the influx is largely gas, the gas nearing the surface through the choke line may be expanding ever more rapidly, particularly in low pressure and delicate well-control situations. This expanding gas may cause an increase in flow rate through the choke and unpredictably raise *SICP*. Third, when gas begins passing through the choke manifold and choke, choke response is likely to change significantly because the gas flows easily through the choke. Because slugs of wellbore fluid will likely accompany the gas, it may be impossible to accurately control *SICP* and *SIDPP* with the choke.

It is therefore important for the choke operator to anti-cipate the need for rapid choke manipulation to try to respond to sudden changes in *SICP* as the kick fluids circulate up the choke line and out of the well. Depending on the severity of the changes and the success in maintaining the desired control pressures, a secondary kick could be admitted into the hole. Personnel should be aware of the possibility and, if necessary, continue kill method circulation long enough to confirm that any secondary kick is removed from the annulus before concluding kill operations.

Bringing the Pump Up to the Kill Rate with a Subsea Stack

During a well-killing operation, a common way to bring the pump up to the kill rate without changing bottomhole pressure is to keep *SICP* constant at the original shut-in value by opening the choke and bringing the pump up to kill-rate speed. When a subsea stack is in use, this procedure may have to be modified because of riser choke-line pressure. One way to modify the procedure is to allow *SICP* to drop by an amount equal to riser choke-line friction pressure as the choke is opened and the pump brought up to the kill rate. If *SICP* is held constant at its original shut-in value, too much back-pressure will be held on the well, which could lead to fracturing and loss of circulation.

Compensating for Low *SICP*

When *SICP* is less than the riser choke-line pressure, it is impossible to open the choke enough to reduce initial and final circulating pressures to the proper value. Choke-line pressure causes bottomhole pressure and thus drill pipe pressure to increase as kill-weight mud approaches the surface. This extra pressure may be sufficient to cause lost returns. Possible solutions include opening the riser kill line, reducing pump kill rate to a slower rate, or filling the choke line with kill-weight mud while keeping the wellbore isolated by use of a lower pipe ram.

Hydrates

In deepwater operations, hydrates can sometimes form in the choke and kill lines when a well is shut in on a gas kick and circulated to the surface. *Hydrates* are a mixture of natural gas and water that form a solid substance. This solid looks and behaves very much like ice. Under the pressures and temperatures encountered in deep water, conditions are favorable for the formation of hydrates. They can plug choke and kill lines and may also interfere with the operation of the BOPs. Plugging or partial plugging of the choke lines as a kick is being circulated could cause bottomhole pressure to get so high that the formation breaks down. Crews should, therefore, be alert to the possibility of hydrate plugging and be prepared to stop pumping if pressures rise too high.

Most operators agree that the best way to deal with hydrates is to prevent or minimize their occurrence. High-salinity drilling muds suppress hydrate formation. In any case, in areas in which hydrate formation is likely, planning should consider the possibility of hydrate formation during well-control operations.

BLOWOUT PREVENTER CONTROL SYSTEM

A subsea BOP control system is more complicated than the system used on surface stacks (fig. 9.2). Because it is more complicated, it can be misused by persons not knowledgeable in its operation. Therefore, it is important that certain procedures be understood and followed. One procedure for operating a typical subsea BOP hydraulically controlled system follows:

1. Before operating a function button or valve, check the readback-pressure to be sure proper pressure is on the system.
2. Press the control button firmly and hold it down until the function light indicates that the control has functioned.
3. Watch the readback-pressure. It should drop and then return to the same, or nearly the same, pressure as before.
4. The flowmeter should show that the approximate operating volume was pumped.
5. Operate the controls one function at a time so that you can see that each control has operated properly.

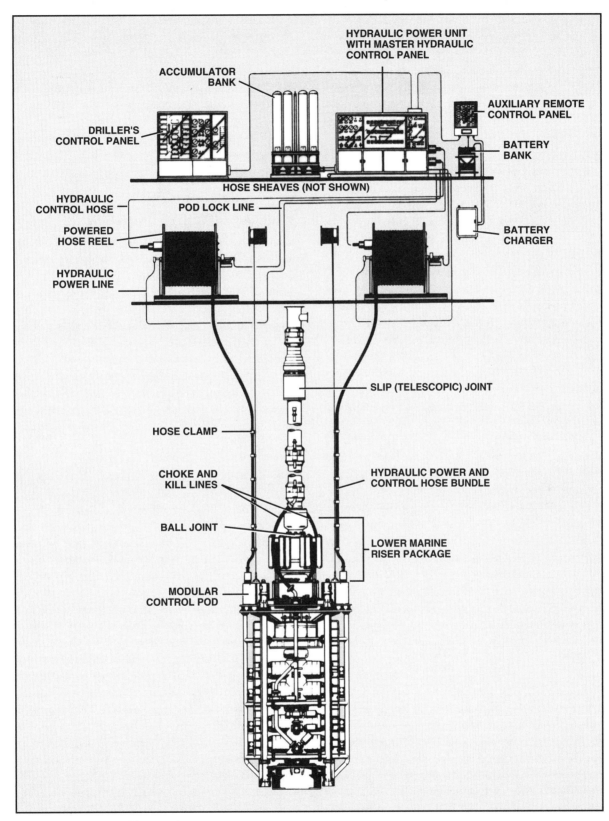

Figure 9.2   Subsea BOP control system

Two special pods that control the operation of the preventers are mounted on the subsea BOP stack. Termed "control pods," one is usually painted yellow and the other blue for purposes of quick identification. If one is not working properly, the other can be selected from the control panel and be used for BOP control. The BOP stack is controlled through one of the pods by means of pilot valves and regulators and, on most rigs, either pod can be retrieved independently of the stack or riser. Each pod has a guiding and orienting mechanism and a hydraulic latch that is operated from the surface.

Whenever a control is operated, the system dumps operating fluid—a soluble oil and water mixture—into the sea. Should the system continue to keep operating, it could empty the surface reserve tank. The flowmeter on the control panel indicates the flow of operating fluid in the system; thus, if the flowmeter does not stop, it is an indication that the system is still dumping fluid. If fluid continues to dump, open and close the functions to switch the control pods, or block the function to stop the loss of hydraulic fluid from the control system. Note that blocking a function removes both opening and closing pressures on the BOP element; it does not lock the element in either position. BOPs in blocked position should be considered to be out of service and alternate BOPs selected as needed for the situation.

Readback-pressure senses the pressure at the subsea accumulator bottles or below the subsea pressure regulator. If readback-pressure does not return to its former value after a function has been actuated, either a valve has not closed and hydraulic fluid is being pumped into the sea, or recharge hydraulic fluid from the surface is not reaching the subsea system. If readback-pressure fails to return to normal, it may be necessary to switch control pods, block the function, or cycle the function to try to clear it.

If a function on the control panel does not operate when the control button is pushed, the problem may be in the surface system. In addition to the obvious action of pushing the button again (many systems have a master button that may need to be pushed at the same time), the main accumulator unit should be checked to see if the control handle is in the proper position. If the control handle is in the proper position, switch the pod to see if a malfunction exists in the control lines. Sometimes cycling the function will clear the lines or control system.

If cycling the function does not solve the problem, the next step depends on rig conditions. During a test or a drill, the function should be cycled several times in an effort to clear debris from a valve or to clear the hang-up. During a well kick, if trouble appears with the upper annular preventer, the lower annular preventer or the pipe rams should be used. Subsea control systems vary from rig to rig, so it is essential for personnel to become familiar with the operating characteristics on their rig.

ANNULAR PREVENTER OPERATION

Each brand of annular preventer has its own operating characteristics and is affected by well pressures and water depth in a different manner. In general, annular BOPs have a maximum operating pressure limit of 1,500 psi, but most operators usually recommend that they be operated at about 800 psi. When using the annular preventer on a floater, a chart should be prepared based on tests of the required closing pressure and the manufacturer's chart. While the packing unit in an annular preventer is a rugged piece of equipment, it can be prematurely worn by putting too much pressure on its closing side. Too much pressure tears the packer and causes a number of problems with annular preventers on subsea BOP stacks. Further, since floaters heave—move up and down—the packing unit can be worn out prematurely when it is closed around the drill stem. Reducing closing pressure to the lowest possible value while maintaining a positive shut in minimizes packer wear. After the well is shut in, many operators recommend that the drill stem be hung off to help reduce wear on the packing unit.

The effect of water depth on the operation of an annular preventer varies with the brand or model of preventer and the weight of mud in the riser. In general, the weight of the mud in the riser tends to resist the closing of the annular preventer so that higher closing pressures are needed as water depth increases or as mud weight increases. In the models that require it, one manufacturer (Hydril) uses the following equation to determine how much more closing pressure is needed as water depth and mud weight increase:

$$CP_i = [(0.+052 \times MW \times D_w) \text{ (Eq. 68)}$$
$$- (0.45 \times D_w)] \div P$$

where

CP_i = closing pressure increase, psi

MW = mud weight, ppg

D_w = water depth, ft

P = a ratio for the preventer used (e.g., 4.74 for Hydril GK 5000, 13⅝).

Other manufacturers use similar equations or have models that require little if any correction for water depth and mud weight in the riser. In any case, closing pressure should be at least equal to the minimum pressure demonstrated to be sufficient to seal the annulus as confirmed by the previous weekly BOP pressure test.

MARINE RISER DISCONNECT

A riser disconnect or a leak in the riser can allow the well to kick or unload. Hydrostatic pressure is, of course, a direct function of mud weight and the height of the column of mud in the hole. In drilling from a floater with a subsea BOP stack, total depth is measured from the kelly bushing to the depth of interest. Should the riser disconnect or fail so that seawater replaces the drilling mud, the hydrostatic pressure in the riser would equal that of seawater. And if the required drilling mud gradient is higher than that of seawater, hydrostatic pressure will be too low to maintain control of the well. Thus, if riser failure occurs, and the mud gradient is greater than the seawater gradient, the well should be shut in and heavier mud circulated to regain control of the well. (The pressure gradient of seawater varies from one location to another. In the Gulf of Mexico, for example, seawater weighs about 8.94 ppg and has a gradient of about 0.465 psi/ft. In the North Sea, seawater weighs about 8.56 ppg and has a gradient of about 0.445 psi/ft.)

KILLING RISER AND TRAPPED GAS
IN THE BOP STACK

After kill-weight mud is circulated to the surface through the choke line, the well is dead, if the choke line is used as a reference point for hydrostatic pressure; however, the well is not dead through the riser. Kill-weight mud must be circulated through the riser to kill the well completely. Before kill-weight mud is circulated through the riser, though, be aware that gas can be trapped in the BOP stack in any area open to the well above the height of the choke outlet used to kill the well.

Typically, a subsea stack is made up of several large preventers mounted one on top of the other, beginning with one or more annular preventers at the very top and ending with several ram preventers installed below. As a result, the stack is often very tall. If procedure calls for the annular preventer to be shut in on a kick, the kick will be circulated through a choke line that is usually located well below the top of the stack. As a result, kick gas can become trapped below the closed preventer. If the annular BOP is opened and gas is trapped below it, the trapped gas could unload the marine riser and escape to the rig floor, where it could burn or explode. Therefore, before kill-weight mud is circulated into the riser, trapped gas must be removed from the BOP stack. The following procedure to remove trapped gas may be used:

1. Keeping the annular BOP closed, close a set of pipe rams that are below the kill and choke lines.

2. Displace kill-weight mud in the stack by pumping water down the kill line and up the choke line; hold back-pressure on the choke equal to the hydrostatic pressure difference at the BOP between the kill mud and the water.

3. When returns are clear water, stop pumping.

4. Close the kill line.

5. Bleed pressure from the choke line to allow water and gas to escape from the line through the choke manifold.

6. Once flow stops, close the diverter, open the fill-up line, fill the hole from the top, and open the annular BOP while taking returns through the choke line.

7. When the well is static, circulate kill mud down the choke and kill lines and up the riser.

Because trapped gas can be a problem, some operators prefer to circulate the kick out of the hole in such a way as to prevent any gas from being trapped in the stack. One method is to shut the well in, hang off the pipe on the ram preventers, and circulate the well through an outlet in the ram preventers. Since the kick is circulated directly out of the stack at the point of closure, no gas can become trapped.

DETECTING KICKS

Because a floating drilling rig moves while on location, it can be difficult to detect some of the basic kick warning signs. Rig movement can mask an increase in return flow and a pit increase. One way in which pit level can be measured accurately is to use several sensors in the pits to sum the outputs and give a more stable reading. As for return-flow sensors, many designs include a paddle that is mounted in the return line. Increased fluid flow depresses the paddle and sends a signal to the driller's console, where flow increase is read as a percentage. The vertical movement, or heave, of a floater and the pumping action of the slip joint can combine to make the return-flow indication questionable. In other words, the return-flow line may run full one moment and dry the next, even when the average flow rate is constant. Today, however, modern integrated and computerized systems can compensate for these fluctuations and give an accurate readout to personnel.

In addition, the use of a drill string compensator often makes it difficult to rapidly spot drilling breaks. Further, information about downhole conditions may be hard to interpret, because the returning mud and cuttings are often jumbled in large-diameter marine risers, where the velocity is quite low. Mud temperature trends are more difficult to analyze because of the heat-exchanger effect of the marine riser in cold seawater. Because of such problems, all systems and devices that can be used for kick detection should be monitored very carefully on a floating rig. Again, modern computerized equipment helps overcome many of the problems associated with rig movements.

Chapter 10

Blowout Prevention Equipment

Today's blowout prevention equipment is rugged, reliable, simple to operate, and widely employed throughout the industry. While the BOP stack is usually the first item thought of when BOP equipment is discussed, other equipment must also be included. Equipment such as chokes, accumulators, pit-volume indicators, gas detectors, and flow detectors make it possible to detect and handle kicks with confidence. When modern equipment is coupled with good well design and thorough training, rig personnel are more able than ever before to control any well.

The BOP stack itself must be able to control high formation pressures and have an internal diameter, or bore, large enough to allow the passage of tools required to drill and complete the well. Because some wells require large-diameter tools and encounter high pressures, large-diameter stacks with high pressure ratings may be employed. Further, since a well that is shut in must be pumped into, that is, circulated with the BOPs closed, outlets to the stack must be provided. Spools located between the ram preventers are sometimes used to provide the outlets; at other times, the outlets in the body of the preventers are used. In any case, when all the requirements of a BOP stack are met, it can be a large, heavy piece of equipment.

Of course, not all blowout prevention equipment is as large and heavy as the stack, but as the pressure rating of any piece of equipment increases, it tends to become heavier, more complex, and less forgiving of operational misuse. Therefore, it is the job of every crew member to be familiar with the equipment and to be aware of its operational limitations.

It is important for crew members to know the rated working pressure of all the equipment used in their rig's well-control system, its wellhead components, and the casing strings in the well. Blowout preventers, other well-control equipment, wellhead equipment, and casing generally have rated working pressures ranging from 2,000 psi to 20,000 psi. Briefly

stated, rated working pressure is the highest pressure that the equipment can reliably withstand. Thus, if a BOP has a rated working pressure of 15,000 psi, a shut-in well could place 15,000 psi on the preventer and the preventer should continue to operate. Sometimes the term maximum allowable pressure is used, which indicates the greatest pressure that may be safely applied to a piece of equipment. Pressure in excess of this amount risks failure and should never be permitted.

STACK ARRANGEMENTS

API RP 53 Blowout Prevention Equipment Systems for Drilling Wells is a document whose purpose is "to provide information that can serve as a guide for installation and testing of blowout prevention equipment systems on land and marine drilling rigs." In addition to giving installation guidelines and equipment requirements for various BOP components, it illustrates example arrangements of surface and subsea blowout preventer stacks with rated working pressures from 2,000 psi to 20,000 psi. Regarding the use of spools, RP 53 points out that, "Choke and kill lines may be connected to side outlets of the BOPs, or to a drilling spool installed below at least one preventer capable of closing on pipe. Utilization of the blowout preventer side outlets reduces the number of stack connections and overall stack height. Typically, drilling spools are not installed on subsea BOPs; however a drilling spool can be used to provide stack outlets (to localize possible erosion in the less expensive spool) and to allow additional space between preventers to facilitate stripping, hang off, and/or shear operations."

Other than the fact that ram preventers are shown below the annular preventer (or preventers), RP 53 does not suggest or specify the position, or arrangement, of pipe ram and blind ram preventers in the stack because various arrangements can be used to control a well successfully. Therefore, ram preventer arrangement depends a great deal on the preferences of the operator or stack designer concerning the advantages and disadvantages of the position of the various rams and side outlets. Most operators acknowledge that virtually any arrangement has advantages and limitations. The responsibility of rig personnel is to understand the capabilities and limitations of the stack on their rig. Because it contains such valuable information, many operators require all crew members to familiarize themselves with RP 53 (available from the American Petroleum Institute, 1220 L Street, NW, Washington, DC 20005).

While operator or designer preferences play a role in subsea stack arrangement, certain requirements dictate the number and placement of ram BOPs and side outlets in the stack. For instance, a subsea BOP stack should provide a means of reentering the well and of circulating it under pressure after pipe is hung off and the location abandoned. Most operators also recommend that the stack provide for at least two sets of hang-off rams below the blind-shear rams. One additional set of rams may be required for casing or smaller drill pipe, and some stacks may need still another set of rams that can be used as a stack handling tool for 5-in. drill pipe.

FLOW PATHS

Regardless of the stack arrangement, well-control personnel should become thoroughly familiar with the arrangement on their rig. Everyone concerned should be able to recognize and identify the flow path of drilling fluid throughout the BOP system during normal drilling operations and during well-control operations, including kill operations. With their particular BOP stack configuration, personnel should be able to identify the shut in, monitoring, and circulation operations that are possible with the

configuration and those that are not. Further, all well-control personnel should be able to select the appropriate BOP in the rig's stack to close on a given tubular. What is more, personnel should be able to point out all the areas that will be exposed to high and low pressure during shut-in and pumping operations. Also important to successful well control is the crew's being able to identify and confirm the configuration and line up of all valves, preventers, and lines when equipment is being pumped through, pressure tested, or shut in. Most operators and contractors routinely train rig personnel on these items and also require them to demonstrate their ability to shut in the well should the primary BOP and equipment fail.

ANNULAR PREVENTERS

The annular preventer is one of the more versatile pieces of equipment in the BOP stack (fig. 10.1). It can close around casing, drill pipe, drill pipe tool joints, drill collars, and the kelly. Most annular preventers are also designed to close on wireline and open hole, although such closing accelerates wear on the preventer's packing element. The annular BOP is usually the first preventer closed when a kick occurs.

Figure 10.1   Installed annular BOP on a land rig

Packing Elements

The rubber or rubberlike packing elements of the annular preventer are subject to wear and abuse. When properly treated, the packer of an annular preventer has a long, reliable life span, even though improper use can damage or destroy it after very few closing cycles. The life of annular preventer packers can be extended by—

1. keeping the closing pressure of the accumulator on the preventer as low as possible;

2. not testing the packer for leaks under high pressure, because high-pressure tests significantly shorten its life;

3. not closing on open hole, because the life of the packer is significantly shortened;

4. not reversing the movement of pipe through a closed packer;

5. storing spare packer elements in a dark, cool room away from electric motors, because the ozone produced by electric motors can cause deterioration of the packer material;

6. careful and correct maintenance procedures.

Closing and Opening Times

Because it is very desirable to keep kick size to a minimum, one of the concerns with annular preventers is the time required for them to close. RP 53 suggests that closing time should not exceed 30 sec for annular preventers with bores smaller than 20 in. and 45 sec for annular preventers with bores 20 in. and larger. One of the factors that determines closing and opening time is the ID of the hydraulic lines running from the accumulator to the preventer. Most operators specify that the lines have a minimum ID of 1½ in. However, some older surface-mounted stacks may have hydraulic lines smaller than 1½ in., or may have restrictions that prevent rapid closing of the annular, because the closing fluid cannot move quickly. While increasing the closing pressure on an annular preventer may make it close faster, a better solution is to use larger hydraulic lines and fittings. Small lines or restrictions may also make the packer less able to flex during stripping operations. Inflexibility can cause excessive packer wear during stripping operations when tool joints are being raised or lowered through a closed BOP.

Packer Flexing

As mentioned earlier, the special regulator valve used to regulate closing pressure on the annular BOP allows fluid to pass back through it so that the packer can flex when a tool joint passes through. This valve must be in good repair, no check valves should be installed ahead of it, and it should not be replaced by an ordinary regulator valve.

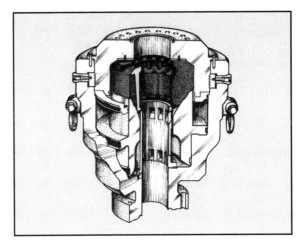

Figure 10.2   Hydril GX annular BOP

A small accumulator bottle can be placed near the annular preventer both on subsea and surface BOP stacks to cut down on packer wear during stripping. Because the bottle is quite close to the preventer, hydraulic operating fluid can easily and rapidly move in and out of the small bottle, thus allowing the packer to flex.

Brands

While various brands of annular preventers share many characteristics, some differences exist. For instance, Hydril preventers have a wedge-shaped contractor piston that closing pressure moves upward to squeeze the packer inward and around the pipe (fig. 10.2). In Shaffer spherical preventers, closing pressure moves a piston upward to push the packer up against a rounded, or spherical, housing (fig. 10.3).

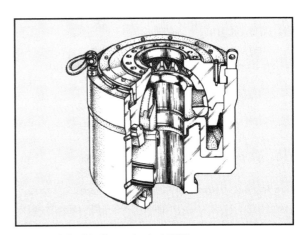

Figure 10.3   Shaffer spherical BOP

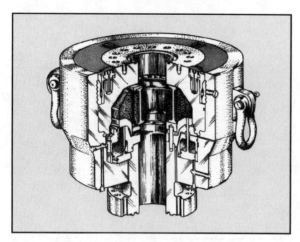

Figure 10.4   Cameron annular BOP

The spherical housing forces the packer inward and around the pipe. In the Cameron annular preventer, closing pressure forces a piston and pusher plate upward to displace a solid rubber doughnut that forces the packer to close inward and around the pipe (fig. 10.4).

Hydril GK

The Hydril GK annular preventer is one of the more common annular preventers in use. It is available in bore sizes from 7¹⁄₁₆ in. to 16¾ in., with working pressures from 2,000 psi to 20,000 psi. The GK's closing pressure requirements vary according to well pressure. As is the case with other annular BOPs, it closes with about 800 psi, but as well pressure increases, closing pressure should be reduced to avoid packer damage. In the GK, well pressure increases the closing force on the packer; in other words, the GK is strongly energized by well pressure. With other annular preventers, closing pressure is either kept the same or increased slightly as well pressure increases. For more detailed information on closing pressure requirements, tables from the manufacturer should be consulted.

Hydril GX

The Hydril GX is similar to the GK, but is designed primarily for high-pressure use offshore or on deep land wells. It is available in three bore sizes: 11 in., 13⅝ in., and 18¾ in. The 11-in. and 13⅝-in sizes are available with either 10,000 psi or 15,000 psi working pressure ratings. It is not as strongly energized as the GK, so adjustment of closing pressure is usually not required, regardless of well pressure.

Hydril GL

Hydril makes the GL annular preventer primarily for subsea service. Like the GX, it is not as strongly energized as the GK, so closing pressure adjustments are not required with this preventer. The GL has a balancing chamber to help compensate for the mud weight in the riser in deep water. Hydril suggests that the compensating chamber be used in waters deeper than 1,200 ft.

Hydril MSP

Hydril MSP annular preventers are usually employed as diverters in areas where shallow gas may be a problem. The MSP 2000 is basically a low-pressure version of the Hydril GK. It is interesting to note that the 29½-in MSP 500 does not have an opening chamber; that is, pressure from hydraulic fluid is not used to open it. Instead, when the MSP 500 is opened, hydraulic fluid is vented from the closing chamber, and the packer opens itself.

Shaffer Spherical Preventer

The Shaffer spherical annular preventer is available in two models: a bolted-cover, or housing, model; and a wedge cover, or housing, model. In bolted-cover models, the upper housing fastens to the lower housing with studs and nuts. Wedge-cover models fasten with locking segments and a locking ring to facilitate change out of the packing element. Wedge-cover models are available with working pressures of 5,000 psi to 15,000 psi, and in bore sizes ranging from 11 in. to 21¼ in. Bolted-cover models are available with working pressures that range from 1,000 psi to 15,000 psi. Bore sizes range from 4¹⁄₁₆ in. to 30 in.

Shaffer specifies that its annular preventers be operated with 1,500 psi closing pressure; however, the company notes that the pressure should be reduced according to operating tables for the preventer if pipe is to be moved. In subsea applications, mud weight in the riser in deep water does not affect the closing pressure of the Shaffer preventer to any great degree; therefore, for deepwater applications, extra closing pressures may not be required.

Cameron Annular Preventer

Cameron annular BOPs are available in bore sizes from 7¹⁄₁₆ in. to 21¼ in. and in pressure ratings from 2,000 psi to 20,000 psi. One of the features of a Cameron annular preventer is a quick-release top that permits relatively fast change out of the packing element. Since it also utilizes a one-piece split lock ring, a visual indicator is provided to show whether the top is locked or unlocked. The operating system is isolated from well pressure, so well pressure does not alter its closing pressure.

Rotating Heads

Although constructed very much like an annular preventer, rotating heads are not primary blowout preventers (fig. 10.5). Instead, they are often used where it is possible to drill underbalanced. In underbalanced drilling, pressure at the wellhead substitutes for part of the hydrostatic head to contain formation pressure. In underbalanced drilling, rotating heads seal around the kelly while it is rotating. A rotating head contains no provision for external control. In operation, the stripper rubber, which is a packing element, contacts the kelly with a pressure-tight seal. The rubber and the bushing assembly rotate with the kelly while the body of the device is stationary.

Rotating heads work well for circulating out low-volume, high-pressure gas kicks while continuing to drill with relatively light mud in the hole. This practice not only permits a faster rate of penetration, but also gains time for drilling that would otherwise be needed to circulate out the gas with the preventers closed.

RAM PREVENTERS

Ram blowout preventers in use today are the result of about eighty years of development. Thus, they are extremely rugged and reliable. As is the case with annular preventers, three companies manufacture the most-popular ram preventers in use: Cameron, Hydril, and Shaffer (fig. 10.6). While differences exist between the brands, they all operate in a similar manner and contain similar components. Regardless

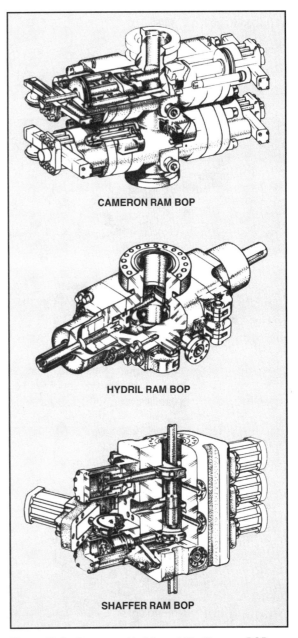

CAMERON RAM BOP

HYDRIL RAM BOP

SHAFFER RAM BOP

Figure 10.6 Cameron, Hydril, and Shaffer ram BOPs

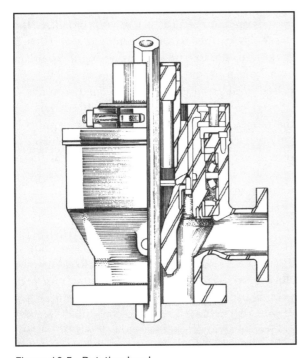

Figure 10.5 Rotating head

of manufacturer, a ram preventer consists of a ram head with extrudable packer material for sealing and a pipe-centering wedge. The head sits on a piston rod, which connects it to the hydraulic chambers and seals.

Closing Pressure

Closing pressure for ram blowout preventers is almost universally specified to be 1,500 psi. However, many operators have observed that, when stripping or moving pipe through the rams, less wear to the packer element will occur if the closing pressure is reduced to about 800 psi.

Types

Three types of ram preventers are available: pipe rams, blind rams, and blind-shear rams (fig. 10.7).

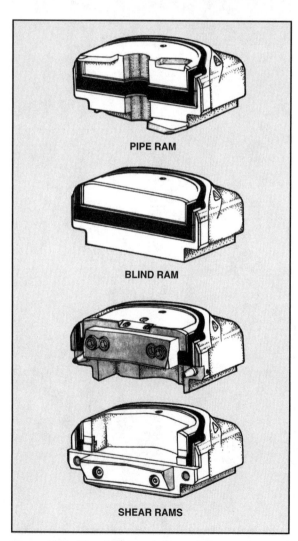

PIPE RAM

BLIND RAM

SHEAR RAMS

Figure 10.7 Pipe, blind, and shear rams

Pipe rams are designed to close on drill pipe (not tool joints), blind rams close on open hole, and blind-shear rams cut, or shear, pipe and form a seal on the resulting open hole. (Blind-shear rams can also be closed to seal open hole.) Blind-shear rams are primarily used in subsea preventer stacks, where it may be necessary for the rig to move off location rapidly because of weather or other factors. Shearing the pipe and sealing the well allows a floating rig to exit the area quickly in an emergency.

Pipe Rams

Two basic types of pipe rams are available: one is manufactured to fit a specific size of pipe; the other is designed to seal on several sizes of pipe or hexagonal kelly (fig. 10.8). Those designed to close on different sizes of pipe are variable-bore rams. Variable-bore pipe rams are able to seal around a range of pipe sizes, for example, 3½-in. pipe, 5-in. pipe, or any size in between. Tests of variable-bore rams have indicated that their performance is comparable to standard pipe rams. One manufacturer suggests that a variable-bore ram could serve as a backup for two sizes of pipe rams, or could serve as a primary ram for one size of pipe and as a backup for another.

Blind and Blind-Shear Rams

When blind rams are closed on open hole with full closing pressure of 1,500 psi, most, if not all, of the packer is extruded to form a complete seal. Thus, if blind rams are to be tested by closing them on an open hole, most operators recommend that they be closed with no more than about 500 psi to avoid damage to the ram packing and possible damage to the ram face. Blind-shear rams are also equipped with a relatively small packer element, which extrudes when well pressure is put under the shear ram. Therefore, extensive pressure testing with shear rams could cause premature packer wear.

Locking Devices

To avoid a possible blowout once rams are closed during a kick, they should remain closed even if closing pressure fails. Therefore, most ram preventers are equipped with either integral or hydraulically actuated, remote-controlled locking devices to ensure that they stay closed (fig. 10.9). Most modern land rigs are equipped with integral or remote locking devices in place of slow, manually

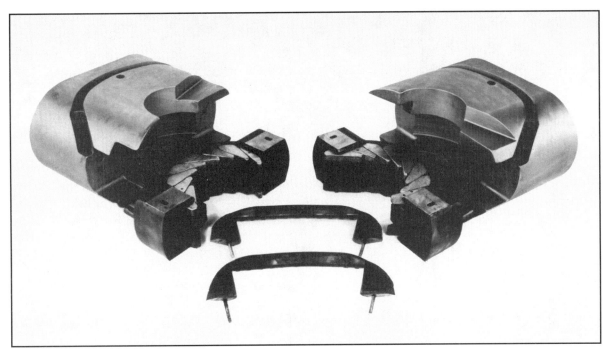

Figure 10.8 Variable-bore rams are designed to close on several sizes of pipe, e.g., 3½ in. to 5 in. drill pipe.

operated locking screws. Integral or remotely controlled locking devices are essential where subsea stacks are in use, in that the only practical way in which the rams can be locked is by integral or remote means.

DRILL STEM VALVES

When a kick is taken during a trip and the well is shut in with the annular or ram BOPs, fluid may flow out of the drill stem. Several types of drill stem valves are available to prevent flow from the drill stem or to stop flow after it has started. They go by several names, such as dart valves, stabbing valves, inside BOPs, drill string safety valves, kelly cocks, and float valves. Regardless of what they are called, however, drill stem valves fall into one of two categories: those that are made up in the drill stem on the surface and those that are placed in the drill stem downhole. In general, drill stem valves that are installed on the surface are termed inside blowout preventers, or safety valves. Those that are made up in the string downhole are termed float, dart, or back-pressure valves.

One of the more widely used drill stem valves is a ball valve that is made up in the drill string on the

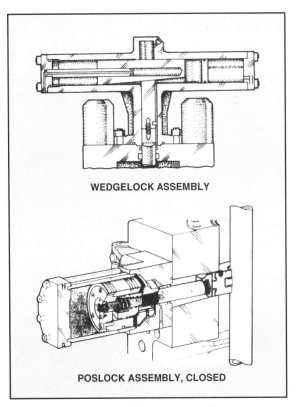

WEDGELOCK ASSEMBLY

POSLOCK ASSEMBLY, CLOSED

Figure 10.9   Cameron Wedgelock and Shaffer Poslock remote-controlled ram locking devices

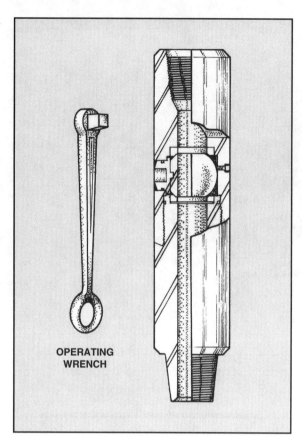

Figure 10.10   Ball-type kelly valve

when the kelly is broken out and fluid is flowing from the drill stem. Once stabbed and made up tightly in the drill string, the wrench is used to close the valve and shut off flow.

Another type of drill stem valve that can be stabbed into the drill pipe is a Gray valve (fig. 10.11). A Gray valve is used only when stripping pipe into the well. It is made up on top of a closed full-opening safety valve already made up in the drill string. After the top assembly and actuating rod are removed from the valve, the check valve closes to keep intruded fluids in the drill stem from acting against the kelly or top drive, which is made up on top of the Gray valve.

Downhole drill stem valves are usually called float, or check, valves because they allow mud to be

surface (fig. 10.10). Typically called a full-opening safety valve, when the ball is in open position, fluid can pass through it in either direction. The valve is opened and closed with a special operating wrench, which should always be stored in a convenient, easily accessible place. When the valve is closed, the ball forms a pressure-tight seal against flow in either direction.

In general, ball valves can be used in three different places in the drill stem above the rotary table. First, they can be used as a kelly cock, or upper kelly valve. When used as a kelly cock, the ball valve is made up above the kelly and below the swivel. When closed, it shuts off kick pressure in the drill stem to protect the swivel and the rotary hose. Second, a ball valve can also be used as a lower kelly valve. Made up between the bottom of the kelly and the top joint of drill pipe, it is often called a drill stem safety valve, or a mud-saver valve. It can be closed when the kelly is raised above the rig floor. Finally, a ball valve can be used as a stabbing valve. When the valve is in its full-open position, it can be stabbed into the drill pipe

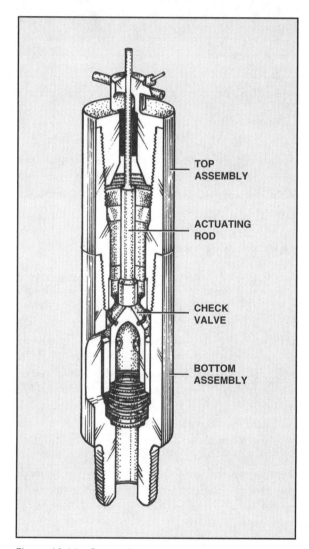

Figure 10.11   Gray valve

pumped through them in a downward direction only; upward flow from a kick closes the valve and shuts off the drill stem. One type of float valve uses a spring-loaded flapper installed inside a special sub (fig. 10.12). The sub, with the flapper inside it, is made up in the drill stem, usually at a point just above the drill collars. Normal circulation keeps the flapper open to allow downward flow of mud. When circulation stops, the flapper closes and prevents any backflow. A vented flapper is available, which helps in obtaining *SIDPP*.

Another downhole check valve is the pump-down type (fig. 10.13). In a pump-down valve, a special landing sub is made up in the drill stem above the drill collars. Once the landing sub is made up in the string, the check valve can be installed in one of three ways. If no backflow is occurring, the check valve is simply dropped into the drill string when the kelly is disconnected. The kelly is then connected and the check valve is pumped into the landing sub.

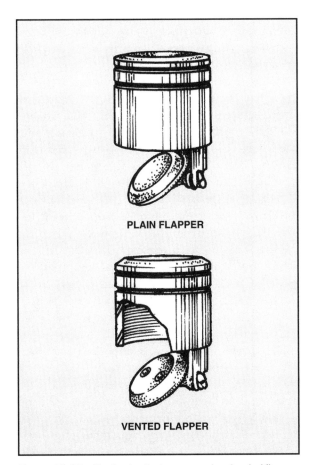

Figure 10.12   Float valve that uses a spring-loaded flapper

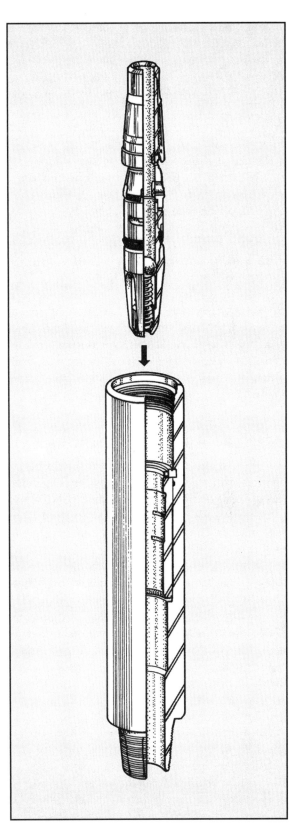

Figure 10.13   Pump-down check valve

If backflow exists or is likely, the check valve can be installed through the lower kelly valve. With the lower kelly valve closed, the kelly is removed and the check valve is placed in the box of the kelly valve. Then, the kelly is stabbed into the kelly valve and over the check valve. Once the kelly is made up, the kelly valve is opened and the check valve is pumped into the landing sub.

Finally, if a kick occurs during a trip and fluid is flowing from the drill stem, a full-opening safety valve can be stabbed into the drill string and made up. The valve is then closed to stop flow. The check valve is then placed in the box of the safety valve and the kelly made up over the safety valve. Pumping is started and the safety valve is opened to send the check valve down to the landing sub.

Regardless of how the check valve is positioned in the landing sub, once it is seated normal circulation can be obtained through it. A spring-loaded ball is forced away from a seat as long as pressure from above is maintained. Upward well pressure, as from a kick, causes the ball to seat and seal to prevent flow into the drill string.

INSPECTING AND TESTING PREVENTERS

All blowout prevention equipment should be inspected and tested on a regular basis. How often to test depends on regulations, company policies, and other factors. For use as a guide, API offers a set of recommended procedures for testing and installing well-control equipment in RP 53.

Crew members should be aware of the maximum safe working pressures of all the equipment (including tubulars, such as drill pipe, casing, and the like) used in controlling a well. They should also be aware of all areas in the system that are exposed to high as well as to low pressures during shut-in and pumping operations. Further, crew members should be able to identify the maximum safe working pressure for the well-control equipment that exists upstream and downstream of the choke on their rig.

Crew members should also keep in mind that in any part of the blowout prevention system, the maximum working pressure of any piece of equipment can be no higher than the lowest rated part in the system. For example, a 10,000-psi choke may be installed in the choke manifold. If, however, the piping in the manifold going to the choke is rated at 5,000 psi, then the choke must also be rated at 5,000 psi working pressure. Obviously, any pressure significantly higher than 5,000 psi may cause the piping to fail, even though the choke may not. Another example: suppose the working pressure of the annular and ram BOPs in a rig's stack is 15,000 psi. Further suppose that the wellhead on which the preventers are mounted is rated at 10,000 psi. What should the rated worked pressure of the BOP stack be? The answer, of course, is that the whole stack must be down rated to 10,000 psi.

Proper installation and maintenance of the well-control equipment is essential to successful well control. As simple as it sounds, many wells have been lost because a crew failed to correctly install equipment. From an action as elementary as installing a ram BOP upside down in the stack to something as complex as incorrectly hooking up a subsea BOP control system, incorrect installation can lead to failure in controlling a well. A good policy when installing equipment is to follow the procedures and recommendations of the drilling contractor, the operating company, and the equipment manufacturer. If any doubt exists, crew members should not hesitate to seek instruction from their supervisor.

Maintenance of well-control equipment is essential to it operating properly and reliably. Virtually all manufacturers publish maintenance and troubleshooting procedures that accompany their equipment into the field. In addition, many contractors and operating companies establish maintenance polices that must also be followed. As equipment is used its parts wear and require replacement. Once a good maintenance procedure is established and followed, crew members should routinely check and replace equipment as required. It is particularly important to install, maintain, and replace, when required, all rings, flanges, and connectors in the well-control system. All these components are subjected to high pressure when the well is shut in and fluids are being pumped. As a part of a quality maintenance program, it is important that crew members regularly and periodically pressure test the equipment. They should also function the BOPs and other well-control equipment regularly and periodically. Certainly one of the best ways in which to ensure that the equipment is operating properly is to try it out before it is needed in an actual well-control situation.

STANDPIPE AND STANDPIPE PRESSURE GAUGE

While not often included in lists of blowout prevention equipment, one of the primary ways in which to read *SIDPP* and pump pressure is the standpipe pressure gauge (fig. 10.14). If the rig layout allows it, the standpipe gauge is often mounted on the rig floor near the driller's position. It is important for crew members to locate this gauge on their own rig and to keep it in good working order. Also, the standpipe itself is important to circulating fluid from the pumps to the drill stem. It, too, should be periodically checked to ensure that it is in good working order.

Figure 10.14 Standpipe pressure gauge

PRESSURE RELIEF VALVES

Like the standpipe and the standpipe pressure gauge, the pressure relief valves installed in the mud pumps are not usually thought of as being part of the well-control system. It is important, however, to bear in mind that when kick fluids are in the drill stem and are therefore exerting pressure in the stem, it is possible for this pressure to exceed the pressure at which the relief valves on the pumps will open. Gas migrating up the drill stem can exert enormous pressures on equipment above the drill stem.

GAS DETECTION EQUIPMENT

Because most kicks have gas associated with them, operators and contractors install gas detection devices to warn personnel of the danger of accumulations on the drill site. Because H_2S is a highly poisonous gas that can be associated with hydrocarbon and other gases, and because hydrocarbon gases are flammable or explosive, H_2S and flammable gas detection equipment is essential in areas where gas may be present or is suspected to be present. Detectors are normally placed at sites on the rig where H_2S and flammable gases are likely to accumulate or come into contact with rig personnel. For example, detectors may be placed in the cellar, near the bell nipple in the shaker room, near the mud tanks, on or near the rig floor, as well as other places on the rig location. Detectors usually have a special sensing device that is placed in an area where gas is likely to be present. The sensing device is electronically connected to an alarm system. When the sensor detects a dangerous gas, it sends a signal to the alarm system, which is usually a combination of bright flashing lights and a loud sound. It is important for the detection equipment to be regularly and periodically inspected and tested to ensure that it is working properly.

DRILLING RECORDERS

Modern drilling rigs usually come equipped with electronic computerized recorders that measure, display, and record, current drilling parameters. Typically, a unit shows the volume of mud in the tanks (pit volume), the flow rate of the mud returning from the hole, the rate of penetration, pump pressure and rate, the mud's weight, and the well's depth. Usually, the equipment can be set to sound a visual or audible alarm if the parameters vary from certain set points. For example, since a drilling break may be a kick sign, modern recorders can alert the driller to a departure from the normal rate of penetration. Similarly, if an increase in the return rate and a pit gain occur, these devices also alert the driller to a possible kick.

ADJUSTABLE CHOKES

While manually adjustable chokes are usually installed in the choke manifold as a backup, the remote-controlled hydraulic drilling choke provides the most efficient and accurate way in which to control a well kick. Chokes are designed to resist abrasion and wear and, when used in conjunction with a remote operating panel on the rig floor or other area, the system provides a valuable control center for killing a well. Many companies manufacture adjustable drilling chokes and remote operating panels, but all of them work on similar principles.

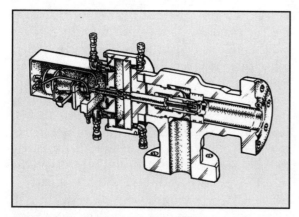

Figure 10.15   Cameron hydraulically actuated drilling choke

Cameron Positive Seal Choke and Panel

The Cameron choke is representative of a number of chokes that use a cylindrical gate-and-seat design. Other manufacturers include Brandt, Swaco, Vetco-Grey, and Willis-McEvoy, to name only a few. They are available in models that have working pressures from 5,000 psi to 20,000 psi, and the choking action occurs when an air-operated hydraulic pump in the control console moves the gate toward or away from the seat (fig. 10.15). A positive seal, or complete shutoff, is achieved when the gate and seat are brought into contact. In most brands, the gate and seat are reversible so that when one side wears, they can be turned around to present new wear surfaces. Some manufacturers offer a nonpositive-seal gate and seat, but if such a choke is in use, it is recommended that an upstream valve be installed so that complete shutoff can be achieved.

The choke panel manufactured by Cameron and other companies is available with a number of features (fig. 10.16). The basic panel contains gauges that display drill pipe and casing pressure (sometimes labeled as standpipe and choke manifold pressure), choke position, rig air pressure supplied to the panel, and hydraulic pressure supplied to the hydraulically actuated choke. Controls include a choke adjustment lever, hydraulic pressure adjustment regulator to control the hydraulic pressure supplied to the choke, and a manually operated emergency pump that permits control of the choke when rig air is lost. A pump stroke counter is available as optional equipment, as are provisions to control standpipe pressure automatically and to open the choke if MASP is reached.

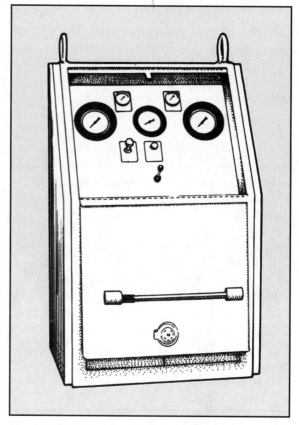

Figure 10.16   Cameron drilling choke control console

Swaco Super Choke, Ultra Choke, and Panel

The Swaco Super Choke and Ultra Choke represent a design that uses two lapped and polished plates of tungsten carbide with half-moon-shaped orifices (fig. 10.17). The plates are rotated by an air-actuated hydraulic pump, which enables adjustment from full open to full close. (Smith-Drilco manufactures a similar choke.) The Swaco Super Choke is designed for 10,000 psi working pressure and the Ultra Choke for 20,000 psi working pressure. The operating handle on the choke panel is connected to a four-way valve that drives the choke open or closed by changing the hydraulic flow. The choke itself is driven by four hydraulic cylinders and a rack-and-pinion system, so if for any reason the hydraulic lines are cut, the choke will stay in its last position. The choke control console, or choke panel, has two direct-reading hydraulic gauges: drill pipe pressure and annulus, or casing, pressure. A pump stroke counter and pump rate meter are provided as standard equipment and can be operated either electrically or pneumatically.

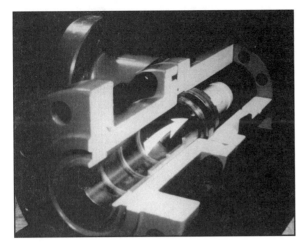

Figure 10.17　Two tungsten carbide plates with half-moon-shaped orifices form an opening for mud flow in the Swaco Super Choke.

There is also a choke position indicator on the panel (fig. 10.18). A panel is available that uses microprocessor technology to control drill pipe or casing pressure automatically as required for basic well-control methods.

Gauge Accuracy

Regardless of the gauge used to monitor *SIDPP*, *SICP*, pump rates, and so on, it is important to bear in mind that all gauges are subject to inaccuracies. In many cases, unless the gauge totally fails to give a reading, minor inaccuracies are not a problem as long as the gauge readings are consistent over its range. Gauge inaccuracies can cause a problem, however, when one gauge completely fails and a backup gauge must be used. The backup gauge may not read the same as the original gauge and well-control errors can occur. Because gauge readings can vary from one gauge to another, many contractors and operators require crew members to test all gauges and counters used in well-control operations and write down any observed discrepancies from one gauge or counter to another.

Gauge inaccuracy can be caused by many things. For example, gauges on a drilling rig are subjected to rough use and it is easy to damage a gauge when the equipment to which it is attached is being rigged up or down. Further, many gauges get their readings from sensors that are remote from the gauge itself. Because of their remote position from the gauge,

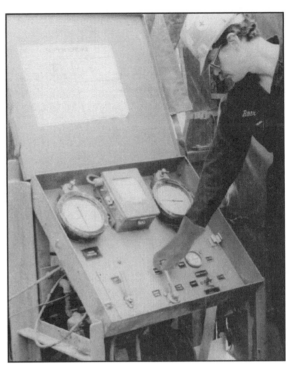

Figure 10.18　Swaco choke control panel

sending devices are used to transmit the sensor's reading to the gauge. Sending devices may be pneumatic, hydraulic, electronic, or a combination. Regardless of how the reading is sent, the sensor can be damaged or fail, or the sending device can be damaged or fail. Such damage or failure can result in erroneous readouts at the gauge.

ACCUMULATOR SYSTEMS

Although blowout preventers date to the early part of the century, it was not until the 1950s that a rapid method of closing them was developed. Early BOPs were closed and opened by means of a manual screw jack; indeed, some manual closing systems are still in use on small rigs. Using a manual method to close blowout preventers is not entirely satisfactory, however, because it is essential to close the well in quickly to keep the kick small. In most cases, a manual operating system is much too slow for effective kick control. Therefore, mud pumps, rig air, and hydraulic-pump closing units were tried as a means of effectively closing preventers, but none proved to be satisfactory. Today, hydraulic accumulator systems combined with recharging pumps are used

Figure 10.19   Pump-accumulator unit for preventer operation

throughout the drilling industry and have proved to meet all requirements (fig. 10.19).

The basic purpose of the accumulator system is to provide a rapid, reliable, and practical way to close the blowout preventers when a kick occurs. Because of the importance of reliability, the system has extra pumps and fluid volume as well as alternate systems. Most current systems use control fluid that is either hydraulic oil or soluble oil and water, stored in 3,000-psi accumulator bottles (fig. 10.20). Enough usable fluid is stored under pressure so that the stack can be closed and opened again by the time the bottles are discharged to 1,200 psi. Air-powered and electric pumps are rigged to recharge the unit automatically as the pressure in the accumulator bottles drops.

Accumulators have a regulator valve that allows the operator to control the amount of closing pressure applied to the BOPs. Generally, it is recommended that the closing pressure be adjusted to maximum operating pressure during normal operations and then readjusted to the manufacturer's recommended operating pressure after the BOP closes.

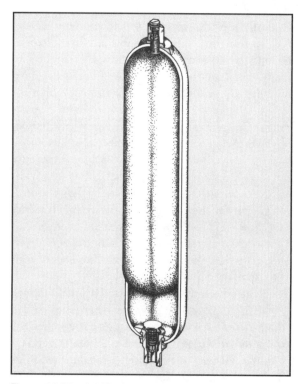

Figure 10.20   3,000-psi accumulator bottle

Adjusting the pressure to a value below maximum not only allows the preventer to maintain a pressure-tight seal on the tubular that is in the preventer, but also prolongs the life of the packing unit, especially in an annular preventer, during the time it is necessary to move pipe through the preventer.

In addition, crew members can make an accumlator unit open or close the BOPs either by manipulating valves on the unit itself or, more commonly, by moving levers or switches on BOP operating panels located remotely from the unit. Often, one remote panel is near the driller's position on the rig floor to allow the driller to quickly operate the BOPs when required. An auxiliary remote operating panel may also be installed on the rig at another remote location. An accumulator isolator valve is also mounted on the remote operating panel. In order to open or close, or function, a BOP, the operator must not only operate the lever or switch that functions the BOP, but also, he or she must actuate the isolator valve on the panel. Unless the isolator valve and the function valve are actuated together, the BOP will not function. The direct controls on the accumlator unit allow crew members to operate the BOPs from a position other than the panel on the rig floor or from the auxiliary panel. On land rigs, the accumulator is usually located some distance from the rig so that should it be necessary to evacuate the rig floor, the BOPs can be opened or closed from the accumulator unit.

Most accumulators also have a bypass valve that, when operated, bypasses the unit's regulator operating valve. Bypassing the regulator operating valve permits from 3,000 to 5,000 psi hydraulic pressure to go directly to the unit's operating manifold for testing or for extreme well-control problems.

Calculating Usable Fluid Volumes

An accumulator bottle contains hydraulic operating fluid and nitrogen. However, only part of the bottle's fluid is available, or usable, to operate the preventers. The reason has to do with the nitrogen precharge pressure and the operating pressures of blowout preventers.

For operating pressures, 1,200 psi is the lower limit. That is, it takes at least 1,200 psi to close an annular preventer on open hole. The upper limit is 1,500 psi. Pressures over 1,500 psi can damage annular preventers. Thus, the working limits range from 1,200 to 1,500 psi.

Using nitrogen, manufacturers first precharged accumulator bottles to 750 psi. Then they added operating fluid until the pressure reached 1,200 psi, the minimum required to close a preventer. They continued to add fluid until the pressure reached the 1,500-psi maximum. This 1,500-psi system worked fine on wells where large BOP stacks were not required; on today's high-pressure wells, however, where large stacks are required, a 1,500-psi system is not practicable.

To understand why, picture an 80-gal bottle with a 750-psi nitrogen precharge. If operating fluid is pumped into this precharged bottle, about 30 gal of fluid bring bottle pressure to 1,200 psi. Only about 10 more gal of fluid bring bottle pressure to 1,500 psi, the upper limit. Thus, the 80-gal bottle contains 40 gal of operating fluid. The rest of the bottle contains nitrogen separated from the operating fluid by a rubber bladder. Of the 40 gal of operating fluid, only 10 gal are available to close the preventer. After the 10 gal of fluid close the BOP, the pressure drops to the 1,200-psi minimum.

Only 10 gal of usable fluid from 40 total gal is not efficient. Thus, since large BOP stacks require relatively large volumes of usable fluid to close, 1,500-psi systems are limited in use.

To avoid this limitation, manufacturers raised the precharge pressure to 1,000 psi and the maximum pressure to 3,000 psi. To see why these two steps solved the problem, again think of an 80-gal bottle. Assume that it is precharged to 1,000 psi. With this amount of precharge, only about 13 gal of operating fluid raises bottle pressure to the 1,200-psi minimum. It takes almost 40 more gal of fluid to raise the pressure from 1,200 psi to 3,000 psi, however. The amount of usable fluid is therefore about 40 gal, a considerable increase over the 10 gal of usable fluid in the 750-psi precharged system. A 3,000-psi system thus provides reasonably large volumes of usable fluid in a relatively compact unit.

Since pressures above 1,500 psi can damage annular BOPs, manufacturers regulate the 3,000-psi system with a special pressure-reducing-and-regulating valve that maintains the pressure at 1,500 psi. Because several sizes of accumulators are available and because they operate at different pressures, the amount of usable fluid varies with each unit. Equations are available for calculating the usable fluid in any system. An example follows:

$$V_u = V_s (P_p \div A_p) (A_p \div R_p - 1) \quad (\text{Eq. } 69)$$

where

V_u = usable volume of fluid, gal

V_s = volume of fluid in the system, gal

P_p = precharge pressure, psi

A_p = total accumulator pressure, psi

R_p = required pressure to close preventer, psi.

As an example, suppose that—

V_s = 80 gal

P_p = 1,000 psi

A_p = 3,000 psi

R_p = 1,200 psi

then,

Vu = 80 (1,000 ÷ 3,000) × (3,000 ÷ 1,200 − 1)

= 80 (0.333) (2.500 − 1)

= 80 (0.333) (1.500)

Vu = 39.960, or 40 gal.

Tables are also available for determining fluid volumes. Table 7 shows stored and usable fluid volumes for accumulators of various sizes. For example, in a fully charged 1,500-psi system with a 750-psi precharge, the amount of stored fluid in an 80-gal accumulator is 40 gal and the usable fluid is only 10 gal.

Similarly, in a fully charged 3,000-psi system with a 1,000-psi precharge, the volume of stored fluid in an 80-gal accumulator is 53.04 gal. The volume of usable fluid is 39.84 gal, or about 40 gal, which is four times the volume of a fully charged 1,500 psi system.

Note that in every case shown on table 7, when accumulator pressure drops to 1,200 psi, no usable fluid is left. Thus, usable fluid is often defined as the volume of fluid stored in the accumulator bottles at pressures above 1,200 psi.

Many operators prefer to use a simple fraction or a decimal to determine usable fluid volumes in an

Table 7

Stored and Usable Accumulator Fluid Volumes

Accumulator Pressure	Accumulator Size									
	10 Gal		20 Gal		27 Gal		40 Gal		80 Gal	
	Stored Fluid	Usable Fluid	Stored Fluid	Usable Fluid	Stored Fluid	Usable Fluid	Stored Fluid	Usable Fluid	Stored Fluid	Usable Fluid
3,000 psi system 1,000 psi precharge										
3,000	6.63	4.98	13.26	9.96	17.90	13.44	26.52	19.92	53.04	39.84
2,500	5.96	4.31	11.92	8.62	16.09	11.63	23.84	17.24	47.68	34.48
2,000	4.96	3.31	9.92	6.62	13.39	8.93	19.84	13.24	39.68	26.48
1,500	3.30	1.65	6.60	3.30	8.91	4.45	13.20	6.60	26.40	13.20
1,200	1.65	0	3.30	0	4.46	0	6.60	0	13.20	0
2,000 psi system 1,000 psi precharge										
2,000	4.96	3.31	9.92	6.62	13.39	8.93	19.84	13.24	39.68	26.48
1,500	3.30	1.63	6.60	3.30	8.91	4.45	13.20	6.60	26.40	13.20
1,200	1.65	0	3.30	0	4.46	0	6.60	0	13.20	0
1,500 psi system 750 psi precharge										
1,500	5.00	1.25	10.00	2.50	13.50	3.37	20.00	5.00	40.00	10.00
1,200	3.75	0	7.50	0	10.13	0	15.00	0	30.00	0

accumulator system. The fraction or decimal depends on system pressures. For a 1,500-psi system with a 750-psi precharge, the fraction is ⅛ and the decimal is 0.125. For a 2,000-psi system with a 1,000-psi precharge, the fraction is ⅓ and the decimal is 0.333. For a 3,000-psi system with a 1,000-psi precharge, the fraction is ½ and the decimal is 0.500.

For example, in a 3,000-psi system precharged to 1,000 psi, the usable fluid in an 80-gal bottle is about 40 gal, since ½ of 80 is 40. Similarly, in a 2,000-psi system precharged to 1,000 psi, usable fluid volume in an 80-gal bottle is about 26.4 gal, since ⅓ of 80 is 26.4 gal. Finally, in a 1,500-psi system precharged to 750 psi, usable fluid volume in an 80-gal bottle is about 10 gal, since ⅛ of 80 is 10.

The reciprocal of the appropriate fraction, which for ½ is 2, for ⅓ is 3, and for ⅛ is 8, is also useful because multiplying it by the usable fluid volume determines the stored fluid volume required by an accumulator.

Determining Fluid Volume Requirements

When operators size an accumulator unit for a particular rig, they must determine the usable fluid volume needed to operate the rig's preventers. Although operators determine usable fluid volumes in several ways, they often use a simple method. First, they find how many gal of fluid are needed to close all the annular and ram BOPs. Then they multiply this volume by 3 to find the usable volume required. This calculation includes a 50-percent safety factor. Volumes needed to close preventers are available from the manufacturer.

As an example, assume a stack with three Cameron 18¾-in., 10,000-psi ram preventers and one Shaffer 18¾-in., 5,000-psi annular preventer. According to the manufacturer, each ram preventer requires 24.9 gal of fluid to close. The annular preventer requires 48.1 gal to close. The three ram preventers in the stack therefore require 74.7 gal of usable fluid to close. Add 74.7 gal to 48.1 gal (the volume needed to close the annular BOP) for a total of 122.8. Multiply 122.8 by 3 to find that 368.4 gal of usable fluid are required for this particular stack.

Once an operator determines usable fluid volume, he or she must determine the stored fluid volume required by the accumulator. (Remember, only part of the accumulator's stored fluid is usable.) Several methods are available, but one way employs the reciprocal of the fraction used to determine usable volume.

For example, in a 3,000-psi accumulator system with a 1,000-psi precharge, the fraction for determining usable fluid volume is ½. The reciprocal of ½ is 2. Therefore, if 368.4 gal of usable fluid are required, multiplying by 2 gives a stored fluid requirement of 736.8 gal. Put another way, 736.8 gal is the volume of stored fluid required to obtain 368.4 gal of usable fluid. The operator's next step is to determine how many bottles are needed to store the required fluid volume. In general, bottles are available in 10-gal increments, such as 10 gal, 20 gal, or 40 gal. Assume 80-gal bottles are in use. In a 3,000 psi system, table 7 shows that an 80-gal bottle stores 53.04 gal of operating fluid. Therefore, the accumulator system requires fourteen 80-gal bottles, since 736.8 ÷ 53.04 = 13.89, or 14 bottles.

In a 2,000-psi system with a 1,000-psi precharge, the fraction is ⅓ and its reciprocal is 3. If the stack requires 368.4 gal of usable fluid, the volume of stored fluid is 3 times the usable volume—1,105.2 gal, or twenty-eight 80-gal bottles. (Remember, an 80-gal bottle does not store 80 gal of operating fluid: nitrogen and a bladder also occupy the bottle. Refer to table 7.) In a 1,500-psi system with a 750-psi precharge, the fraction is ⅛ and its reciprocal is 8. So, if the stack needs 368.4 gal of usable fluid, the stored-fluid volume is 8 times the usable volume—2,947.2 gal, or seventy-four 80-gal bottles. Note that a 74-bottle accumulator is probably much too bulky to be practicable. Even if an operator used 110-gal bottles, a 1,500-psi accumulator system would need 41.

Determining Fluid Volumes for Subsea Accumulators

In floating operations, many operators and contractors prefer to use subsea accumulator bottles to improve BOP response times. If operators mount bottles on or near the subsea stack, they calculate usable and stored fluid volumes in much the same way as they calculate the volumes for bottles on the surface. They must, however, account for the effect of the seawater's hydrostatic pressure on the subsea accumulator bottles. Seawater pressure on the bottles, unlike atmospheric pressure, can be significant.

To account for the seawater's hydrostatic pressure, operators merely add its hydrostatic pressure to all the pressures involved in the calculations.

Since they must account for seawater's hydrostatic pressure, it is convenient to use an equation for determining the stored volume requirement. One equation that can be used follows:

$$V_s = V_u (P_p \div R_p) - (P_p \div A_p) \quad \text{(Eq. 70)}$$

where

V_s = stored fluid volume, gal
V_u = usable fluid volume, gal
P_p = precharge pressure, psi
R_p = required closing pressure, psi
A_p = total accumulator pressure, psi.

Note that to use the equation, the required usable fluid volume (Vu) must be determined in advance.

Assume that a rig is drilling in 1,500 ft of water and that the water's hydrostatic pressure gradient is 0.465 psi/ft. In this case, 698 psi must be added to all pressures, because 1,500 × 0.465 = 698 psi. Also assume that the required usable fluid volume is 100 gal. Further assume that precharge pressure is 1,000 psi, required closing pressure is 1,200 psi, and total accumulator pressure is 3,000 psi. Thus, if—

V_u = 100 gal
P_p = 1,698 psi (1,000 psi + 698 psi)
R_p = 1,898 psi (1,200 psi + 698 psi)
A_p = 3,698 psi (3,000 psi + 698 psi),

then,

V_s = 100 ÷ (1,698 ÷ 1,898) - (1,698 ÷ 3,698)
= 100 ÷ (0.895 - 0.459)
= 100 ÷ 0.436
V_s = 229.35, or 230 gal.

In this case, the subsea accumulator bottles must store 230 gal of hydraulic operating fluid to achieve the required 100 gal of usable fluid. Because of sea-water's hydrostatic pressure, the subsea accumulator bottles in the example require 30 more gal of operating fluid than are required for a comparable surface system.

Maintaining Accumulator Systems

Most accumulator manufacturers recommend that maintenance work be performed on the accumulator system at least every 30 days; most operators recommend that more frequent checks be made under extreme conditions or in bad weather. One manufacturer recommends the following checks as a minimum:

1. Clean and wash the air strainer.

2. Fill the lubricator with 10-weight oil.
3. Check the air pump packing. It should be loose enough to allow rod lubrication but not loose enough to allow dripping.
4. Check the electric pump packing.
5. Remove and clean the suction strainers (located on the suctions of both air and electric pumps).
6. If the electric pump has a chain drive, check the oil bath, which should be kept full of oil. Also check the pump for water.
7. Check the fluid volume in the reservoir to see that it is at operating level.
8. Remove and clean the high-pressure hydraulic strainers.
9. Lubricate the four-way operating valves. Use the zerk on the mounting bracket and the oil cup for the piston rod.
10. Clean the air filter on the regulator line.

Maintaining Nitrogen Precharge

An important factor in the operation of the accumulator is the 1,000-psi nitrogen precharge in the bottles. If the bottles lose their charge completely, the accumulator will stop working. Since nitrogen has a tendency to leak away or be lost over time, at different rates with each bottle, it is necessary to check each bottle individually. The charge in the bottles should be kept near the 1,000-psi precharge operating pressure. The charge on the bottles should also be checked before each well is drilled. To check the nitrogen precharge, the following recommended procedure may be used.

1. Shut off air to the air pumps and power to the electric pump.
2. Close the accumulator shutoff valve.
3. Open the bleeder valve and bleed the fluid back into the main reservoir.
4. After the fluid has bled down, close the bleeder valve.
5. Check the pressure on the bottles with a charging and gauging assembly. Recharge with nitrogen only if the reading is below 1,000 psi. (Note that precharges in excess of 1,000 psi may reduce the capabilities of the accumulator.)
6. Open the accumulator shutoff valve.
7. Turn on air and power. The unit should recharge automatically.

Performing Accumulator Tests

Most operators recommend that accumulator performance be tested at the time the blowout preventers are installed. Thereafter, they often recommend that a testing drill be conducted one or more times each week and that the drill be conducted from the remote control every other week. The purpose of the drill is to minimize the causes of equipment failure, such as improperly made-up lines, loss of nitrogen precharges, leaking seal gaskets, and leaking lines and fittings; to ensure quick initial closure times; and to check for the possibility of continued operation in the event of pump failure. A testing drill may be conducted in the following manner:

1. Position a drill pipe joint in the stack, making sure that a tool joint is clear of the rams to be used.
2. Record the initial accumulator pressure.
3. Turn off the accumulator pumps.
4. Simultaneously close the annular preventer and the pipe rams that fit the pipe. Do not close the blind rams or pipe rams of a different size from the OD of the test drill pipe.
5. Record the time taken for each preventer to close.
6. Record final accumulator pressure.
7. Turn on the accumulator pumps.
8. Open all preventer control valves; set the control valves in the open position during normal operations, not in the neutral position. Setting the valves in the open position keeps opening pressure on the preventers and ensures that they remain open until closure is required.

Most operators agree that the equipment is in satisfactory condition if blowout preventers with bores of 20 in. or less close completely within 30 sec, and if those with bores of more than 20 in. close completely within 45 sec. As for final accumulator pressure, this pressure should be compared with the pressure obtained with the initial test of the system when it was installed and the bottle precharges were known to be correct. Any significant discrepancy could indicate that some precharge gas has leaked out and that the system requires service.

PIT-LEVEL AND PIT-VOLUME INDICATORS

Since one of the best signs of a kick is an increase in pit volume, many operators require the installation of pit-level and pit-volume indicators on every well.

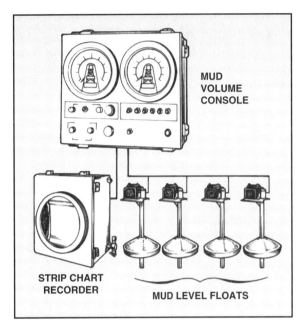

Figure 10.21   In this pit-volume indicator, floats in each pit transmit pit-level deviations to a console and recorder.

Such indicators continuously monitor the level of drilling mud in the mud pits. Pit-level and pit-volume indicators are available from several manufacturers, but most operate on similar principles. Most indicators consist of a series of floats in each of the pits (fig. 10.21). The floats sense the level in each pit and transmit a totalized average to a recorder and alarm. The recorder usually consists of a gauge and chart mounted near the driller's position on the rig floor.

Using air pressure or electronic signals to indicate and record the mud level and volume in the pits, modern pit-volume instruments are calibrated in total barrels or cubic feet of fluid. They have low- and high-level alarms that sound a warning and actuate lights when selected limits are reached, as might occur as a result of lost circulation or a well kick. Recorder charts are clock-driven and furnish a continuous record of the amount of mud in the pits. The 1-bbl indicator quickly notes any change of mud volume and can also be used to measure the volume of fluid required to fill the hole when pipe is pulled.

MUD-RETURN INDICATORS

Mud-return indicators have some limitations in that they may not always give accurate readings on floating drilling rigs, but they can be valuable in assisting the rig crew in noting an increase in return flow of drilling mud from the well. Because the entry of formation fluid into the hole causes an increase in the rate of returns from the well, mud-return indicators may give the first positive sign that formation fluid has entered the well.

Many different brands of return indicators are available, but most operate with some type of flow sensor that is mounted in the return line from the bell nipple to the mud pits. The associated meter continuously indicates the amount of mud coming out of the well. One return indicator compares input volume by pump strokes to output volume, which is gauged by a flow sensor in the return line from the well (fig. 10.22). If the volumes are not the same, indication of an increase or decrease in flow is given immediately. The sensor is capable of detecting increases or

decreases with 99 percent accuracy. While mud is being circulated, the sensor gauges the mud flow in terms of percentage points on the indicator dial. The device also includes a pump-stroke counter that can be used to total the number of strokes of either pump in a given period of time. In addition, the unit can indicate the number of strokes required to fill the hole after a given number of stands have been pulled.

GAS DETECTORS

Gas detectors monitor and detect changes in hydrocarbon gas entrained in the mud returning from the hole. Since gas is often one of the fluids in a kick, an increase in gas may indicate that formation fluids are entering the hole. Using a sensor mounted in or near the possum belly of the shale shaker, signals are sent to a device that can be mounted near the driller's position. Should a change occur, the device warns the driller of the change and indicates that steps may need to be taken to see if a kick has occurred. Many rigs operate with on-site service company personnel who monitor gas sensing and recording equipment and assist drilling personnel in interpreting trends in gas reading.

MUD TANKS AND MIXING EQUIPMENT

The rig's mud tanks (or pits as they used to be called) are a vital part of well control. While the number, size, and arrangement of the mud tanks on almost every rig is different, virtually all rigs have a suction tank, a return tank, one or more working tanks, and several other tanks for special purposes. The suction tank is the tank out of which the mud pumps take mud that goes into the hole. The return tank is the tank into which mud returning from the well flows. Working tanks are tanks generally located between the return and suction tanks; their number and size vary and depend on the hole's geometry, formation characteristics, and the like. The volume or capacity of the mud tanks determines how much kill-weight mud can be prepared and pumped into the well. A rig's mud mixing equipment is a factor in the speed with which kill-weight mud can be prepared. The larger the mixing equipment, the faster weighting material can be added to the mud. In cases where deep holes are being drilled and high formation pressures are expected, operators and contractors normally size the tanks and the mixing equipment to be able to handle the large volumes of mud anticipated.

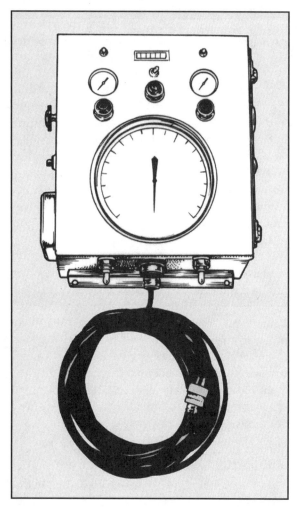

Figure 10.22   Pump-stroke counter and mud-flow sensor

Figure 10.23   Trip tank

TRIP TANKS

One of the best ways to ensure that the hole is taking the proper amount of fill-up mud when pipe is being tripped from the hole is to use a trip tank (fig. 10.23). Trip tanks usually have a capacity of 10 bbl to 40 bbl and have ½-bbl gauge markers that can be seen from the rig floor. When the tank is empty, it is filled from the active mud pit. Filling from the active pit permits a double check to be made of the fluid needed to fill the hole. In some instances, two tanks are provided; one is lined up with the hole and the other is shut off to be filled. This arrangement eliminates the need to stop pulling pipe while the tank is being filled. Trip tanks may be elevated so that gravity is the only force needed to run mud from the tank to the hole; or they can be at ground level and a small centrifugal pump used to transfer mud from tank to hole. In general, trip tanks should be built so that small changes in mud level in the tank are easily seen.

MUD-GAS SEPARATORS

A mud-gas separator is essential for handling a gas kick. It provides a means of safely venting gas away from the rig and makes it possible to save the liquid mud (fig. 10.24). Various separator designs can be

Figure 10.24   Mud-gas separator

employed, but many of them are constructed from a length of large-diameter pipe with interior baffles to slow down the mud-gas stream. A U-tube arrangement at the bottom permits mud to flow to the shaker pit while maintaining a fluid seal to hold gas in the upper part of the separator. Some mud-gas separators on rigs are modified production separators and have a mechanical float valve to control the fluid head of mud in the separator. No matter how the separator is built, pressure loss in the gas vent line at the top of the separator must be less than the hydrostatic pressure of the mud column in the bottom of the separator for the gas to be vented a safe distance from the rig. All vent lines should be large enough to keep pressure losses to a minimum and should be anchored firmly to prevent movement when large volumes of gas are blown off.

The separator is located downstream from the choke manifold so that flow from the well can be directed through the separator when necessary. A mud-gas separator can be overloaded both in terms of volume and pressure. A separator of a given size can handle only so much gas before it fails to do its job. Further, a separator of a given pressure rating can handle only so much pressure before it ruptures. Therefore, a bypass around the separator should be provided so that flow from the choke discharge line can be directed to a burn pit or over the side if the capacity of the separator is exceeded or if the vessel ruptures during a well-killing operation. Further, flanged openings should be provided near the bottom to permit the separator to be cleaned.

DEGASSERS

Degassers are available in many designs. Some work by means of a vacuum tank, others incorporate centrifugal force, still others use a combination of centrifugal force and vacuum. In vacuum degassers, pressure below that of the atmosphere is maintained inside a vessel (fig. 10.25). Mud enters the degassing vessel through an inlet pipe because of the low pressure in the vessel. Once mud is in the tank, it flows across inclined flat surfaces, creating thin layers so

Figure 10.25   Vacuum-degasser installation

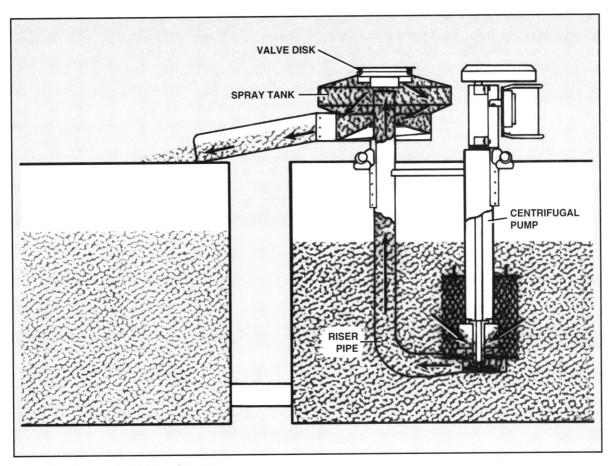

Figure 10.26 | Centrifugal force degasser

the gas bubbles can separate from the mud. Degassed mud falls to the bottom of the vessel and travels through an outlet pipe into the next pit. A hydraulic jet is placed in the downspout to pull mud from the vessel despite the vacuum. This jet is operated by fluid pressure from an auxiliary pump on the rig.

One type of degasser that employs centrifugal force consists of a submerged, centrifugal pump and a relatively small, elevated spray tank on which no vacuum is held. The pump picks up the gas-contaminated mud, moves it into a riser pipe, sprays it past an adjustable valve disk, and into the spray tank (fig. 10.26). As the mud hits the wall of the tank, small gas bubbles surface and separate from the mud. The degassed mud flows to the base of the spray tank and into the next mud pit by means of a trough. If desired or necessary, a gas disposal vent system is available that allows the gas to be discharged away from the rig. One vacuum-and-centrifugal design draws mud into the degasser tank by a vacuum pump (fig. 10.27).

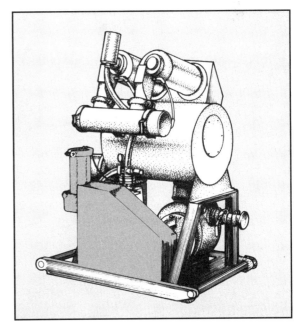

Figure 10.27 | Vacuum-and-centrifugal pump degasser

The mud is spun in centrifugal degassing tubes where high g forces are obtained. Mud flowing through the degassing tube inlets is dispersed in thin layers on the tube walls, where the vacuum in the tank allows gas to break out of the mud. The lighter gas bubbles rise to the surface of the mud, where they burst. The heavier mud is sprayed onto a turbulence hood, where any remaining gas breaks out. Gas is discharged through a blower to atmosphere or to a flare line as required. Degassed mud returns to the system by means of a centrifugal discharge pump.

TOP-DRIVE SYSTEMS

Top-drive systems have won wide acceptance within the industry. They can speed up drilling operations, make the floor crew's job easier, and allow the driller to rotate the drill stem up or down while in the hole. What is more, three-joint stands of pipe can be added during connections to further speed the operation. From a well-control point of view, top drives present no special problems when it comes to kick detection and shutting in the well. Because no kelly is required when drilling with top drives, only drill pipe lengths should be taken into account when spacing out the drill string for proper shut in using pipe rams. Virtually all top-drive units are supplied with inside blowout preventers (IBOPs), which the driller can open or close from his position on the rig floor. Nevertheless, if a kick is taken during a trip, when the top drive is not made up to the string, crew members must stab a full opening safety valve into the top joint of drill pipe hanging in the rotary. After the valve is closed, many contractors and operators recommend making up a pup joint or a single joint between the safety valve and the top drive. This short joint makes the top-drive unit much more accessible to the rig floor. After the joint and top drive are made up, the safety valve is opened.

COILED TUBING UNITS

A coiled tubing unit is a portable, compact device that eliminates making up and breaking out drill pipe or tubing connections (fig. 10.28). It uses a continuous length of ½-in. to 2-in. OD tubing that is stored on a reel. A tubing injector on the unit moves the pipe into or out of the well. Contrary to appearances, the reel on which the tubing is stored does not move the pipe. The reel moves only when the tubing moves in or out of the well.

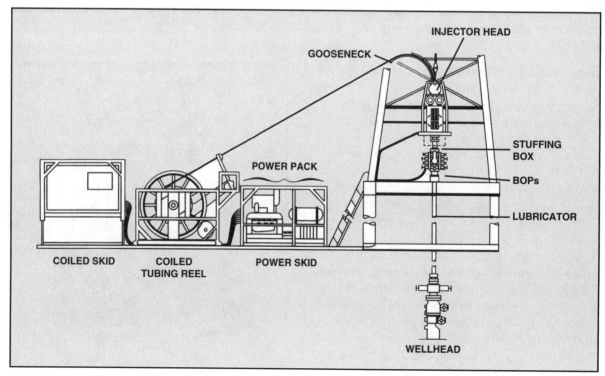

Figure 10.28 Coiled tubing unit

Coiled tubing units may be used for sand washing, jetting, spot acidizing, mud washing, cementing, paraffin removal, and cleanout. In certain well-control situations, such as when a large hole or washout occurs in a drill string, coiled tubing can be snubbed into the string and circulation established to regain control of bottomhole pressure. Coiled tubing can also be used in sand consolidation projects, as well as for light drilling and milling.

Coiled tubing has a number of advantages over conventional units. For example, (1) no connections are required; (2) better clearances inside most tools are provided; (3) no open-ended pipe exists at floor level; (4) no connections, which could leak, are present; (5) coiled tubing is lighter than conventional tubing strings; (6) coiled tubing requires less time to run and pull; (7) it is less expensive; (8) coiled tubing units require smaller crews than conventional units; (9) coiled tubing can reduce formation damage; (10) with coiled tubing, the well does not have to be killed; (11) coiled tubing lasts longer; and (12) coiled tubing units are very mobile and compact.

Coiled tubing units also have disadvantages, for example, (1) only light drilling can be done; (2) tubing sizes are limited; (3) tubing collapse pressures are relatively low; and (4) coiled tubing has a relatively low yield strength. The disadvantage of not being able to rotate the pipe has been partly overcome by downhole rotating tools, which are fluid-powered motors, that can be used for light drilling or milling.

Power Pack

The hydraulic plant, or power pack, is the heart of a coiled tubing unit. It consists of two main sections: the prime mover and a hydraulic system.

Prime Mover

Diesel engines rather than electric motors are preferred as prime movers because of the potential of fire or explosion when electric motors are used around a well site. Diesel engines are sized according to the type, weight, and length of the tubing used in the unit, as well as to the horsepower demands of the hydraulic system. In most cases, 6- or 8-cylinder engines are used.

Hydraulic System

Engine-driven pumps generate hydraulic power to operate a unit. Oil filters and coolers condition the hydraulic fluid and protect components from inter-

nal damage. Although hydraulic systems vary from manufacturer to manufacturer, most consist of five basic circuits:

1. Main power circuit—provides fluid power to operate the tubing injector;
2. BOP power circuit—provides fluid power for the operation of the blowout preventers. A hydraulic accumulator should be incorporated at some point in this circuit to provide an emergency supply of pressurized fluid, should power pack failure occur;
3. Reel power circuit—provides fluid power for the operation of the tubing reel and crane (on crane-equipped coiled tubing units);
4. Pilot circuit—provides fluid power to the valve network that hydraulically controls the primary components of the system; and
5. Auxiliary circuit—standby fluid power for emergencies and for operating nonstandard accessories.

These circuits can be activated individually. Further, circuit pressure can be monitored and adjusted, and the direction of flow of hydraulic fluid controlled.

Tubing Injector

The tubing injector, or injector head, imparts upward and downward movement to the work string. The injector head is held in position over the well with a hoisting device such as a crane or a mast support mechanism. Using a crane allows more flexibility because it is not necessary to maneuver a truck body near the well. On the other hand, a mast support mechanism allows tubing to remain threaded through the injector head during transport. Being able to leave the tubing threaded reduces rig up time. The direction of pipe movement is controlled from the operating console. A tubing injector consists of hydraulic motors and a pipe transport mechanism. Force required to move the pipe comes from hydraulic motors that are mounted on injector side plates.

The pipe transport mechanism uses friction to hold and direct the pipe. The mechanism is a series of sprockets around which two independent chains—one to the left and one to the right—are wrapped. Each chain is three links wide, and the outer links of each are connected to the outer links of the other to form a continuous chain. The inner links are steel gripper blocks with semicircular faces that correspond to the diameter of the work string.

Each of the two upper sprockets is connected to one of the hydraulic motors and drives a chain. The motors are assigned opposite hydraulic flow, which causes one to turn clockwise while the other turns counterclockwise. Since the motors face away from each other, they run in the same direction; that is, both turn the chains toward or away from the wellbore. The sprockets are arranged so that the chains face each other as they revolve down the middle of the injector. The semicircular gripper blocks hold the pipe in place in the chain.

The energy needed to force and hold opposing gripper blocks together is generated by backup plates, or skates, mounted behind the chains. They constrict inward when hydraulic pressure is applied and pinch the opposing gripper block faces together to generate friction to grip and hold the pipe. As the blocks grip and hold the pipe, they move up or down. This motion carries the pipe in the direction in which the gripper blocks are traveling in the injector assembly.

Safety features prevent runaway conditions, in which the work string might run in or out of the hole uncontrollably. Direct control is accomplished by use of flow-control valves. These valves do not allow hydraulic flow in the injector motors when hydraulic operating pressure is absent or abruptly lost. Loss of pressure load-locks the injector and prevents pipe travel until operating pressure returns. Further, most units have mechanical braking systems to provide additional safety. When rigging up the injector, check to make sure it is firmly secured and cannot move.

Tubing Reel

The tubing reel assembly consists of two independent components—a reel mechanism, and a level-wind mechanism—that work together to store the work string. Each is controlled from the console and driven by a hydraulic motor. The reel mechanism maintains tension on the work string during operation to prevent unused pipe from backlashing or uncoiling. The level-wind mechanism distributes and spools the pipe onto the reel in an orderly fashion.

When the unit is operating, the reel's motor provides tension by pulling slightly against the injector. This tension, coupled with the level-wind, maximizes pipe storage capacity on the reel. What is

more, tension minimizes pipe damage from overlap kinking and delays from tangled or crossed pipe wraps.

The reel is mounted on an axle that turns on two bearings. At one end, the axle is fitted with a circulating swivel. One end of the swivel connects to the work string on the reel; the opposite end of the circulating swivel is fitted with a pump manifold secured to the reel support skid. Pumping operations can, therefore, be conducted without interruption while running pipe in and out of the hole.

Stripper Assembly

The tubing stripper assembly is attached to the lower structure of the injector assembly, as closely as possible to the lower gripping area of the chain. Attaching the stripper assembly close to the lower gripping area prevents the work string from deviating from its proper path when well pressure exerts upward force on the pipe.

Two semicircular rubber stripper seal elements fit together in the cylindrical bore of the stripper body. They surround the work string and are retained by a cap, which is secured by two pins, that fits into the top of the stripper body.

An energizer, or hydraulic piston, below the seal elements is forced upward hydraulically to make contact with the seal elements. A taper on the end of the seal element corresponds to a taper on the energizer. When force is exerted on the tapers, the seal elements are forced together, creating a pressure-tight seal around the work string. Wellbore force may energize the stripper assembly, thereby assisting in an effective seal in some types of stripper preventers.

Blowout Preventers

The blowout preventers are placed below the stripper assembly. A common arrangement, from top to bottom, for coiled tubing units is blind rams, cutter rams, slip rams, and pipe rams. The rams are actuated hydraulically from the control console; they also can be operated manually in the event of hydraulic failure.

Certain pressure applications, wellbore conditions, or safety procedures may warrant additional preventers. A flow tee or drill cross may also be incorporated to prevent flow-cut damage to the BOPs when high-pressure or abrasive fluids are used.

Operating Console

The operating console consists of controls and various gauges. It provides a protective shelter for personnel and sensitive components and serves as a central location for all the critical controls. Almost every function of the unit can be controlled and regulated from the console.

The following items can be monitored inside the operating console: (1) wellhead pressure; (2) hydraulic stripping rubber pressure; (3) hydraulic skate, or backup plate, pressure; (4) blowout preventer actuating pressure; (5) tubing reel operating pressure; (6) tubing injector operating pressure; (7) pipe depth; (8) air system pressure; and (9) auxiliary circuit pressure.

The following are controlled from the operating console: (1) hydraulic stripping rubber pressure; (2) hydraulic skate, or backup pad, pressure; (3) BOP operation; (4) tubing reel tension; (5) tubing reel direction; (6) level-wind direction; (7) tubing injector operating pressure; (8) tubing injector direction (in or out of the hole or stop); (9) BOP accumulator actuation; (10) power pack start and stop; and (11) power pack engine throttle speed.

Accessories and Auxiliary Equipment

Several pieces of auxiliary equipment are available on coiled tubing units:

1. Tubing depth counter—used to measure the work string depth. It may be mechanical, a magnetic transducer, or a microprocessor.

2. Hydraulic crane—often incorporated into a unit when another method of lifting is not available at the location.

3. Freestanding injector bases—used for crane-free rigging up where overhead space, crane capacity, or weight prohibits normal rig-up. The injector base may incorporate self-elevating hydraulic cylinders for height adjustment when rigging up and include hydraulic winches for skidding the assembly across the location for positioning.

4. Fluid pumping unit—provides circulation or pressure, whether liquid or nitrogen is used as a fluid.

5. Liquid pump—has an intake located in or selected from the desired tanks. Fluid is drawn from a tank, through the pump, to a manifold and lines, where it may be mixed with nitrogen.

It then goes to the circulating swivel on the tubing reel, where it connects to the tubing.

6. Tanks and equipment for mixing and storing fluids typically consist of two or three compartments and are available in various capacities. The suction of the tank may be on the side several inches from the bottom to prevent sediment from entering the suction line; or suction may be on the bottom.

7. Centrifugal pumps, a mixing hopper, and a filtering unit are often included.

EQUIPMENT LIMITATIONS

The primary source of information for equipment testing and limitations is the manufacturer's specifications and recommendations. Also keep in mind that normal wear from the routine use and testing of equipment eventually results in malfunctions. Moreover, while manufacturers' recommendations are usually based on the best information they can obtain, each operating company and contractor should establish its own testing, inspection, and maintenance program.

To ensure proper equipment performance, regularly test and visually inspect the equipment. Routine equipment testing should include applying the correct pressure, if required, and then operating the equipment. For example, when testing an annular preventer, apply the proper operating pressure and then observe whether the preventer opens and closes properly. Be aware that in the outer continental shelf of the United States, and in other areas that fall under the jurisdiction of the Minerals Management Service, the MMS has established pressure limitations for BOP testing. These limitations and regulations should be adhered to.

Visual inspections of equipment for wear, improper sealing, corrosion, and other problems should be conducted periodically. Examples of items that can be checked visually include the following:

1. casing wear;
2. frozen valves;
3. valve erosion caused by the valve's being partially open;
4. use of incorrect material in the closing elements of a BOP;
5. incorrect assembly;
6. inadequate tightening of joints so that they fail to seal;

7. worn rings and gaskets that fail to seal;
8. stripped bolt and stud threads;
9. pitting of rings and grooves;
10. excessive wear of BOP bores;
11. cracked welds;
12. failed bonnet seals;
13. cracked and hardened packing;
14. clogged weep holes;
15. pitted or cracked flanges;
16. bolts and nuts that fail to meet specifications;
17. improper test procedures; and
18. welds that do not conform to the American Society of Testing and Materials Boiler and Pressure Vessel Code, Section IX, Welding Specifications.

Chapter 11

Organizing and Directing Well-Control Operations

During well-control operations, it is imperative that operators and contractors carefully organize rig personnel so that everyone knows where to be and what to do. Station and duty assignments and knowledge of the operator's and contractor's well-control policies and procedures are essential parts of successful kick control and blowout prevention. During well-control drills, the effectiveness of the crew's organization should be a part of the drill. In addition, general well-control procedures should be posted as part of emergency safety station bills for the rig. In some cases, station bills may be required by a regulatory body, such as the U.S. Coast Guard.

ORGANIZING CONSIDERATIONS

When organizing the crew for well control, the following points should be raised by supervisory personnel on the rig:

1. Who is directly responsible for well-control operations, the contractor or the operator?
2. If the toolpusher or operator's representative is designated to operate the choke, who is responsible for duties that the toolpusher or representative would normally perform if not operating the choke?
3. Have the rig supervisors participated in all well-control drills, and do they understand who is responsible for various actions?
4. How are communications with the office to be handled?
5. Does every person on the crew thoroughly understand where to be and what to do?

The crew and the rig are the responsibility of the toolpusher, but he or she must work closely with the operator to ensure that the crew has a clear understanding of applicable well-control procedures and policies.

The drilling crew's reaction time to a kick plays a major role in getting the well properly shut in to prevent a blowout; however, the crew must also give proper attention to equipment. Blowout prevention equipment is primarily intended for emergencies and, as such, it is useful only if it is adequate for the pressures involved, only if it is in good operating condition, and only if it is used correctly. Ensuring that the BOP system is in good condition and adequate for its job involves the operating company, the drilling contractor, and the drilling crews. If just one person fails to do his or her part, the good work of everyone else—for example, those who designed and installed the mud program, the casing, the wellhead equipment, and the blowout preventers—may be canceled out. The company representative, the toolpusher, and the drillers are key persons in performing well-control duties; however, every person on the crew should know how to operate the basic preventer equipment and be alert for the signs of a well kick. Trial operation of the preventers and thorough pressure testing ensure that the equipment is in good operating condition and ready for use.

DRILLS

Blowouts can be avoided if people on the job are familiar with the warnings of a kick and understand control procedures. Time is important in killing any well; therefore, early recognition of the signs and prompt and proper action for control are essential. A vital part of any crew's training is practice in closing the preventers under simulated kick conditions. Practice drills should be initiated without warning and at unexpected times. The drilling crew should be trained by means of detailed instruction and repeated drills to ensure that it can detect well flow quickly and can close in the well promptly. A properly trained floor crew should be expected to handle a kick in such a manner that they do not make the situation worse before the toolpusher or company representative can take charge.

Many types of drills may be required by different operators, depending on whether the rig is on land or offshore, whether a surface or subsea stack is in use, or other factors. Examples of different drills are a choke drill, hang-off drill, stripping drill, and pit drills.

Pit Drills

One of the most important drills is probably the pit drill, because it is a basic procedure that all crew members must know and understand. The primary objectives of a pit drill are to train rig personnel in the importance of constantly being aware of mud level in the pits and of being able to react quickly to shut the well in at the first indication of a kick. Pit drills should be conducted both when the bit is on bottom and drilling and when a trip is being made. To conduct a pit drill when the bit is on bottom and drilling, the following set of steps has been used:

1. The toolpusher or company representative raises the pit-level sensor to simulate a rise in pit level.
2. Upon recognition of the simulated kick, the driller sounds the alarm.
3. The driller then picks up the kelly above the rotary table, being sure that no tool joints are in the ram BOPs.
4. The driller shuts down the pump.
5. The driller opens the remote-controlled choke-line (HCR) valve.
6. The driller closes the annular preventer.
7. The driller closes the remote choke.

8. The driller records the time (as measured from the time the simulated pit gain was recognized) required to shut the well in, in the driller's report.

When a trip is being made, the following pit drill has been used:

1. The toolpusher or company representative manipulates the flow sensor or trip tank level indicator to simulate a kick.
2. Upon recognition of the simulated kick, the driller sounds the alarm.
3. Driller and crew set the top tool joint on the slips.
4. The rotary helpers (floorhands) stab a fully open safety valve in the drill pipe.
5. The driller opens the remote-controlled choke-line (HCR) valve.
6. The rotary helpers make up and close the drill pipe safety valve.
7. The driller closes the annular preventer.
8. The driller records the time since the simulation started to conduct the steps in the driller's report.

H₂S Drills

Since hydrogen sulfide (H_2S) can be present in a kick, and since it is a highly toxic gas, crew training in H_2S procedures is necessary in areas where H_2S is expected. Crew training involves awareness of the danger of the gas, familiarization with contingency plans, use of H_2S detection equipment, use and maintenance of breathing equipment, and several other factors. Once the crew has been trained, drills for H_2S emergencies should be carried out on a frequent basis. Some companies require them at least once a week, more often if conditions warrant. Many companies also require that records be kept on the date of drills and the personnel who participated. The following suggested actions may be taken.

1. Upon hearing or seeing an H_2S alarm, all personnel don air-breathing equipment. Assigned individuals check the breathing-air supply valves for the piped-air system. The driller takes the necessary precautions as indicated by the company's contingency plan.
2. Bug blowers are made operational and all open flames are extinguished.

3. The buddy system is implemented and all crew members act on directions from the supervisor.
4. Nonessential personnel on the location don breathing equipment and move off location.
5. Gates to the location entrance are closed and patrolled, and appropriate H_2S warning signs are displayed.

After the drill, the H_2S contingency plan regarding the notification of local authorities and the alerting of residents near the location of the possible need to evacuate the area should be discussed. Appendix A gives more details about planning for drilling in H_2S zones and gives suggested H_2S guidelines for offshore and land locations.

Hydrogen Sulfide Concerns

HYDROGEN SULFIDE CONSIDERATIONS

I. Planning for hydrogen sulfide (H_2S)

    A. A contingency plan should be drawn up when H_2S may be a problem; the contingency plan should include the following:

        1. Information on the physical effects of exposure to H_2S and sulfur dioxide (SO_2)

        2. Safety and training procedures to be followed and safety equipment that will be used

        3. Procedures for operations when H_2S conditions exist

            a. Prealarm condition

            b. Moderate danger to life

            c. Extreme danger to life

        4. Responsibilities and duties of personnel under each operating condition

        5. Designation of briefing areas

        6. Evacuation plan

        7. Governmental agency or agencies to be notified in case of extreme emergency

        8. A list of emergency medical facilities and law enforcement agencies, including addresses and telephone numbers

        9. Equipment layout and location construction

    B. A study should be made of the geology of the area, including history of adjacent wells, to predict the zone in which H_2S may be encountered

    C. The drilling mud program should provide adequate well control for the area; consideration should also be given to the use of an H_2S scavenger or inhibitor in the mud to reduce the reaction of H_2S on the drill stem and related equipment

    D. Information about the area and known field conditions, including temperatures, pressures, proposed well depth, and H_2S concentrations, should be obtained and taken into consideration

    E. In a prespud meeting, the operator should review the drilling program with the drilling and service contractors and outline each party's responsibilities in the event H_2S is encountered

    F. All personnel should be fully trained and all H_2S-related equipment should be in place either 1,000 ft above the expected zone or 1 week in advance of expected encountering of the zone

II. Well-control surveillance and training

    A. During drilling operations, continued surveillance should be maintained of mud conditions, pump pressures, and H_2S-monitoring equipment

B. Operating company and drilling contractor personnel should be trained in well-control procedures and techniques; records should be maintained on the dates and type of training that was carried out for all personnel; personnel should also be fully trained in operating the blowout prevention equipment, and the equipment should be tested periodically

III. General operating procedures

A. During trips or fishing operations, every effort should be made to pull a dry drill string while maintaining control of the well; if it is necessary to pull the string wet after an H_2S zone is penetrated, increased monitoring of the work area should be provided and protective breathing equipment should be worn; ammonia or hydrogen peroxide may be used to oxidize H_2S and thus reduce H_2S hazards during trips

B. When an H_2S zone has been penetrated and the well is being circulated bottoms-up when not drilling (such as during cementing or logging operations, or after incurring a drilling break), protective breathing apparatus should be worn by personnel in the work area in case H_2S is circulated to the surface; breathing equipment should be donned at least 30 min prior to bottoms-up time

C. During coring operations in H_2S zones, personal protective breathing equipment should be donned 10 stands to 20 stands in advance of retrieving the core barrel; breathing equipment should continue to be worn, especially while cores are being removed from the barrel; cores to be transported should be sealed and marked for the presence of H_2S

IV. Well testing

A. Well testing should be performed with a minimum number of personnel in the immediate vicinity of the rig floor and test equipment; drill stem tests should be conducted only during daylight hours; inhibitors should be used to minimize drill pipe H_2S corrosion problems during drill stem testing

B. Prior to any test, a special safety meeting should be held with all personnel who will be present during the test to emphasize the use of personal protective breathing equipment, first-aid procedures, and the H_2S contingency plan

C. During the test, use of all H_2S detection equipment should be intensified; all gases from the test should be flared and, since flared H_2S produces SO_2, SO_2 precautions should also be taken

D. No-smoking rules should be rigorously enforced, since H_2S is flammable and explosive

E. During all tests, warning signs indicating, "Danger—Poison Gas," should be displayed

H₂S GUIDELINES FOR OFFSHORE OPERATIONS

I. Drilling equipment

A. Blowout prevention equipment selected for H_2S service should have the chemical composition, heat treatment, hardness, and trim to comply with current metallurgical specifications defined by the American Petroleum Institute (API) and the National Association of Corrosion Engineers (NACE)

B. BOP remote-control equipment should be tested, and all drills and training on such equipment should be conducted in accordance with the 30 Code of Federal Regulations (CFR), Subpart D, Drilling Operations §250.50-§250.68, as published in the *Federal Register*, vol. 53, no. 63, and Oil and Gas and Sulphur Operations in the Outer Continental Shelf; Training, 30 CFR, Subpart O. The location of BOP remote-control units should be in accordance with API's *Recommended Practices for Safe Drilling of Wells Containing Hydrogen Sulfide (API RP 49)*, and installation of such equipment should be in accordance with the manufacturer's specifications

C. Flange, bonnet cover, bolting, and nut material for H_2S use should meet the requirements described in *API Specification for Wellhead Equipment (API Spec 6A)*

D. Piping, flanges, valves, fittings, and discharge, or flare, lines used in the choke-manifold assembly should contain metals and seals recommended by API and NACE

E. Degassers should be of a rated capacity to remove H_2S effectively from contaminated drilling fluid circulated back to the surface, should have explosion-proof electrical components, and should be arranged so that extracted gas can be routed to an area for flaring

F. Mud-gas separators, or gas busters, used to extract H_2S from drilling fluids should be tied into a vent line for burning so that gas is not released into the air near the rig

G. Flare lines should be installed from the degasser, choke manifold, and mud-gas separator in accordance with recommendations in *API RP 49*; all flare lines should be equipped with constant or automatic ignition devices

H. Because drill pipe comes into direct contact with H_2S in the wellbore, the composition of the pipe should be such that hydrogen embrittlement or sulfide stress cracking (SSC) is minimized; methods of minimizing hydrogen embrittlement and SSC can be found in *API RP 49*

I. The mud logging unit and equipment should be installed on the drilling site in accordance with the recommendations in *API RP 49*

J. Electrical equipment used on drilling units engaged in H_2S operations should comply with U.S. Coast Guard (USCG), MMS, or foreign regulatory requirements

K. Venting systems on drilling units with H_2S exposure should be in compliance with USCG and MMS requirements

L. Enclosed engine rooms should comply with USCG and MMS requirements

M. Any internal-combustion engine exhausting into a hazardous zone should be equipped with spark arresters

II. Monitoring equipment

A. A fixed H_2S monitor and detection unit that sets off audible and visual alarms upon detecting 10 ppm H_2S in the air should be mounted on each drilling unit; fixed systems should have a central readout panel located in an area where it can be monitored constantly

B. Sensor heads and channels for fixed H_2S monitoring systems should be located at the—

1. Bell nipple

2. Diverter opening

3. Driller's console

4. Mud tanks

5. Shale shaker

6. Ventilation system of the living quarters

7. Lower hull pump or ballast room that can be entered through hatchways on the main deck

8. Mud pump room

9. Barge engine room

C. Portable gas-detection devices, capable of reading a minimum of 10 ppm of H_2S, should be used to make frequent inspections of all areas with poor ventilation; persons making the inspection should wear protective breathing equipment

D. Personal detection devices should be available for use by all working personnel

E. At least one portable detection instrument should be available for the detection of SO_2

III.   Alarm systems

A. Warning devices should be installed at various places on the rig so that personnel can react when H_2S is encountered; the meaning of each warning signal should be made known to all personnel on or near the drilling unit

B. An audible warning should be a yelping-type electronic siren; the audible unit should be explosion-proof if in a hazardous area and should be connected to the fixed H_2S-sensing system

C. Rotating or flashing amber warning lights should be used as a visual warning; they should also be explosion-proof if in a hazardous area

D. All warning systems connected to the fixed H_2S-sensing system should be set to actuate when 10 ppm H_2S is detected at any or all of the sensor heads; when 10 ppm H_2S is encountered, the contingency program should go into effect

E. All personnel on the rig should be trained to react properly to the audible and visual warning system; designated essential personnel should move to assigned stations or job functions after donning their breathing apparatus; nonessential personnel should move to the designated safe briefing area, upwind from the rig floor, after donning their breathing apparatus

F. Warning lights on offshore rigs and platforms should be shielded from outboard view so they are not confused with navigational aids and lights

G. H_2S warning alarms should be located at the—

1.   Driller's console (audible and visual)

2.   Engine room (audible and visual)

3.   Mud room (audible)

4.   Living quarters (audible at each level)

5.   Central area of each structural level (audible and visual)

IV.   Personal protective equipment

A. All personnel whose presence is required on a drilling location where H_2S may be encountered should have properly certified breathing equipment

B. Designated essential personnel should be assigned a pressure-demand work unit with a 5-min escape, or egress, cylinder; a minimum of 25 ft of airline hose should be provided from the air source to the mask; individuals wearing the work unit should not exceed a distance of 250 ft from the source of breathing air; all airline couplings should be of corrosion-resistant material and should be designed so that they cannot be connected to any source of air or oxygen other than breathing air; with the work unit donned, the wearer should be able to turn on the escape cylinder valve, disconnect from the air hose, and move to a designated safe area or briefing station; persons working with masks on should have safety lines attached to their bodies and a buddy system should be utilized

C. Nonessential personnel should have available or be assigned an approved escape unit; escape units can be a 30-min pressure-demand, self-contained breathing apparatus, a 5-min pressure-demand emergency breathing apparatus, or any properly certified breathing apparatus that allows an individual to reach a gas-free area; designated escape routes and briefing areas should be familiar to all personnel

D. All personnel should be trained to use the breathing apparatus on the rig site; written procedure concerning the breathing apparatus should be posted and be available to all personnel on the rig site

E. Protective breathing units should be stored at locations where they can be donned quickly and readily available to all personnel on the rig; suggested locations are—

1. Rig floor

2. Derrick monkeyboard

3. Mud logging unit

4. Shale shaker

5. Mud and cement pump rooms

6. Crew quarters

7. Toolpusher's and drilling foreman's office

8. Each designated briefing area

9. Heliport

10. Standby boats

11. Work boats

12. Helicopters

H₂S GUIDELINES FOR LAND OPERATIONS

I. Drilling equipment

A. Blowout prevention equipment selected for H_2S service should have the chemical composition, heat treatment, hardness, and trim to comply with current metallurgical specifications defined by API and NACE

B. Remote-controlled BOP equipment should be located in accordance with the suggestions in *API RP 49*

C. Flange, bonnet cover, bolting, and nut material for H_2S use should meet the requirements described in *API Spec 6A*

D. Piping, flanges, valves, fittings, and discharge, or flare, lines used in the choke-manifold assembly should contain metals and seals recommended by API and NACE

E. Degassers should be of a rated capacity to remove H_2S effectively from contaminated drilling fluid circulated back to the surface, should have explosion-proof electrical components, and should be arranged so that extracted gas can be routed to an area for flaring

F. Mud-gas separators, or gas busters, used to extract H_2S from drilling fluids should be tied into a vent line for burning so that gas is not released into the air near the rig

G. Flare lines should be installed from the degasser, choke manifold, and mud-gas separator in accordance with recommendations in *API RP 49*; all flare lines should be equipped with constant or automatic ignition devices

H. Because drill pipe comes into direct contact with H_2S in the wellbore, the composition of the pipe should be such that hydrogen embrittlement or sulfide stress cracking (SSC) is minimized; methods of minimizing hydrogen embrittlement and SSC can be found in *API RP 49*

I. The mud logging unit and equipment should be installed on the drilling site in accordance with the recommendations in *API RP 49*

J. Electrical equipment used on drilling rigs engaged in H_2S operations should comply with the National Electric Code

L. Venting systems on weatherized drilling rigs affixed with permanent partitions should be provided with a sufficient ventilation system for the removal of H_2S accumulations. Generators and electrical panels should be placed in accordance with *API RP 49*

M. Mud tanks should be located in accordance with *API RP 49*

N. Any internal-combustion engine exhausting into a hazardous area should be equipped with spark arresters

II. Monitoring and detection equipment should be installed on any rig known or expected to be operating in an H_2S area; should the need for specific continuous monitoring or detection instruments be deemed necessary, it is recommended that this equipment be installed 1,000 ft prior to entry into the zone

CORROSION AND H_2S

Hydrogen embrittlement occurs when hydrogen atoms are produced on a steel surface by the reaction of iron and H_2S. Some of the atomic hydrogen migrates through the grain boundaries of the steel and recombines into hydrogen molecules. The larger molecules cannot escape from the steel and thus increase internal stresses, which cause blistering or hydrogen embrittlement. The time to failure decreases as the amount of hydrogen absorbed, applied stress, and strength level of the material increase. Sulfide stress cracking (SSC) is a brittle failure that occurs when steel is exposed to moist hydrogen sulfide. SSC is thought to be a form of hydrogen embrittlement. Most corrosion specialists agree that, for SSC to occur, H_2S, water, high-strength steel, and applied or residual stress must be present.

When H_2S is in contact with casing in an electrolyte such as water-base drilling mud, both sulfide ions and atomic hydrogen are present. Tests have shown that the sulfide ion reduces the rate at which hydrogen atoms combine to form diatomic molecules outside the metal. Since fewer diatomic hydrogen molecules form outside the metal, more atomic hydrogen atoms migrate into the steel and increase the tendency for blistering or embrittlement to occur. If no sulfide ions are present in the electrolyte, most of the hydrogen atoms combine into diatomic molecules outside the steel, and little or no blistering or embrittlement occurs.

The NACE *Standard MR-01-75* lists several acceptable steels for tubular goods that will be exposed directly to H_2S. The API casing grades that NACE lists as acceptable for all temperatures are (1) H-40, (2) J-55, (3) K-55, (4) C-75, and (5) L-80. For temperatures over 150° F, NACE lists (1) N-80, (2) C-95, and (3) proprietary grades with 110 Ksi or less maximum yield strength. For temperatures over 175° F, NACE lists (1) H-40, (2) N-80, (3) P-105, (4) P-110, and (5) proprietary grades with 110 Ksi minimum to 140 Ksi maximum yield strength.

Reference Tables

Table B.1
Specifications of API Drill Pipe

Nominal Size	Weight (ppf)	OD (in.)	ID (in.)	Capacity (bbl/ft)	Displacement (bbl/ft)
2⅜*	4.80	2.375	2.000	.00389	.00167
2⅜	6.65	2.375	1.815	.00320	.00230
2⅞*	6.45	2.875	2.469	.00592	.00233
2⅞*	8.35	2.875	2.323	.00524	.00290
2⅞	10.40	2.875	2.151	.00449	.00397
3½*	8.50	3.500	3.063	.00911	.00290
3½*	11.20	3.500	2.900	.00817	.00380
3½	13.30	3.500	2.764	.00742	.00503
3½	15.50	3.500	2.602	.00658	.00587
4	14.00	4.000	3.340	.01084	.00500
4½	12.75	4.500	4.000	.01554	.00467
4½	13.75	4.500	3.958	.01522	.00549
4½	16.60	4.500	3.826	.01422	.00648
4½	20.00	4.500	3.640	.01287	.00776
5	19.50	5.000	4.276	.01776	.00750
5½	21.90	5.500	4.778	.02218	.00770
5½	24.70	5.500	4.670	.02119	.00870
5⁹⁄₁₆*	19.00	5.563	4.975	.02404	.00650
5⁹⁄₁₆*	22.20	5.563	4.859	.02294	.00760
5⁹⁄₁₆*	25.25	5.563	4.733	.02176	.00880
6⅝*	22.20	6.625	6.065	.03573	.00750
6⅝*	25.20	6.625	5.965	.03456	.00870
6⅝*	31.90	6.625	5.761	.03224	.01100
7⅝*	29.25	7.625	6.969	.04718	.01010
8⅝*	40.00	8.625	7.825	.05948	.01400

* Not API Standard

Table B.2
Specifications of Drill Collars

OD (in.)	ID (in.)	Capacity (bbl/ft)	Displacement (bbl/ft)
3.125	1.2500	.00151	.00796
3.500	1.5000	.00218	.00971
3.750	1.5000	.00218	.01147
4.000	2.0000	.00388	.01165
4.125	2.0000	.00388	.01264
4.250	2.0000	.00388	.01366
4.500	2.2500	.00491	.01475
4.750	2.2500	.00491	.01700
5.000	2.2500	.00491	.01936
5.250	2.2500	.00491	.02185
5.500	2.2500	.00491	.02446
5.750	2.2500	.00491	.02719
6.000	2.2500	.00491	.03005
6.250	2.2500	.00491	.03002
6.250	2.2125	.00468	.03026
6.500	2.2500	.00491	.03612
6.500	2.8125	.00768	.03335
6.750	2.8125	.00768	.03657
7.000	2.8125	.00768	.03991
7.250	2.8125	.00768	.04337
7.500	2.8125	.00768	.04695
7.750	2.8125	.00768	.05066
8.000	2.8125	.00768	.05448
8.000	3.0000	.00874	.05342
8.250	3.0000	.00874	.05737
8.500	3.0000	.00874	.06144
8.750	3.0000	.00874	.06563
9.000	3.0000	.00874	.06994
9.250	3.0000	.00874	.07437
9.500	3.0000	.00874	.07892
9.750	3.0000	.00874	.08360
10.000	3.0000	.00874	.08839
11.000	3.0000	.00874	.10880
11.250	3.0000	.00874	.11420

Table B.3
Annular Volume and Height between Hole and Drill Collars

Drill Collar OD (in.)	Hole Size (in.)	Bbl per Linear Ft	Linear Ft per Bbl
4	5⅝	.0152	65.8
4¾		.0088	113.4
4	6¼	.0224	44.6
4¾		.0160	62.5
5		.0137	73.2
4¾	6¾	.0224	44.6
5		.0200	48.5
4¾	7⅞	.0383	26.1
5		.0360	27.8
6	8½	.0352	28.4
6¼		.0323	31.0
6½		.0292	34.2
6	8¾	.0394	25.4
6½		.0334	29.9
7		.0268	37.3
6½	9¾	.0513	19.5
7		.0448	22.3
6½	10⅝	.0686	14.6
7		.0621	16.1
8		.0475	21.1
9		.0309	32.4
7½	12¼	.0911	11.0
8		.0836	12.0
7½	15	.1639	6.1
8		.1564	6.4
9		.1399	7.2
10		.1214	8.2
11		.1010	9.9

Table B.4
Capacity of Hole

Diameter of Hole (in.)	Gal per Linear Ft	Linear Ft per Gal	Bbl per Linear Ft	Linear Ft per Bbl
2	.1632	6.1275	.0039	257.3536
⅛	.1842	5.4278	.0044	227.9672
¼	.2055	4.8415	.0049	203.3411
⅜	.2301	4.3452	.0055	182.5000
½	.2550	3.9216	.0061	164.7063
⅝	.2811	3.5570	.0067	149.3934
¾	.3085	3.2410	.0073	136.1209
⅞	.3372	2.9653	.0080	124.5416
3	.3672	2.7233	.0087	114.3794
⅛	.3984	2.5098	.0095	105.4120
¼	.4309	2.3205	.0103	97.4593
⅜	.4647	2.1518	.0111	90.3738
½	.4998	2.0008	.0119	84.0338
⅝	.5361	1.8652	.0128	78.3383
¾	.5737	1.7429	.0137	73.2028
⅞	.6126	1.6323	.0146	68.5562
4	.6528	1.5319	.0155	64.3384
⅛	.6942	1.4404	.0165	60.4982
¼	.7369	1.3569	.0175	56.9918
⅜	.7809	1.2805	.0186	53.7816
½	.8262	1.2104	.0197	50.8353
⅝	.8727	1.1458	.0208	48.1245
¾	.9205	1.0863	.0219	45.6250
⅞	.9696	1.0313	.0231	43.3153
5	1.0200	.9804	.0243	41.1766
⅛	1.0716	.9332	.0255	39.1924
¼	1.1245	.8892	.0268	37.3484
⅜	1.1787	.8484	.0281	35.6314
½	1.2342	.8102	.0294	34.0302
⅝	1.2909	.7746	.0307	32.5346
¾	1.3489	.7413	.0321	31.1354
⅞	1.4082	.7101	.0335	29.8245
6	1.4688	.6808	.0350	28.5948
⅛	1.5306	.6533	.0364	27.4396
¼	1.5937	.6275	.0379	26.3530
⅜	1.6581	.6031	.0395	25.3297
½	1.7238	.5801	.0410	24.3648
⅝	1.7907	.5584	.0426	23.4541
¾	1.8589	.5379	.0443	22.5935
⅞	1.9284	.5186	.0459	21.7793

Table B.4
Capacity of Hole
(continued)

Diameter of Hole (in.)	Gal per Linear Ft	Linear Ft per Gal	Bbl per Linear Ft	Linear Ft per Bbl
7	1.9992	.5002	.0476	21.0085
1/8	2.0712	.4828	.0493	20.2778
1/4	2.1445	.4663	.0511	19.5846
3/8	2.2191	.4506	.0528	18.9263
1/2	2.2950	.4357	.0546	18.3007
5/8	2.3721	.4216	.0565	17.7055
3/4	2.4505	.4081	.0583	17.1390
7/8	2.5302	.3952	.0602	16.5993
8	2.6112	.3830	.0622	16.0846
1/8	2.6934	.3713	.0641	15.5935
1/4	2.7769	.3601	.0661	15.1245
3/8	2.8617	.3494	.0681	14.6764
1/2	2.9478	.3392	.0702	14.2479
5/8	3.0351	.3295	.0723	13.8380
3/4	3.1237	.3201	.0744	13.4454
7/8	3.2136	.3112	.0765	13.0693
9	3.3048	.3026	.0787	12.7088
1/8	3.3972	.2944	.0809	12.3630
1/4	3.4909	.2865	.0831	12.0311
3/8	3.5859	.2789	.0854	11.7124
1/2	3.6822	.2716	.0877	11.4063
5/8	3.7797	.2646	.0900	11.1119
3/4	3.8785	.2578	.0923	10.8288
7/8	3.9786	.2513	.0947	10.5564
10	4.0800	.2451	.0971	10.2941
1/8	4.1826	.2391	.0996	10.0415
1/4	4.2865	.2333	.1021	9.7981
3/8	4.3917	.2277	.1046	9.5634
1/2	4.4982	.2223	.1071	9.3371
5/8	4.6059	.2171	.1097	9.1187
3/4	4.7149	.2121	.1123	8.9079
7/8	4.8252	.2072	.1149	8.7043
11	4.9368	.2026	.1175	8.5076
1/8	5.0496	.1980	.1202	8.3174
1/4	5.1637	.1937	.1229	8.1336
3/8	5.2791	.1894	.1257	7.9559
1/2	5.3958	.1853	.1285	7.7839
5/8	5.5137	.1814	.1313	7.6174
3/4	5.6329	.1775	.1341	7.4561
7/8	5.7534	.1738	.1370	7.3000

Table B.4
Capacity of Hole
(continued)

Diameter of Hole (in.)	Gal per Linear Ft	Linear Ft per Gal	Bbl per Linear Ft	Linear Ft per Bbl
12	5.8752	.1702	.1399	7.1487
⅛	5.9982	.1667	.1428	7.0021
¼	6.1225	.1633	.1458	6.8599
⅜	6.2481	.1600	.1488	6.7220
½	6.3750	.1569	.1518	6.5883
⅝	6.5031	.1538	.1548	6.4584
¾	6.6325	.1508	.1579	6.3324
⅞	6.7632	.1479	.1610	6.2101
13	6.8952	.1450	.1642	6.0912
⅛	7.0284	.1423	.1673	5.9757
¼	7.1629	.1396	.1705	5.8635
⅜	7.2987	.1370	.1738	5.7544
½	7.4358	.1345	.1770	5.6484
⅝	7.5741	.1320	.1803	5.5452
¾	7.7137	.1296	.1837	5.4448
⅞	7.8546	.1273	.1870	5.3472
14	7.9968	.1251	.1904	5.2521
⅛	8.1402	.1228	.1938	5.1596
¼	8.2849	.1207	.1973	5.0694
⅜	8.4309	.1186	.2007	4.9817
½	8.5782	.1166	.2042	4.8961
⅝	8.7267	.1146	.2078	4.8128
¾	8.8765	.1127	.2113	4.7316
⅞	9.0276	.1108	.2149	4.6524
15	9.1800	.1089	.2186	4.5752
⅛	9.3336	.1071	.2222	4.4999
¼	9.4885	.1054	.2259	4.4264
⅜	9.6447	.1037	.2296	4.3547
½	9.8022	.1020	.2334	4.2848
⅝	9.9609	.1004	.2372	4.2165
¾	10.1209	.0988	.2410	4.1498
⅞	10.2822	.0973	.2448	4.0847
16	10.4448	.0957	.2487	4.0211
¼	10.7737	.0928	.2565	3.8984
½	11.1078	.0900	.2645	3.7811
¾	11.4469	.0874	.2725	3.6691
17	11.7912	.0848	.2807	3.5620
¼	12.1405	.0824	.2891	3.4595
½	12.4950	.0800	.2975	3.3614
¾	12.8545	.0778	.3061	3.2673

Table B.4
Capacity of Hole
(continued)

Diameter of Hole (in.)	Gal per Linear Ft	Linear Ft per Gal	Bbl per Linear Ft	Linear Ft per Bbl
18	13.2192	.0756	.3147	3.1772
¼	13.5889	.0736	.3235	3.0908
½	13.9638	.0716	.3325	3.0078
¾	14.3437	.0697	.3415	2.9281
19	14.7288	.0679	.3507	2.8516
¼	15.1189	.0661	.3600	2.7780
½	15.5142	.0645	.3694	2.7072
¾	15.9145	.0628	.3789	2.6391
20	16.3200	.0613	.3886	2.5735
¼	16.7305	.0598	.3983	2.5104
½	17.1462	.0583	.4082	2.4495
¾	17.5669	.0569	.4183	2.3909
21	17.9928	.0556	.4284	2.3343
¼	18.4237	.0543	.4387	2.2797
½	18.8598	.0530	.4490	2.2270
¾	19.3009	.0518	.4595	2.1761
22	19.7472	.0506	.4702	2.1269
¼	20.1985	.0495	.4809	2.0794
½	20.6550	.0484	.4918	2.0334
¾	21.2265	.0474	.5028	1.9890
23	21.5831	.0463	.5139	1.9460
¼	22.0549	.0453	.5251	1.9043
½	22.5317	.0444	.5365	1.8640
¾	23.0137	.0435	.5479	1.8250
24	23.5007	.0426	.5595	1.7872
¼	23.9929	.0417	.5713	1.7505
½	24.4901	.0408	.5831	1.7150
¾	24.9925	.0400	.5951	1.6805
25	25.4999	.0392	.6071	1.6471
¼	26.0125	.0384	.6193	1.6146
½	26.5301	.0377	.6317	1.5831
¾	27.0529	.0370	.6441	1.5525
26	27.5807	.0363	.6567	1.5228
¼	28.1137	.0356	.6694	1.4939
½	28.6517	.0349	.6822	1.4659
¾	29.1949	.0343	.6951	1.4386

Table B.4
Capacity of Hole
(continued)

Diameter of Hole (in.)	Gal per Linear Ft	Linear Ft per Gal	Bbl per Linear Ft	Linear Ft per Bbl
27	29.7431	.0336	.7082	1.4121
¼	30.2965	.0330	.7213	1.3863
½	30.8549	.0324	.7346	1.3612
¾	31.4185	.0318	.7481	1.3368
28	31.9871	.0313	.7616	1.3130
¼	32.6509	.0307	.7753	1.2899
½	33.1397	.0302	.7890	1.2674
¾	33.7237	.0297	.8029	1.2454
29	34.3127	.0291	.8170	1.2240
¼	34.9069	.0286	.8311	1.2032
½	35.5061	.0282	.8454	1.1829
¾	36.1105	.0277	.8598	1.1631
30	36.7199	.0272	.8743	1.1438
¼	37.3345	.0268	.8889	1.1250
½	37.9541	.0263	.9037	1.1066
¾	38.5789	.0259	.9185	1.0887
31	39.2087	.0255	.9335	1.0712
¼	39.8437	.0251	.9487	1.0541
½	40.4837	.0247	.9639	1.0375
¾	41.1289	.0243	.9793	1.0212
32	41.7791	.0239	.9947	1.0053
¼	42.4345	.0236	1.0103	.9898
½	43.0949	.0232	1.0261	.9746
¾	43.7604	.0229	1.0419	.9598
33	44.4311	.0225	1.0579	.9453
¼	45.1068	.0222	1.0740	.9311
½	45.7877	.0218	1.0902	.9173
¾	46.4736	.0215	1.1065	.9037
34	47.1647	.0212	1.1230	.8905
¼	47.8608	.0209	1.1395	.8775
½	48.5621	.0206	1.1562	.8649
¾	49.2684	.0203	1.1731	.8525
35	49.9799	.0200	1.1900	.8403
¼	50.6964	.0197	1.2071	.8285
½	51.4181	.0194	1.2242	.8168
¾	52.1448	.0192	1.2415	.8054

Table B.5
Capacity of Casing

OD (in.)	Weight (ppf)	ID (in.)	Capacity (bbl/ft)
4½	9.50	4.090	.0163
4½	11.60	4.000	.0155
4½	13.50	3.920	.0149
4½	15.10	3.826	.0142
5	11.50	4.560	.0202
5	13.00	4.494	.0196
5	15.00	4.408	.0189
5*	21.00	4.154	.0168
5½	13.00	5.044	.0247
5½	14.00	5.012	.0244
5½ *	15.00	4.974	.0240
5½	15.50	4.950	.0238
5½	17.00	4.892	.0232
5½	20.00	4.778	.0222
5½	23.00	4.670	.0212
6⅝	17.00	6.135	.0366
6⅝	20.00	6.049	.0355
6⅝ *	22.00	5.989	.0348
6⅝	24.00	5.921	.0341
7	17.00	6.538	.0415
7	20.00	6.456	.0405
7*	22.00	6.398	.0398
7	23.00	6.366	.0394
7*	24.00	6.336	.0390
7	26.00	6.276	.0383
7*	28.00	6.214	.0375
7	29.00	6.184	.0371
7*	30.00	6.154	.0368
7	32.00	6.094	.0361
7*	34.00	6.040	.0354
7	35.00	6.004	.0350
7	38.00	5.920	.0340
7*	40.00	5.836	.0331
7⅝	20.00	7.125	.0493
7⅝	24.00	7.025	.0479
7⅝	26.40	6.969	.0472
7⅝	29.70	6.875	.0459
7⅝	33.70	6.765	.0445
7⅝	39.00	6.625	.0426
8⅝	24.00	8.097	.0637
8⅝	28.00	8.017	.0624
8⅝	32.00	7.921	.0609
8⅝	36.00	7.825	.0595
8⅝ *	38.00	7.775	.0587

* Not API Standard

Table B.5
Capacity of Casing

OD (in.)	Weight (ppf)	ID (in.)	Capacity (bbl/ft)
8⅝	40.00	7.725	.0580
8⅝*	43.00	7.651	.0569
8⅝	44.00	7.625	.0565
8⅝	49.00	7.511	.0548
9⅝	29.30	9.063	.0798
9⅝	32.30	9.001	.0787
9⅝	36.00	8.921	.0773
9⅝*	38.00	8.877	.0765
9⅝	40.00	8.835	.0758
9⅝	43.50	8.755	.0745
9⅝	47.00	8.681	.0732
9⅝	53.50	8.535	.0708
10¾	32.75	10.192	.1009
10¾*	35.75	10.136	.0998
10¾	40.50	10.050	.0981
10¾	45.50	9.950	.0962
10¾	51.00	9.850	.0943
10¾*	54.00	9.784	.0930
10¾	55.50	9.760	.0925
10¾	60.70	9.660	.0906
10¾	65.70	9.560	.0888
11¾	38.00	11.150	.1208
11¾	42.00	11.084	.1193
11¾	47.00	11.000	.1175
11¾	54.00	10.880	.1150
11¾	60.00	10.772	.1127
13⅜	48.00	12.715	.1571
13⅜	54.50	12.615	.1546
13⅜	61.00	12.515	.1521
13⅜	68.00	12.415	.1497
13⅜	72.00	12.347	.1481
13⅜*	83.00	12.175	.1440
16	55.00	15.376	.2297
16	65.00	15.250	.2259
16*	70.00	15.198	.2244
16	75.00	15.124	.2222
16	84.00	15.010	.2189
18⅝*	78.00	17.855	.3097
18⅝*	87.50	17.755	.3062
18⅝*	88.50	17.655	.3028
20*	90.00	19.166	.3568
20	94.00	19.124	.3553

* Not AP¾ Standard

Table B.6
Displacement of Duplex Mud Pumps
90% Efficiency
(bbl/s)

Liner Size (in.)	Stroke Length (in.)								
	8	10	12	14	15	16	18	20	22
4	0.034	0.044	0.053	0.062					
4¼	0 040	0.050	0.060	0.070					
4½	0.044	0.056	0.067	0.078	0.084	0.087	0.090	0.099	
4¾	0.050	0.062	0.075	0.087	0.093	0.096	0.102	0.113	
5	0.055	0.069	0.083	0.096	0.103	0.110	0.115	0.138	
5¼	0.061	0.076	0.091	0.106	0.114	0.121	0.137	0.152	
5½	0.067	0.083	0.099	0.115	0.121	0.127	0.142	0.167	
5¾	0.073	0.091	0.109	0.128	0.137	0.146	0.164	0.182	
6	0.079	0.099	0.119	0.139	0.146	0.153	0.172	0.198	
6¼	0.086	0.108	0.130	0.151	0.161	0.168	0.189	0.215	
6½	0.093	0.116	0.141	0.164	0.176	0.182	0.205	0.228	
6¾	0.100	0.126	0.152	0.178	0.188	0.198	0.223	0.247	
7	0.108	0.135	0.164	0.192	0.202	0.214	0.241	0.267	
7¼	0.115	0.145	0.174	0.201	0.217	0.231	0.259	0.288	
7½		0.155	0.186	0.217	0.232	0.248	0.279	0.310	0.342
7¾			0.199	0.232	0.248	0.265	0.299	0.331	0.365
8		0.176	0.212	0.247	0.264	0.282	0.317	0.355	0.413
8½			0.225	0.263	0.281	0.300	0.338	0.375	0.439

Table B.7
Displacement of Triplex Mud Pumps
100% Efficiency
(bbl/s)

Liner Size (in.)	Stroke Length (in.)								
	7	7½	8	8½	9	9¼	10	11	12
3	0.015	0.016	0.017	0.019	0.020	0.020	0.022	0.024	0.026
3¼	0.018	0.019	0.021	0.022	0.023	0.024	0.026	0.028	0.031
3½	0.021	0.022	0.024	0.025	0.027	0.028	0.030	0.033	0.036
3¾	0.024	0.026	0.027	0.029	0.031	0.032	0.034	0.038	0.041
4	0.027	0.030	0.031	0.033	0.035	0.036	0.039	0.043	0.047
4¼	0.031	0.033	0.035	0.037	0.040	0.040	0.044	0.048	0.053
4½	0.035	0.037	0.039	0.042	0.044	0.045	0.049	0.054	0.059
4¾	0.038	0.041	0.044	0.047	0.049	0.051	0.055	0.060	0.066
5	0.043	0.045	0.048	0.052	0.055	0.056	0.061	0.067	0.073
5¼	0 047	0.050	0.054	0.057	0.060	0.062	0.067	0.074	0.080
5½	0.051	0.055	0.059	0.062	0.066	0.068	0.074	0.081	0.088
5¾	0.056	0.060	0.064	0.068	0.072	0.074	0.080	0.088	0.096
6	0.061	0.065	0.070	0.074	0.079	0.081	0.087	0.096	0.105
6¼	0.066	0.071	0.076	0.081	0.085	0.088	0.095	0.104	0.114
6½	0.072	0.077	0.082	0.087	0.092	0.095	0.103	0.113	0.123
6¾	0.077	0.083	0.089	0.094	0.100	0.102	0.110	0.121	0.133
7	0.083	0.089	0.095	0.101	0.107	0.110	0.119	0.131	0.143
7½	0.096	0.103	0.109	0.116	0.123	0.126	0.137	0.150	0.164

Table B.8
Mud Weight Increase in Pounds per Gallon
Required to Balance a Kick

True Vertical Depth (ft)	Shut-In Drill Pipe Pressure (psi)													
	100	200	300	400	500	600	700	800	900	1,000	1,100	1,200	1,300	1,400
3,000	0.6	1.3	1.9	2.5	3.2	3.8	4.5	5.1	5.8	6.4	7.1	7.7	8.3	9.0
4,000	0.5	1.0	1.4	1.9	2.4	2.9	3.4	3.8	4.3	4.8	5.3	5.8	6.3	6.7
5,000	0.4	0.8	1.2	1.5	1.9	2.3	2.7	3.1	3.5	3.9	4.2	4.6	5.0	5.4
6,000	0.3	0.6	1.0	1.3	1.6	1.9	2.3	2.6	2.9	3.2	3.5	3.9	4.2	4.5
7,000	0.3	0.6	0.8	1.1	1.4	1.7	1.9	2.2	2.5	2.8	3.0	3.3	3.6	3.9
8,000	0.2	0.5	0.7	1.0	1.2	1.4	1.7	1.9	2.2	2.4	2.6	2.9	3.1	3.4
9,000	0.2	0.4	0.6	0.9	1.1	1.3	1.5	1.7	1.9	2.1	2.4	2.6	2.8	3.0
10,000	0.2	0.4	0.6	0.8	1.0	1.2	1.3	1.5	1.7	1.9	2.1	2.3	2.5	2.7
11,000	0.2	0.4	0.5	0.7	0.9	1.1	1.2	1.4	1.6	1.8	1.9	2.1	2.3	2.5
12,000	0.2	0.3	0.5	0.6	0.6	1.0	1.1	1.3	1.4	1.6	1.8	1.9	2.1	2.2
13,000	0.1	0.3	0.4	0.6	0.7	0.9	1.0	1.2	1.3	1.5	1.6	1.8	1.9	2.1
14,000	0.1	0.3	0.4	0.6	0.7	0.8	0.9	1.1	1.2	1.4	1.5	1.7	1.8	1.9
15,000	0.1	0.3	0.4	0.5	0.6	0.7	0.9	1.0	1.2	1.3	1.4	1.5	1.7	1.8
16,000	0.1	0.2	0.4	0.5	0.6	0.7	0.8	1.0	1.1	1.2	1.3	1.4	1.6	1.7
17,000	0.1	0.2	0.3	0.5	0.6	0.7	0.8	0.9	1.0	1.1	1.2	1.4	1.5	1.6
18,000	0.1	0.2	0.3	0.4	0.5	0.6	0.8	0.9	1.0	1.1	1.2	1.3	1.4	1.5
19,000	0.1	0.2	0.3	0.4	0.5	0.6	0.7	0.8	0.9	1.0	1.1	1.2	1.3	1.4
20,000	0.1	0.2	0.3	0.4	0.5	0.6	0.7	0.8	0.9	1.0	1.1	1.2	1.3	1.3

Table B.9
Annular Pressure Loss, Ppg

Hole Diameter (in.)	Drill Pipe OD (in.)	Yield Point (lb/100 ft²)												
		2	4	6	8	10	12	14	16	18	20	22	24	26
12¼	5	0.1	0.1	0.1	0.1	0.1	0.1	0.2	0.2	0.2	0.2	0.3	0.3	0.3
9⅞	5	0.1	0.1	0.1	0.1	0.2	0.2	0.2	0.3	0.3	0.4	0.4	0.4	0.4
8¾	4½	0.1	0.1	0.2	0.2	0.2	0.2	0.3	0.3	0.4	0.4	0.4	0.5	0.5
7⅞	4½	0.1	0.1	0.2	0.2	0.3	0.3	0.4	0.4	0.5	0.5	0.6	0.6	0.6
6¾	4½	0.1	0.2	0.2	0.3	0.4	0.5	0.5	0.6	0.7	0.7	0.8	0.9	1.0
6¼	3½	0.1	0.1	0.2	0.2	0.3	0.4	0.4	0.5	0.5	0.6	0.7	0.7	0.8
6	2⅞	0.1	0.1	0.2	0.2	0.3	0.3	0.3	0.4	0.5	0.5	0.6	0.7	0.7

Table B.10
Approximate Gain in Barrels of Mud per
100 Barrels from Increasing Mud Weight

Present Mud Weight (ppg)	Desired Mud Weight (ppg)								
	10	11	12	13	14	15	16	17	18
9	5	10	15	20	25	30	30	45	50
10		5	10	15	20	25	30	40	50
11			5	10	15	20	25	35	40
12				5	10	15	20	30	35
13					5	10	15	20	30
14						5	10	15	25
15							5	10	20
16								5	10
17									5

Table B.11
Approximate Number of 100-Pound Sacks of Barite
Required to Increase Weight of 100 Barrels of Mud

Present Mud Weight (ppg)	Desired Mud Weight (ppg)								
	10	11	12	13	14	15	16	17	18
9	50	120	190	270	350	440	540	650	780
10		60	130	200	280	370	460	570	700
11			64	130	210	300	390	490	600
12				67	140	220	310	410	520
13					70	150	230	330	430
14						70	150	240	350
15							80	160	260
16								80	170
17									90

Table B.12
Casing Burst Values

Size (in.)	Weight (ppf)	Grade	Burst (psi)	80% Burst (psi)
8⅝	32	K55	3,930	3,144
	36	K55	4,460	3,568
	36	N80	6,490	5,192
	36	S95	7,710	6,168
	40	N80	7,300	5,840
	40	S95	8,670	6,936
	40	P110	10,040	8,032
	44	N80	8,120	6,496
	44	S95	9,640	7,712
	44	P110	11,160	8,928
	49	N80	9,040	7,232
	49	S95	10,740	8,592
	49	P110	12,430	9,944
9⅝	36	K55	3,520	2,816
	40	K55	3,950	3,160
	40	N80	5,750	4,600
	40	S95	6,820	5,456
	43.5	N80	6,330	5,064
	43.5	S95	7,510	6,008
	43.5	P110	8,700	6,960
	47	N80	6,870	5,496
	47	S95	8,150	6,520
	47	P110	9,440	7,552
	53.5	N80	7,930	6,344
	53.5	S95	9,410	7,528
	53.5	P110	10,900	8,720
	58.4	S95	10,280	8,224
	58.4	S105	10,280	8,224
	61.1	S95	10,800	8,640
	61.1	S105	10,800	8,640
10¾	40.5	K55	3,130	2,504
	45.5	K55	3,580	2,864
	51	K55	4,030	3,224
	51	N80	5,860	4,688
	51	S95	6,960	5,568
	51	P110	8,060	6,448
	55.5	N80	6,450	5,160
	55.5	S95	8,860	7,088
	60.7	S95	8,430	6,744
	60.7	P110	9,760	7,808
	65.7	S95	9,200	7,360
	65.7	P110	10,650	8,520

Table B.12
Casing Burst Values
(continued)

Size (in.)	Weight (ppf)	Grade	Burst (psi)	80% Burst (psi)
11 ¾	47	K55	3,070	2,456
	54	K55	3,560	2,848
	60	K55	4,010	3,208
	60	N80	5,830	4,664
	60	S95	6,920	5,536
	65	S95	7,560	6,048
	71	S95	8,230	6,584
	71.8	S95	8,150	6,520
13⅜	48	H40	1,730	1,384
	54.5	K55	2,730	2,184
	61	K55	3,090	2,472
	72	N80	5,380	4,304
	72	S95	6,390	5,112
	72	P110	7,400	5,920
	80.7	S80	4,170	3,336
	80.7	S95	7,210	5,768
	85	N80	6,360	5,088
	86	S95	7,770	6,216
16	65	H40	1,640	1,312
	75	K55	2,630	2,104
	84	K55	2,980	2,384
20	94	K55	2,110	1,688
	133	K55	3,060	2,448

Table B.13
Pressure Gradients for Various Weight Fluids

Ppg	Psi/Ft	Ppg	Psi/Ft	Ppg	Psi/Ft	Ppg	Psi/Ft	Ppg	Psi/Ft
0.0	.0000	4.0	.2076	8.0	.4152	12.0	.6228	16.0	.8304
0.1	.0052	4.1	.2128	8.1	.4204	12.1	.6280	16.1	.8356
0.2	.0104	4.2	.2180	8.2	.4256	12.2	.6332	16.2	.8408
0.3	.0156	4.3	.2232	8.3	.4308	12.3	.6384	16.3	.8460
0.4	.0208	4.4	.2284	8.33	.4328	12.4	.6436	16.4	.8512
0.5	.0260	4.5	.2336	8.4	.4360	12.5	.6488	16.5	.8564
0.6	.0311	4.6	.2387	8.5	.4412	12.6	.6539	16.6	.8615
0.7	.0363	4.7	.2439	8.6	.4463	12.7	.6591	16.7	.8667
0.8	.0415	4.8	.2491	8.7	.4515	12.8	.6643	16.8	.8719
0.9	.0467	4.9	.2543	8.8	.4567	12.9	.6695	16.9	.8771
1.0	.0519	5.0	.2595	8.9	.4619	13.0	.6747	17.0	.8823
1.1	.0571	5.1	.2647	9.0	.4671	13.1	.6799	17.1	.8875
1.2	.0623	5.2	.2699	9.1	.4723	13.2	.6851	17.2	.8927
1.3	.0675	5.3	.2751	9.2	.4775	13.3	.6903	17.3	.8979
1.4	.0727	5.4	.2803	9.3	.4827	13.4	.6955	17.4	.9031
1.5	.0779	5.5	.2855	9.4	.4879	13.5	.7007	17.5	.9083
1.6	.0830	5.6	.2906	9.5	.4931	13.6	.7058	17.6	.9134
1.7	.0883	5.7	.2958	9.6	.4982	13.7	.7110	17.7	.9186
1.8	.0934	5.8	.3010	9.7	.5034	13.8	.7162	17.8	.9238
1.9	.0986	5.9	.3062	9.8	.5086	13.9	.7214	17.9	.9290
2.0	.1038	6.0	.3114	9.9	.5138	14.0	.7266	18.0	.9342
2.1	.1090	6.1	.3166	10.0	.5190	14.1	.7318	18.1	.9394
2.2	.1142	6.2	.3218	10.1	.5242	14.2	.7370	18.2	.9446
2.3	.1194	6.3	.3270	10.2	.5294	14.3	.7422	18.3	.9498
2.4	.1246	6.4	.3322	10.3	.5346	14.4	.7474	18.4	.9550
2.5	.1298	6.5	.3374	10.4	.5398	14.5	.7526	18.5	.9602
2.6	.1349	6.6	.3425	10.5	.5450	14.6	.7577	18.6	.9653
2.7	.1401	6.7	.3477	10.6	.5501	14.7	.7629	18.7	.9705
2.8	.1453	6.8	.3529	10.7	.5553	14.8	.7681	18.8	.9757
2.9	.1505	6.9	.3581	10.8	.5605	14.9	.7733	18.9	.9809
3.0	.1557	7.0	.3633	10.9	.5657	15.0	.7785	19.0	.9861
3.1	.1609	7.1	.3685	11.0	.5709	15.1	.7837	19.1	.9913
3.2	.1661	7.2	.3737	11.1	.5761	15.2	.7889	19.2	.9965
3.3	.1713	7.3	.3789	11.2	.5813	15.3	.7941	19.3	1.0017
3.4	.1765	7.4	.3841	11.3	.5865	15.4	.7993	19.4	1.0069
3.5	.1817	7.5	.3893	11.4	.5917	15.5	.8045	19.5	1.0121
3.6	.1868	7.6	.3944	11.5	.5969	15.6	.8096	19.6	1.0172
3.7	.1920	7.7	.3996	11.6	.6020	15.7	.8148	19.7	1.0224
3.8	.1972	7.8	.4048	11.7	.6072	15.8	.8200	19.8	1.0276
3.9	.2024	7.9	.4100	11.8	.6124	15.9	.8252	19.9	1.0328
				11.9	.6176			20.0	1.0380

Table B.14
Buoyancy Factors for Open-ended Steel Pipe in Various Weight Fluids

Ppg	Buoyancy Factor	Ppg	Buoyancy Factor	Ppg	Buoyancy Factor
6.0	.9100	10.6	.8410	15.3	.7705
6.1	.9085	10.7	.8395	15.4	.7690
6.2	.9070	10.8	.8380	15.5	.7675
6.3	.9055	10.9	.8365	15.6	.7660
6.4	.9040	11.0	.8350	15.7	.7645
6.5	.9025	11.1	.8335	15.8	.7630
6.6	.9010	11.2	.8305	15.9	.7615
6.7	.8995	11.3	.8305	16.0	.7600
6.8	.8980	11.4	.8290	16.1	.7585
6.9	.8965	11.5	.8275	16.2	.7570
7.0	.8950	11.6	.8260	16.3	.7555
7.1	.8935	11.7	.8245	16.4	.7540
7.2	.8920	11.8	.8230	16.5	.7525
7.3	.8905	11.9	.8215	16.6	.7510
7.4	.8890	12.0	.8200	16.7	.7495
7.5	.8875	12.1	.8185	16.8	.7480
7.6	.8860	12.2	.8170	16.9	.7465
7.7	.8845	12.3	.8155	17.0	.7450
7.8	.8830	12.4	.8140	17.1	.7435
7.9	.8815	12.5	.8125	17.2	.7420
8.0	.8800	12.6	.8110	17.3	.7405
8.1	.8785	12.7	.8095	17.4	.7390
8.2	.8770	12.8	.8080	17.5	.7375
8.3	.8755	12.9	.8065	17.6	.7360
8.33	.8749	13.0	.8050	17.7	.7345
8.4	.8740	13.1	.8035	17.8	.7330
8.5	.8725	13.2	.8020	17.9	.7315
8.6	.8710	13.3	.8005	18.0	.7300
8.7	.8695	13.4	.7990	18.1	.7285
8.8	.8680	13.5	.7975	18.2	.7270
8.9	.8665	13.6	.7960	18.3	.7255
9.0	.8650	13.7	.7945	18.4	.7240
9.1	.8635	13.8	.7930	18.5	.7225
9.2	.8620	13.9	.7915	18.6	.7210
9.3	.8605	14.0	.7900	18.7	.7195
9.4	.8590	14.1	.7885	18.8	.7180
9.5	.8575	14.2	.7870	18.9	.7165
9.6	.8560	14.3	.7855	19.0	.7150
9.7	.8545	14.4	.7840	19.1	.7135
9.8	.8530	14.5	.7825	19.2	.7120
9.9	.8515	14.6	.7810	19.3	.7105
10.0	.8500	14.7	.7795	19.4	.7090
10.1	.8485	14.8	.7780	19.5	.7075
10.2	.8470	14.9	.7765	19.6	.7060
10.3	.8455	15.0	.7750	19.7	.7045
10.4	.8440	15.1	.7735	19.8	.7030
10.5	.8425	15.2	.7720	19.9	.7015
				20.0	.7000

Table B.15
Operating Characteristics of Annular Blowout Preventers

CAMERON

MODEL OR TYPE	B.O.P. SIZE (IN.)	WORKING PRESSURE (MAX.PSI)	VERT. BORE (IN.)	HYDRAULIC CONTROL (MAX.PSI)	GALS. TO CLOSE	GALS. TO OPEN	PACKOFF OPEN HOLE (MIN.PSI)
D	6	5,000	7 1/16	3,000	1.69	1.39	N.A.
D	7 1/16	10,000	7 1/16	3,000	2.94	2.55	N.A.
D	10	5,000	11	3,000	5.65	4.69	N.A.
D	11	10,000	11	3,000	10.15	9.06	N.A.
D	13 5/8	5,000	13 5/8	3,000	12.12	10.34	N.A.
D	13 5/8	10,000	13 5/8	3,000	18.10	16.15	N.A.

NL SHAFFER

MODEL OR TYPE	B.O.P. SIZE (IN.)	WORKING PRESSURE (MAX.PSI)	VERT. BORE (IN.)	HYDRAULIC CONTROL (MAX.PSI)	GALS. TO CLOSE	GALS. TO OPEN	PACKOFF OPEN HOLE (MIN.PSI)
Spherical BOP	6	3,000	7 1/16	1,500	4.57	3.21	
	6	5,000	7 1/16	1,500	4.57	3.21	
	7 1/16	10,000	7 1/16	1,500	17.11	13.95	
	8	3,000	9	1,500	7.23	5.03	
	8	5,000	9	1,500	11.05	8.72	
	10	3,000	11	1,500	11.00	6.78	
	10	5,000	11	1,500	18.67	14.59	VARIABLE
	11	10,000	11	1,500	30.58	24.67	
	12	3,000	13 5/8	1,500	23.50	14.67	
	13 5/8	5,000	13 5/8	1,500	23.58	17.41	
	13 5/8	10,000	13 5/8	1,500	51.24	42.68	
	16 3/4	5,000	16 3/4	1,500	33.26	25.61	
	18 3/4	5,000	18 3/4	1,500	48.16	37.61	
	20	2,000	21 1/4	1,500	32.59	16.92	
	21 1/4	5,000	21 1/4	1,500	61.37	47.76	

HYDRIL

MODEL OR TYPE	B.O.P. SIZE (IN.)	WORKING PRESSURE (MAX.PSI)	VERT. BORE (IN.)	HYDRAULIC CONTROL (MAX.PSI)	GALS. TO CLOSE	GALS. TO OPEN	PACKOFF OPEN HOLE (MIN.PSI)
GK	6	3,000	7 1/16	1,500	2.42	1.90	1,000
GK	6	5,000	7 1/16	1,500	3.28	2.81	1,000
GK	8	3,000	8 15/16	1,500	3.68	2.90	1,050
GK	8	5,000	8 15/16	1,500	5.81	4.93	1,150
GK	10	3,000	11	1,500	6.32	4.71	1,150
GK	10	5,000	11	1,500	8.34	6.78	1,150
GK	12	3,000	13 5/8	1,500	9.66	7.60	1,150
GK	13 5/8	5,000	13 5/8	1,500	15.28	12.04	1,150
GK	16	2,000	16 3/4	1,500	14.81	10.65	1,150
GK	16	3,000	16 3/4	1,500	17.86	13.43	1,150
GK	16 3/4	5,000	16 3/4	1,500	24.40	16.94	1,150
GK	18	2,000	17 7/8	1,500	17.93	12.28	1,110
GK	7 1/16	10,000	7 1/16	1,500	8.06	6.02	1,150
GK	9	10,000	9	1,500	13.52	10.17	1,150
GK	11	10,000	11	1,500	21.34	16.04	1,150
GK	13 5/8	10,000	13 5/8	1,500	29.35	20.96	1,150
GL	13 5/8	5,000	13 5/8	1,500	16.80	16.80	1,300
GL	16 3/4	5,000	16 3/4	1,500	28.73	28.73	1,300
GL	18 3/4	5,000	18 3/4	1,500	37.4	37.4	1,300
GL	21 1/4	5,000	21 1/4	1,500	49.3	49.3	1,300
MSP	6	2,000	7 1/16	1,500	2.42	1.68	1,000
MSP	8	2,000	8 15/16	1,500	3.89	2.51	1,050
MSP	10	2,000	11	1,500	6.32	4.45	1,150
MSP	20	2,000	20 3/4	1,500	26.39	16.09	1,100
MSP	20	2,000	21 1/4	1,500	26.39	16.09	1,100
MSP	29 1/2	500	29 1/2	1,500	60.0**	0	1,500

Table B.16
Operating Characteristics of
Ram Blowout Preventers

CAMERON

MODEL OR TYPE	B.O.P. SIZE (IN.)	WORKING PRESSURE (MAX.PSI)	VERT. BORE (IN.)	HYDRAULIC OPERATOR PSI	GALS. TO CLOSE	GALS. TO OPEN	CLOSE RATIO	OPEN RATIO
U	6	3,000	7 1/16	1500/5000	1.22	1.17	6.9:1	2.3:1
U	6	5,000	7 1/16	1500/5000	1.22	1.17	6.9:1	2.3:1
U Shear	6	5,000	7 1/16	1500/5000	1.54	1.48	6.9:1	2.3:1
U	7 1/16	10,000	7 1/16	1500/5000	1.22	1.17	6.9:1	2.3:1
U	7 1/16	15,000	7 1/16	1500/5000	1.22	1.17	6.9:1	2.3:1
U	10	3,000	11	1500/5000	3.31	3.16	7.3:1	2.5:1
U	10	5,000	11	1500/5000	3.31	3.16	7.3:1	2.5:1
U Shear	10	5,000	11	1500/5000	4.23	4.03	7.3:1	2.5:1
U	11	10,000	11	1500/5000	3.31	3.16	7.3:1	2.5:1
U Shear	11	10,000	11	1500/5000	4.23	4.03	7.3:1	2.5:1
U	11	15,000	11	1500/5000	5.54	5.42	9.9:1	2.2:1
U	12	3,000	13 5/8	1500/5000	5.54	5.20	7.0:1	2.3:1
U	13 5/8	5,000	13 5/8	1500/5000	5.54	5.20	7.0:1	2.3:1
U Shear	13 5/8	5,000	13 5/8	1500/5000	6.78	6.36	7.0:1	2.2:1
U	13 5/8	10,000	13 5/8	1500/5000	5.54	5.20	7.0:1	2.3:1
U Shear	13 5/8	10,000	13 5/8	1500/5000	6.78	6.36	7.0:1	2.2:1
U	13 5/8	15,000	13 5/8	1500/5000	11.70	11.28	6.6:1	8.9:1
U	16 3/4	3,000	16 3/4	1500/5000	10.16	9.45	6.8:1	2.3:1
U	16 3/4	5,000	16 3/4	1500/5000	10.16	9.45	6.8:1	2.3:1
U Shear	16 3/4	5,000	16 3/4	1500/5000	12.03	11.19	6.8:1	1.9:1
U	16 3/4	10,000	16 3/4	1500/5000	12.03	11.19	6.8:1	2.3:1
U	18 3/4	10,000	18 3/4	1500/5000	24.88	22.99	7.4:1	3.6:1
U	20	2,000	20 3/4	1500/5000	8.11	7.61	7.0:1	1.3:1
U	20	3,000	20 3/4	1500/5000	8.11	7.61	7.0:1	1.3:1
U Shear	20	3,000	20 3/4	1500/5000	9.35	8.77	7.0:1	1.2:1
U	21 1/4	2,000	21 1/4	1500/5000	8.11	7.61	7.0:1	1.3:1
U Shear	21 1/4	2,000	21 1/4	1500/5000	9.35	8.77	7.0:1	1.2:1
U	21 1/4	7,500	21 1/4	1500/5000	20.41	17.78	5.5:1	3.0:1
U Shear	21 1/4	7,500	21 1/4	1500/5000	23.19	20.20	5.5:1	2.3:1
U	21 1/4	10,000	21 1/4	1500/5000	26.54	24.14	7.2:1	4.1:1
U Shear	21 1/4	10,000	21 1/4	1500/5000	30.15	27.42	7.2:1	3.1:1
U	26	2,000	26 3/4	1500/5000	10.50	9.84	7.0:1	1.0:1
U	26	3,000	26 3/4	1500/5000	10.50	9.84	7.0:1	1.0:1
Type F with Type L Oper.	6	3,000	7 1/16	250/1,500	3.97	3.46	VARIABLE	4.9:1
	6	5,000	7 1/16	250/1,500	3.97	3.46		4.9:1
	7	10,000	7 1/16	250/1,500	3.97	3.46		4.9:1
	7	15,000	7 1/16	250/1,500	3.97	3.46		4.9:1
	8	3,000	9	250/1,500	6.85	6.19		3.44:1
	8	5,000	9	250/1,500	6.85	6.19		3.44:1
	10	3,000	11	250/1,500	6.85	6.19		3.44:1
	10	5,000	11	250/1,500	6.85	6.19		3.44:1
	11	10,000	11	250/1,500	6.85	6.19		3.44:1
	12	3,000	13 5/8	250/1,500	10:30	9.38		2.3:1
	14	5,000	13 5/8	250/1,500	10:30	9.38		2.3:1
	16	2,000	16 3/4	250/1,500	11.71	10.66		2.3:1
	16	3,000	16 3/4	250/1,500	11.71	10.66		2.3:1
	20	2,000	20 1/4	250/1,500	11.71	10.66		2.3:1
	20	3,000	20 1/4	250/1,500	11.71	10.66		2.3:1
Type F with Type H Opr.	6	3,000	7 1/16	1,000/5,000	0.52	1.05	VARIABLE	1.5:1
	6	5,000	7 1/16	1,000/5,000	0.52	1.05		1.5:1
	7	10,000	7 1/16	1,000/5,000	0.52	1.05		1.5:1
	7	15,000	7 1/16	1,000/5,000	0.52	1.05		1.5:1
	8	3,000	9	1,000/5,000	0.90	1.80		1:1
	8	5,000	9	1,000/5,000	0.90	1.80		1:1
	10	3,000	11	1,000/5,000	0.90	1.80		1:1
	10	5,000	11	1,000/5,000	0.90	1.80		1:1
	11	10,000	11	1,000/5,000	0.90	1.80		1:1
	12	3,000	13 5/8	1,000/5,000	1.52	2.70		2/3:1
	14	5,000	13 5/8	1,000/5,000	1.52	2.70		2/3:1
	16	2,000	16 3/4	1,000/5,000	1.73	3.08		2/3:1
	16	3,000	16 3/4	1,000/5,000	1.73	3.08		2/3:1
	20	2,000	20 1/4	1,000/5,000	1.73	3.08		2/3:1
	20	3,000	20 1/4	1,000/5,000	1.73	3.08		2/3:1

CAMERON (Continued)

MODEL OR TYPE	B.O.P. SIZE (IN.)	WORKING PRESSURE (MAX.PSI)	VERT. BORE (IN.)	HYDRAULIC OPERATOR PSI	GALS. TO CLOSE	GALS. TO OPEN	CLOSE RATIO	OPEN RATIO
U-Blind Ram with Shear Booster	13 5/8	5,000	13 5/8	1,500/2,500	11.6	10.9	14:1	2.3:1
	13 5/8	10,000	13 5/8	1,500/2,500	11.6	10.9	14:1	2.3:1
	16 3/4	3,000	16 3/4	1,500/2,500	10.8	11.7	9:1	1.4:1
	16 3/4	5,000	16 3/4	1,500/2,500	10.8	11.7	9:1	1.4:1
	20	2,000	20 3/4	1,500/2,500	16.8	15.7	14:1	1.2:1
	20	3,000	20 3/4	1,500/2,500	16.8	15.7	14:1	1.2:1
QRC	6	3,000	7 1/16	1,500/3,000	0.81	0.95	7.75:1	1.5:1
QRC	6	5,000	7 1/16	1,500/3,000	0.81	0.95	7.75:1	1.5:1
QRC	8	3,000	9	1,500/3,000	2.36	2.70	9.05:1	1.83:1
QRC	8	5,000	9	1,500/3,000	2.36	2.70	9.05:1	1.83:1
QRC	10	3,000	11	1,500/3,000	2.77	3.18	9.05:1	1.21:1
QRC	10	5,000	11	1,500/3,000	2.77	3.18	9.05:1	1.21:1
QRC	12	3,000	13 5/8	1,500/3,000	4.42	5.10	8.64:1	1.07:1
QRC	16	2,000	16 3/4	1,500/3,000	6.0	7.05	8.64:1	0.62:1
QRC	18	2,000	17 3/4	1,500/3,000	6.0	7.05	8.64:1	0.62:1
QRC	20	2,000	17 3/4	1,500/3,000	6.0	7.05	8.64:1	0.62:1
SS	6	3,000	7 1/16	1,500/3,000	0.8	0.7	3.8:1	1:1
SS	6	5,000	7 1/16	1,500/3,000	0.8	0.7	3.8:1	1:1
SS	8	3,000	9	1,500/3,000	1.5	1.3	3.9:1	1:1
SS	8	5,000	9	1,500/3,000	1.5	1.3	3.9:1	1:1
SS	10	3,000	11	1,500/3,000	1.5	1.3	3.9:1	1:1
SS	10	5,000	11	1,500/3,000	1.5	1.3	3.9:1	1:1
SS	12	3,000	13 5/8	1,500/3,000	2.9	2.5	3.7:1	1:1
SS	14	5,000	13 5/8	1,500/3,000	2.9	2.5	3.7:1	1:1
Type F with Type W$_x$ Opr.	6	3,000	7 1/16	500/1,500	1.5	2.3	VARIABLE	4.5:1
	6	5,000	7 1/16	500/1,500	1.5	2.3		4.5:1
	7	10,000	7 1/16	500/1,500	1.5	2.3		4.5:1
	7	15,000	7 1/16	500/1,500	1.5	2.3		4.5:1
	8	3,000	9	500/1,500	2.8	3.7		2.5:1
	8	5,000	9	500/1,500	2.8	3.7		2.5:1
	10	3,000	11	500/1,500	2.8	3.7		2.5:1
	10	5,000	11	500/1,500	2.8	3.7		2.5:1
	11	10,000	11	500/1,500	2.8	3.7		2.5:1
	12	3,000	13 5/8	500/1,500	4.1	5.3		2:1
	14	5,000	13 5/8	500/1,500	4.1	5.3		2:1
	16	2,000	16 3/4	500/1,500	5.0	6.0		2:1
	16	3,000	16 3/4	500/1,500	5.0	6.0		2:1
	20	2,000	20 1/4	500/1,500	5.0	6.0		2:1
	20	3,000	20 1/4	500/1,500	5.0	6.0		2:1
Type F with Type W Opr.	6	3,000	7 1/16	500/1,500	2.3	3.05	VARIABLE	4.5:1
	6	5,000	7 1/16	500/1,500	2.3	3.05		4.5:1
	7	10,000	7 1/16	500/1,500	2.3	3.05		4.5:1
	7	15,000	7 1/16	500/1,500	2.3	3.05		4.5:1
	8	3,000	9	500/1,500	3.7	4.6		2.5:1
	8	5,000	9	500/1,500	3.7	4.6		2.5:1
	10	3,000	11	500/1,500	3.7	4.6		2.5:1
	10	5,000	11	500/1,500	3.7	4.6		2.5:1
	11	10,000	11	500/1,500	3.7	4.6		2.5:1
	12	3,000	13 5/8	500/1,500	6.8	8.1		2:1
	14	5,000	13 5/8	500/1,500	6.8	8.1		2:1
	16	2,000	16 3/4	500/1,500	7.6	9.1		2:1
	16	3,000	16 3/4	500/1,500	7.6	9.1		2:1
	20	2,000	20 1/4	500/1,500	7.6	9.1		2:1
	20	3,000	20 1/4	500/1,500	7.6	9.1		2:1

Table B.16
Operating Characteristics of Ram Blowout Preventers
(continued)

HYDRIL

MODEL OR TYPE	B.O.P. SIZE (IN.)	WORKING PRESSURE (MAX.PSI)	VERT. BORE (IN.)	HYDRAULIC OPERATOR PSI	GALS. TO CLOSE	GALS. TO OPEN	CLOSE RATIO	OPEN RATIO
Manual Lock	6	3,000	7 1/16	750/3,000	1.2	1.3	5.32:1	N.A.
Automatic Lock	6	3,000	7 1/16	600/3,000	1.5	1.3	6.75:1	N.A.
Manual Lock	6	5,000	7 1/16	1,175/3,000	1.2	1.3	5.32:1	N.A.
Automatic Lock	6	5,000	7 1/16	950/3,000	1.5	1.3	6.75:1	N.A.
Manual Lock	10	3,000	11	550/3,000	3.3	3.2	6.00:1	N.A.
Automatic Lock	10	3,000	11	500/3,000	3.8	3.2	6.85:1	N.A.
Manual Lock	10	5,000	11	850/3,000	3.3	3.2	6.0:1	N.A.
Automatic Lock	10	5,000	11	750/3,000	3.8	3.2	6.85:1	N.A.
Manual Lock Pipe	11	10,000	11	1,050/3,000	11.8	11.8	10.2:1	N.A.
Manual Lock Shear	11	10,000	11	1,050/3,000	11.8	11.8	10.2:1	N.A.
Automatic Lock Pipe	11	10,000	11	1,050/3,000	12.9	11.8	10.56:1	N.A.
Automatic Lock Shear	11	10,000	11	1,050/3,000	12.9	11.8	10.56:1	N.A.
Manual Lock Pipe	12	2,000	13 5/8	700/3,000	5.4	4.9	4.75:1	N.A.
Manual Lock Shear	12	2,000	13 5/8	400/3,000	11.5	11.2	10.14:1	N.A.
Automatic Lock Pipe	12	2,000	13 5/8	700/3,000	5.9	4.9	5.2:1	N.A.
Automatic Lock Shear	12	2,000	13 5/8	400/3,000	12.0	11.2	10.56:1	N.A.
Manual Lock Pipe	13 5/8	5,000	13 5/8	1,050/3,000	5.4	4.9	4.75:1	N.A.
Manual Lock Shear	13 5/8	5,000	13 5/8	600/3,000	11.5	11.2	10.14:1	N.A.
Automatic Lock Pipe	13 5/8	5,000	13 5/8	1,050/3,000	5.9	4.9	5.2:1	N.A.
Automatic Lock Shear	13 5/8	5,000	13 5/8	600/3,000	12.0	11.2	10.56:1	N.A.
Manual Lock Pipe	13 5/8	10,000	13 5/8	1,050/3,000	11.8	11.8	10.2:1	N.A.
Manual Lock Shear	13 5/8	10,000	13 5/8	1,050/3,000	11.8	11.8	10.2:1	N.A.
Automatic Lock Pipe	13 5/8	10,000	13 5/8	1,050/3,000	12.9	11.8	10.56:1	N.A.
Automatic Lock Shear	13 5/8	10,000	13 5/8	1,050/3,000	12.9	11.8	10.56:1	N.A.
Manual Lock Pipe	16 3/4	10,000	16 3/4	1,050/3,000	15.0	14.1	10.2:1	N.A.
Manual Lock Shear	16 3/4	10,000	16 3/4	1,050/3,000	15.0	14.1	10.2:1	N.A.
Automatic Lock Pipe	16 3/4	10,000	16 3/4	1,050/3,000	15.6	14.1	10.56:1	N.A.
Automatic Lock Shear	16 3/4	10,000	16 3/4	1,050/3,000	15.6	14.1	10.56:1	N.A.
Manual Lock Pipe	18 3/4	10,000	18 3/4	1,050/3,000	16.4	15.6	10.2:1	N.A.
Manual Lock Shear	18 3/4	10,000	18 3/4	1,050/3,000	16.4	15.6	10.2:1	N.A.
Automatic Lock Pipe	18 3/4	10,000	18 3/4	1,050/3,000	17.1	15.6	10.56:1	N.A.
Automatic Lock Shear	18 3/4	10,000	18 3/4	1,050/3,000	17.1	15.6	10.56:1	N.A.

HYDRIL (Continued)

MODEL OR TYPE	B.O.P. SIZE (IN.)	WORKING PRESSURE (MAX.PSI)	VERT. BORE (IN.)	HYDRAULIC OPERATOR PSI	GALS. TO CLOSE	GALS. TO OPEN	CLOSE RATIO	OPEN RATIO
Manual Lock Pipe	20	2,000	21 1/4	500/3,000	8.1	7.2	4.74:1	N.A.
Manual Lock Shear	20	2,000	21 1/4	300/3,000	17.2	16.3	10.14:1	N.A.
Automatic Lock Pipe	20	2,000	21 1/4	1,050/3,000	8.9	7.2	5.2:1	N.A.
Automatic Lock Shear	20	2,000	21 1/4	1,050/3,000	18.0	16.3	10.56:1	N.A.
Manual Lock Pipe	20	3,000	20 3/4	500/3,000	8.1	7.2	4.74:1	N.A.
Manual Lock Shear	20	3,000	20 3/4	300/3,000	17.2	16.3	10.14:1	N.A.
Automatic Lock Pipe	20	3,000	20 3/4	1,050/3,000	8.9	7.2	5.2:1	N.A.
Automatic Lock Shear	20	3,000	20 3/4	1,050/3,000	18.0	16.3	10.56:1	N.A.

NL SHAFFER

MODEL OR TYPE	B.O.P. SIZE (IN.)	WORKING PRESSURE (MAX.PSI)	VERT. BORE (IN.)	HYDRAULIC OPERATOR PSI	GALS. TO CLOSE	GALS. TO OPEN	CLOSE RATIO	OPEN RATIO
LWS with Manual Lock	4 1/16	10,000	4 1/16	1,500/3,000	.59	.52	8.45:1	4.74:1
	6	5,000	7 1/16	1,500/3,000	1.19	.99	4.45:1	1.82:1
	7 1/16	10,000	7 1/16	1,500/3,000	6.35	5.89	10.63:1	19.40:1
	7 1/16	15,000	7 1/16	1,500/3,000	6.35	5.89	10.63:1	19.40:1
	8	5,000	9	1,500/3,000	2.58	2.27	5.57:1	3.00:1
	10	3,000	11	1,500/3,000	1.74	1.45	4.45:1	1.16:1
	10	5,000	11	1,500/3,000	2.98	2.62	5.57:1	2.09:1
	11	10,000	11	1,500/3,000	8.23	7.0	7.11:1	3.44:1
	12	3,000	13 5/8	1,500/3,000	4.35	5.30	8.16:1	1.74:1
	13 5/8	5,000	13 5/8	1,500/3,000	4.35	5.30	8.16:1	1.74:1
	13 5/8	10,000	13 5/8	1,500/3,000	11.56	10.52	10.85:1	3.48:1
	16 3/4	5,000	16 3/4	1,500/3,000	13.97	12.71	10.85:1	3.61:1
	20	2,000	21 1/4	1,500/3,000	5.07	4.46	5.57:1	.78:1
	20	3,000	21 1/4	1,500/3,000	5.07	4.46	5.57:1	.78:1
LWP Type	6	3,000	7 1/16	1,500/3,000	.55	.51	4:1	2.5:1
	8	3,000	9	1,500/3,000	.77	.68	4:1	1.81:1
SL and LWS Poslock	11*	10,000	11	1,500/3,000	8.23	7.0	7.11:1	3.44:1
	13 5/8	5,000	13 5/8	1,500/3,000	4.35	5.30	8.16:1	1.74:1
	13 5/8*	5,000	13 5/8	1,500/3,000	11.56	10.52	10.85:1	3.48:1
	13 5/8*	10,000	13 5/8	1,500/3,000	11.56	10.52	10.85:1	3.48:1
	16 3/4	5,000	16 3/4	1,500/3,000	13.97	12.60	11.85:1	2.45:1
	16 3/4*	10,000	16 3/4	1,500/3,000	14.47	12.50	7.11:1	2.06:1
	18 3/4*	10,000	18 3/4	1,500/3,000	15.30	13.21	7.11:1	1.83:1
	20	2,000	21 1/4	1,500/3,000	7.80	6.86	8.16:1	1.15:1
	20*	2,000	21 1/4	1,500/3,000	16.88	15.35	10.85:1	2.52:1
	20	3,000	21 1/4	1,500/3,000	7.80	6.86	8.16:1	1.15:1
	20*	3,000	21 1/4	1,500/3,000	16.88	15.35	10.85:1	2.52:1
	21 1/4*	10,000	21 1/4	1,500/3,000	16.05	13.86	7.11:1	1.63:1
Types B & E	6	3,000	7 1/16	1,500/3,000	2.75	2.3	6:1	2.57:1
	6	5,000	7 1/16	1,500/3,000	2.75	2.3	6:1	2.57:1
	8	3,000	9	1,500/3,000	2.75	2.3	6:1	1.89:1
	8	5,000	9	1,500/3,000	2.75	2.3	6:1	1.89:1
	10	3,000	11	1,500/3,000	3.25	2.7	6:1	1.51:1
	10	5,000	11	1,500/3,000	3.25	2.7	6:1	1.35:1
	12	3,000	13 5/8	1,500/3,000	3.55	2.9	6:1	1.14:1
	14	5,000	13 5/8	1,500/3,000	3.55	2.9	6:1	1.14:1
	16	2,000	15 1/2	1,500/3,000	3.65	3.0	6:1	1.05:1

Table B.17
Drill Pipe Collapse Pressure
New Drill Pipe

OD size, in.	Nominal Weight, lb/ft	Collapse Pressure, psi			
		E-75	X-95	G-105	S-135
2⅜	4.85	11,040	13,980	15,460	19,070
	6.65	15,600	19,760	21,840	28,080
2⅞	6.85	10,470	12,930	14,010	17,060
	10.40	16,510	20,910	23,110	29,720
3½	9.50	10,040	12,060	13,050	15,780
	13.30	14,110	17,880	19,760	25,400
	15.50	16,770	21,250	23,480	30,190
4	11.85	8,410	9,960	10,700	12,650
	14.00	11,350	14,380	15,900	20,170
	15.70	12,900	16,340	18,050	23,210
4½	13.75	7,200	8,400	8,950	10,310
	16.60	10,390	12,750	13,820	16,800
	20.00	12,960	16,420	18,150	23,330
	22.82	14,810	18,770	20,740	26,670
5	16.25	6,970	8,090	8,610	9,860
	19.50	10,000	12,010	12,990	15,700
	25.60	13,500	17,100	18,900	24,300
5½	19.20	6,070	6,930	7,300	8,120
	21.90	8,440	10,000	10,740	12,710
	24.70	10,460	12,920	14,000	17,050
6⅝	25.20	4,810	5,300	5,480	6,040

Table B.18
Drill Pipe Collapse Pressure
Premium Used Drill Pipe

OD size, in.	Nominal Weight, lb/ft	Collapse Pressure, psi			
		E-75	X-95	G-105	S-135
2⅜	4.85	8,550	10,150	10,900	12,920
	6.65	13,380	16,950	18,730	24,080
2⅞	6.85	7,670	9,000	9,620	11,210
	10.40	14,220	18,020	19,910	25,600
3½	9.50	7,100	8,270	8,800	10,102
	13.30	12,020	15,220	16,820	21,630
	15.50	14,470	18,330	20,260	26,050
4	11.85	5,730	6,490	6,820	7,470
	14.00	9,040	10,780	11,610	13,870
	15.70	10,910	13,820	15,180	18,630
4½	13.75	4,710	5,170	5,340	5,910
	16.60	7,550	8,850	9,460	10,990
	20.00	10,980	13,900	15,340	18,840
	22.82	12,660	16,030	17,120	22,780
5	16.25	4,510	4,920	5,060	5,670
	19.50	7,070	8,230	8,760	10,050
	25.60	11,460	14,510	16,040	20,540
5½	19.20	3,760	4,140	4,340	4,720
	21.90	5,760	6,530	6,860	7,520
	24.70	7,670	9,000	9,620	11,200
6⅝	25.20	2,930	3,250	3,350	3,430

Table B.19
Drill Pipe Collapse Pressure
Class 2 Used Drill Pipe

OD size, in.	Nominal Weight, lb/ft	Collapse Pressure, psi			
		E-75	X-95	G-105	S-135
2⅜	4.85	6,020	6,870	7,240	8,030
	6.65	11,480	14,540	16,080	20,630
2⅞	6.85	5,270	5,900	6,150	6,610
	10.40	12,260	15,520	17,160	22,060
3½	9.50	4,790	5,270	5,450	6,010
	13.30	10,250	12,420	13,450	16,310
	15.50	12,480	15,810	17,480	22,470
4	11.85	3,620	4,020	4,210	4,550
	14.00	6,440	7,410	7,850	8,840
	15.70	8,560	10,150	10,910	12,930
4½	13.75	2,960	3,290	3,400	3,480
	16.60	5,170	5,770	6,010	6,490
	20.00	8,660	10,280	11,050	13,120
	22.82	10,830	13,710	14,950	18,320
5	16.25	2,850	3,150	3,240	3,300
	19.50	4,760	5,230	5,410	5,970
	25.60	9,420	11,270	12,160	14,590
5½	19.20	2,440	2,610	2,650	2,650
	21.90	3,640	4,040	4,230	4,580
	24.70	5,260	5,890	6,140	6,610
6⅝	25.20	1,870	1,900	1,900	1,900

Table B.20
Drill Pipe Collapse Pressure
Class 3 Used Drill Pipe

OD size, in.	Nominal Weight, lb/ft	Collapse Pressure, psi			
		E-75	X-95	G-105	S-135
2⅜	4.85	4,260	4,590	4,810	5,350
	6.65	10,030	12,050	13,040	15,760
2⅞	6.85	3,600	4,010	4,190	4,530
	10.40	10,800	13,680	14,880	18,230
3½	9.50	3,230	3,650	3,790	4,000
	13.30	8,040	9,480	10,160	11,930
	15.50	11,010	13,950	15,420	18,960
4	11.85	2,570	2,790	2,840	2,850
	14.00	4,630	5,070	5,230	5,810
	15.70	6,490	7,480	7,920	8,940
4½	13.75	2,090	2,170	2,170	2,170
	16.60	3,520	3,930	4,110	4,420
	20.00	6,580	7,590	8,040	9,100
	22.82	8,960	10,680	11,500	13,710
5	16.25	1,990	2,050	2,050	2,050
	19.50	3,210	3,630	3,770	3,960
	25.60	7,250	8,460	9,020	10,410
5½	19.20	1,640	1,640	1,640	1,640
	21.90	2,580	2,810	2,860	2,870
	24.70	3,600	4,000	4,190	4,520
6⅝	25.20	1,170	1,170	1,170	1,170

Equations

The equations that appear in the text are listed alphabetically by type below. The equation number assigned in the text is given in parentheses.

ACCUMULATOR USABLE FLUID

$$V_u = V_s (P_p \div A_p)(A_p \div R_p - 1) \qquad \text{(Eq. 69)}$$

where

V_u = usable volume of fluid, gal
V_s = volume of fluid in the system, gal
P_p = precharge pressure, psi
A_p = total accumulator pressure, psi
R_p = required pressure to close preventer, psi

ANNULAR VOLUME

$$AV = (ID_h^2 - OD_{dp}^2) \div 1{,}029.4 \qquad \text{(Eq. 39)}$$

where

AV = annular volume, bbl/ft
ID_h = ID of open or cased hole, in.
OD_{dp} = OD of drill pipe or drill collars, in.

BIT-TO-SURFACE STROKES

$$BSS = [(D_{h2} - D_{dc}^2) \div 1{,}029.4 \times (L_{dc} \div PD)]$$
$$+ \qquad \text{(Eq. 44)}$$
$$[(D_h^2 - D_{dp}^2) \div 1{,}029.4 \times (L_{dp} \div PD)]$$

where

BSS = number of bit-to-surface pump strokes
D_h = diameter of hole, in.
D_{dc} = OD of drill collars, in.
L_{dc} = length of drill collars, ft
PD = pump displacement, bbl/s
D_{dp} = OD of drill pipe, in.
L_{dp} = length of drill pipe, ft

BUOYANCY FACTOR

$$BF \ = \ (65.5 - MW) \div 65.5 \tag{Eq. 61}$$

where

BF = buoyancy factor, dimensionless

MW = mud weight, ppg.

CAPACITY

$$C_h \ = \ ID^2 \div 1{,}029.4 \tag{Eq. 40}$$

where

C = capacity of drill pipe, drill collars, or cased or open hole, bbl/ft

ID = ID of drill pipe, drill collars, or cased or open hole, in.

CHOKE-AND-KILL-LINE

$$V_{ckl} \ = \ ID^2 \div 1{,}029.4 \tag{Eq. 43}$$

where

V_{ckl} = choke line or kill line volume, bbl/ft

ID^2 = choke line or kill line ID, in.

$$S_{ckl} \ = \ ID^2 \div 1{,}029.4 \times (L_{ckl} \div PD) \tag{Eq. 46}$$

where

S_{ckl} = number of strokes to displace the choke line or the kill line

ID^2 = inside diameter of the choke line or the kill line, in.

L_{ckl} = length of the choke line or the kill line, ft

PD = pump displacement, bbl/s

$$T_{ckl} \ = \ S_{ckl} \div SPM \tag{Eq. 47}$$

where

T_{ckl} = time to displace fluid from the choke line or the kill line, min

S_{ckl} = number of strokes to displace the choke line or the kill line

SPM = number of pump stokes, spm.

CIRCULATING PRESSURE

$$NCP \ = \ OCP \times NMW \div OMW \tag{Eq. 22}$$

where

NCP = new circulating pressure, psi

OCP = old circulating pressure, psi

NMW = new mud weight, ppg

OMW = old mud weight, ppg.

CLOSING PRESSURE

$$CP_i = [(0.052 \times MW \times D_w) - (0.45 \times D_w)] \div P \qquad \text{(Eq. 68)}$$

where

CP_i = closing pressure increase, psi

MW = mud weight, ppg

D_w = water depth, ft

P = a ratio for the preventer used.

DENSITY OF INFLUX

$$SICP = FP - HP - P_f \qquad \text{(Eq. 20)}$$

where

$SICP$ = shut-in casing pressure, psi

FP = formation pressure, psi

HP = hydrostatic pressure from top of kick to surface, psi

P_f = pressure of the kick fluid, psi.

$$D_i = MW - [(SICP - SIDDP) \div (L \times 0.052)] \qquad \text{(Eq. 21)}$$

where

D_i = density of influx, ppg

MW = original mud weight, ppg

$SICP$ = shut-in casing pressure, psi

$SIDPP$ = shut-in drill pipe pressure, psi

L = influx height in the annulus, ft.

DROP IN BOTTOMHOLE PRESSURE

$$BHP = SICP + (F_b \times P_{bbl}) \qquad \text{(Eq. 59)}$$

where

BHP = amount bottomhole pressure is reduced after fluid is bled, psi

$SICP$ = amount of pressure drop in SICP after fluid is bled, psi

F_b = amount of fluid bled, bbl

P_{bbl} = pressure exerted by mud, psi/bbl.

EQUIVALENT CIRCULATING DENSITY

$$ECD = MW + (APL \div 0.052 \div TVD) \qquad \text{(Eq. 5)}$$

where

ECD = equivalent circulating density, ppg

MW = mud weight, ppg

APL = annular pressure loss, psi

TVD = true vertical depth, ft.

EQUIVALENT MUD WEIGHT

$$EMW_{cs} = (P_{cl} \div 0.052 \div D_{cs}) + MW \qquad \text{(Eq. 67)}$$

where

EMW_{cs} = equivalent mud weight at casing shoe, ppg

P_{cl} = choke-line friction pressure, psi

D_{cs} = depth of casing shoe, ft

MW = mud weight in kill and choke lines

FINAL CIRCULATING PRESSURE

$$FCP = KRP \times (KWM \div OMW) \qquad \text{(Eq. 48)}$$

where

FCP = final circulating pressure, psi

KRP = kill-rate pressure, psi

OMW = old mud weight, ppg

KWM = kill-weight mud, ppg.

FLOW RATE

$$Q = 0.007 \times md \times \Delta p \times L \div \mu \times 1n\ (R_e \div R_w)\ 1,440 \qquad \text{(Eq. 6)}$$

where

Q = flow rate, bbl/min

md = permeability, md

Δp = pressure differential, psi

L = length of section open to wellbore, ft

μ = viscosity of intruding gas, cp

R_e = radius of drainage, ft

R_w = radius of wellbore, ft.

FLUID HEIGHT GAIN

$$h = L \times (C_{dp} + D_{dp}) \div AV \qquad \text{(Eq. 64)}$$

where

h = height gain, ft

L = length of pipe stripped, ft

C_{dp} = drill pipe or drill collar capacity, bbl/ft

D_{dp} = drill pipe or drill collar displacement, bbl/ft

AV = annular volume, bbl/ft.

FORMATION PRESSURE

$$FP = HP + SIDPP \qquad \text{(Eq. 23)}$$

where

FP = formation pressure, psi

HP = hydrostatic pressure, psi

$SIDPP$ = shut-in drill pipe pressure, psi.

GAS MIGRATION

$$R_{gm} = \Delta SICP \div MG \qquad \text{(Eq. 31)}$$

where

R_{gm} = rate of gas migration, ft/h

$\Delta SICP$ = change in SICP after 1 h, psi

MG = mud gradient, psi/ft.

GAS PRESSURE

$$P_1 V_1 \div T_1 = P_2 V_2 \div T_2 \qquad \text{(Eq. 28)}$$

where

P_1 = formation pressure, psi

P_2 = hydrostatic pressure at any depth in the wellbore, psi

V_1 = original pit gain, bbl

V_2 = gas volume at surface, bbl

T_1 = temperature of formation fluid, ° R

T_2 = temperature at the surface, ° R.

$$P_1 V_1 \div T_1 Z_1 = P_2 V_2 \div T_2 Z_2 \qquad \text{(Eq. 29)}$$

where

Z_1 = compressibility factor under pressure in formation, dimensionless

Z_2 = compressibility factor at the surface, dimensionless.

$$P_1 V_1 = P_2 V_2 \qquad \text{(Eq. 30)}$$

where

P_1 = formation pressure, psi

P_2 = hydrostatic pressure, psi

V_1 = original pit gain, bbl

V_2 = gas volume at surface, bbl.

HOLE CAPACITY

$$C_h = ID^2 \div 1,029.4 \qquad \text{(Eq. 39)}$$

where

C_h = hole capacity, bbl/ft

ID = inside diameter of the hole (either open or cased), in.

HYDROSTATIC PRESSURE

$$HP = C \times MW \times TVD \qquad \text{(Eq. 1)}$$

where

HP = hydrostatic pressure, psi

C = a constant (value depends on unit used to express mud weight)

MW = mud weight, ppg or other units

TVD = true vertical depth, ft.

$$HP = MG \times TVD \qquad \text{(Eq. 3)}$$

where

HP = hydrostatic pressure, psi

MG = mud gradient, psi/ft

TVD = true vertical depth, ft.

$$HP_1 = 0.052 \times EMW \times TVD_{cs} \qquad \text{(Eq. 52)}$$

$$HP_2 = 0.052 \times MWH \times TVD_{cs} \qquad \text{(Eq. 53)}$$

$$STP = HP_1 - HP_2 \qquad \text{(Eq. 54)}$$

where

HP_1 = hydrostatic pressure, psi

EMW = equivalent mud weight, ppg

TVD_{cs} = true vertical depth of casing shoe, ft

HP_2 = hydrostatic pressure, psi

MWH = mud weight in hole, ppg

STP = surface test pressure, psi.

$$P_{bbl} = (MW \times 0.052) \div AV \qquad \text{(Eq. 58)}$$

where

P_{bbl} = psi/bbl of mud

MW = mud weight, ppg

AV = annular capacity, bbl/ft.

INITIAL CIRCULATING PRESSURE

$$ICP = KRP + SIDPP \qquad \text{(Eq. 33)}$$

where

ICP = initial circulating pressure, psi

KRP = kill-rate pressure, psi

$SIDPP$ = shut-in drill pipe pressure, psi.

MAXIMUM PIT GAIN

$$MPG_{gk} = 4\sqrt{(P \times V \times C) \div W} \qquad \text{(Eq. 27)}$$

where

MPG_{gk} = maximum pit gain from a gas

P = formation pressure, psi

V = original pit gain, bbl

C = annulus capacity, bbl/ft

W = mud weight to kill well, ppg.

MAXIMUM SURFACE PRESSURE

$$MSP_{gk} = 0.2 \sqrt{(P \times V \times W) \div C}$$ (Eq. 26)

where

MSP_{gk} = maximum surface pressure from a gas kick, psi
P = formation pressure, psi
V = original pit gain, bbl
W = mud weight to kill well, ppg
C = annulus capacity at surface, bbl/ft.

MUD GRADIENT

$$MG = MW \times C$$ (Eq. 2)

where

MG = mud gradient, psi/ft of depth
MW = mud weight, ppg
C = a constant (value depends on unit used to express mud weight).

MUD TANK VOLUME

$$V_{mt} = L \times W \times H \div 5.614$$ (Eq. 56)

where

V_{mt} = volume of mud tank, bbl
L = length of mud tank, ft
W = width of mud tank, ft
H = height of mud tank, ft.

MUD VOLUME INCREASE

$$V_i = N \div 14.9$$ (Eq. 57)

where

V_i = volume of increase, bbl
N = number of sacks of barite added to the system.

MUD WEIGHT

$$MW = MG \div 0.052$$ (Eq. 4)

where

MW = mud weight, ppg
MG = mud gradient, psi/ft.

$$MW = BHP_i \div 0.052 \div TVD$$ (Eq. 13)

where

MW = mud weight, ppg
BHP_i = desired increase in bottomhole pressure, psi
TVD = true vertical depth, ft.

$$TM \;=\; YP \div 11.7\,(D_h - D_p) \tag{Eq. 14}$$

where

TM = trip margin, ppg

YP = yield point, lb/100 ft^2

D_h = diameter of the hole, in.

D_p = outside diameter (OD) of pipe.

$$MWI \;=\; SIDPP \div TVD \div 0.052 \tag{Eq. 24}$$

where

MWI = mud-weight increase, ppg

$SIDPP$ = shut-in drill pipe pressure, psi

TVD = true vertical depth, ft.

$$KWM \;=\; OMW + MWI \tag{Eq. 25}$$

where

KWM = kill-weight mud, ppg

OMW = old mud weight, ppg

MWI = mud-weight increase, ppg.

$$sx \;=\; 1{,}470\,(W_2 - W_1) \div (35 - W_2) \tag{Eq. 55}$$

where

sx = sacks of barite to add per 100 bbl of mud

W_2 = desired mud weight, ppg

W_1 = initial mud weight, ppg.

PENETRATION RATE

$$R \div N \;=\; a\,(W^d \div D) \tag{Eq. 17}$$

where

R = penetration rate, ft/hr

N = rotary speed, rpm

a = a constant, dimensionless

W = weight on bit, lb

d = exponent in general drilling equation, dimensionless

D = hole diameter or bit size, in.

$$d \;=\; \log\,(R \div 60N) \div \log\,(12W \div 1{,}000D) \tag{Eq. 18}$$

where

d = d exponent, dimensionless

R = penetration rate, ft/hr

N = rotary speed, rpm

W = weight on bit, 1,000 lb

D = bit size, in.

$$d_c = d\,(MW_1 \div MW_2) \qquad \text{(Eq. 19)}$$

where

d_c = corrected d exponent

MW_1 = normal mud weight, 9.0 ppg

MW_2 = actual mud weight used, ppg.

PRESSURE LOSS

$$PL_w = MG \times (DP_c + DP_d) \div C_c - (DP_c + DP_d) \qquad \text{(Eq. 7)}$$

where

PL_w = pressure lost for each foot of pipe pulled wet, psi

MG = mud gradient, psi/ft

DP_c = capacity of drill pipe, bbl/ft

DP_d = displacement of drill pipe, bbl/ft

C_c = capacity of casing or hole, bbl/ft.

$$PL_d = (MG \times DP_d) \div (C_c - DP_d) \qquad \text{(Eq. 8)}$$

where

PL_d = pressure lost for each ft of pipe pulled dry, psi

MG = mud gradient, psi/ft

DP_d = displacement of drill pipe, bbl/ft

C_c = capacity of casing or hole, bbl/ft.

$$V_{dc1} = OD_{dc1}^2 \div 1,029.4 \qquad \text{(Eq. 9)}$$

$$V_{dp1} = OD_{dp1}^2 \div 1,029.4 \qquad \text{(Eq. 10)}$$

where

V_{dc1} = volume displaced by drill collars with check valve installed, bbl/ft

OD_{dc1}^2 = outside diameter of drill collars, in.

V_{dp1} = volume displaced by drill pipe with check valve installed, bbl/ft

$$V_{dc2} = OD_{dc2}^2 - ID_{dc2}^2 \div 1,029.4 \qquad \text{(Eq. 11)}$$

$$V_{dp2} = OD_{dp2}^2 - ID_{dp2}^2 \div 1,029.4 \qquad \text{(Eq. 12)}$$

where

V_{dc2} = volume of drill collars without check valve installed, bbl/ft

OD_{dc2}^2 = outside diameter drill collars, in.

ID_{dc2}^2 = inside diameter of drill collars, in.

V_{dp2} = volume of drill pipe without check valve installed, bbl/ft

OD_{dp2}^2 = outside diameter of drill pipe, in.

ID_{dp2}^2 = inside diameter of drill pipe, in.

PRESSURE TO BREAK GEL STRENGTH

$$P_{gs} = (\gamma \div 300 \div d)\, L \qquad \text{(Eq. 15)}$$

where

P_{gs} = pressure required to break gel strength, psi
γ = 10-min gel strength of drilling fluid, lb/100 ft^2
d = inside diameter (ID) of drill pipe, in.
L = length of drill pipe, ft.

$$P = L\,(\gamma \div 300d) \qquad \text{(Eq. 49)}$$

where

P = pressure required to overcome mud's gel strength, psi
L = length of drill stem, ft
γ = 10-min gel strength of mud, lb/100 ft^2
d = ID of drill pipe, in.

$$P = L\,[\gamma \div 300\,(D_h - D_p)] \qquad \text{(Eq. 50)}$$

where

P = pressure required to overcome mud's gel strength, psi
L = length of drill stem, ft
γ = 10-min gel strength of mud, lb/100 ft^2
Dh = diameter of hole, in.
Dp = OD of drill pipe, in.

PUMP DISPLACEMENT

$$PD = 0.000162 \times L\,[(2 \times D^2) - d^2]\,0.90 \qquad \text{(Eq. 37)}$$

where

PD = pump displacement, bbl/s
L = stroke length, in.
D = piston diameter, in.
d = rod diameter, in.

$$PD = 0.000243 D^2 L \qquad \text{(Eq. 38)}$$

where

PD = pump displacement, bbl/s
D = piston diameter, in.
L = stroke length, in.

PUMP SPEED

$$P_2 = P_1 \times SPM_2{}^2 \div SPM_1{}^2 \qquad \text{(Eq. 32)}$$

where

P_1 = original pump pressure at SPM_1, psi
P_2 = reduced or increased pump pressure at SPM_2, psi
SPM_1 = original pump rate, spm
SPM_2 = reduced or increased pump rate, spm .

REDUCTION IN BOTTOMHOLE PRESSURE

$$P \ = \ (MG \div AV)\, PVI \qquad \text{(Eq. 16)}$$

where

P = reduction in bottomhole pressure, psi

MG = mud gradient, psi/ft

AV = annular volume, bbl/ft

PVI = pit volume increase, bbl

RISER VOLUME

$$V_r \ = \ (ID_r^2 - OD_{dp}^2) \div 1{,}029.4 \qquad \text{(Eq. 41)}$$

where

V_r = riser volume, bbl/ft

ID_r = ID of riser, in.

OD_{dp} = OD of drill pipe or drill collars, in.

$$V_r \ = \ ID^2 \div 1{,}029.4 \qquad \text{(Eq. 42)}$$

where

V_r = riser volume, bbl/ft

ID^2 = riser ID, in.

SHUT-IN CASING PRESSURE

$$SICP \ = \ FP - HP - P_f \qquad \text{(Eq. 20)}$$

or

$$SICP \ = \ (\text{mud}_{\text{grad}})\, \text{Influx}_{\text{grad}} - HT_{\text{influx}} + SIDPP$$

where

$SICP$ = shut-in casing pressure, psi

FP = formation pressure, psi

HP = hydrostatic pressure from top of kick to surface, psi

P_f = pressure of the kick fluid, psi

HT = height.

$$\Delta SICP \ = \ PV_c \times P_{bbl} + SICP_i \qquad \text{(Eq. 63)}$$

where

$\Delta SICP$ = new casing pressure, psi

PV_c = pit-volume change, bbl

P_{bbl} = pressure per bbl of mud, psi

$SICP_i$ = initial shut-in casing pressure, psi.

STORED VOLUME

$$V_s = V_u \div (P_p \div R_p) - (P_p \div A_p)$$ (Eq. 70)

where

V_s = stored fluid volume. gal

V_u = usable fluid volume, gal

P_p = precharge pressure, psi

R_p = required closing pressure, psi

A_p = total accumulator pressure, psi

SURFACE TEST PRESSURE

$$STP = (EMW - MWH)\, 0.052 \times TVD_{cs}$$ (Eq. 51)

where

STP = surface test pressure, psi

EMW = equivalent mud weight, ppg

MWH = mud weight in hole, ppg

TVD_{cs} = true vertical depth of casing shoe, ft.

SURFACE-TO-BIT STROKES

$$SBS_{dp} = C_{dp} \times L_{dp} \div PD$$ (Eq. 34)

where

SBS_{dp} = number of surface-to-bit strokes to displace mud in drill pipe

C_{dp} = capacity of drill pipe, bbl/ft

L_{dp} = length of drill pipe, ft

PD = pump displacement, bbl/s.

$$SBS_{dc} = C_{dc} \times L_{dc} \div PD$$ (Eq. 35)

where

SBS_{dc} = number of surface-to-bit strokes to displace mud in drill collars

C_{dc} = capacity of drill collars, bbl/ft

L_{dc} = length of drill collars, ft

PD = pump displacement, bbl/s.

SURFACE-TO-BIT TIME

$$SBT = SBS \div SPM$$ (Eq. 36)

where

SBT = surface-to-bit time, min

SBS = number of surface-to-bit strokes

SPM = pump rate, spm.

$$BST = BSS \div SPM \qquad \text{(Eq. 45)}$$

where

  BST = bit-to-surface time, min

  BSS = number of bit-to-surface strokes

  SPM = number of pump strokes, spm.

TRIP MARGIN

$$TM = YP \div 11.7\,(D_h - D_p) \qquad \text{(Eq. 14)}$$

where

  TM = trip margin, ppg

  YP = yield point, lb/100 ft^2

  D_h = diameter of the hole, in.

  D_p = outside diameter (OD) of pipe, in.

WATER CUSHION

$$WC_{ft} = P_d \div F_g \qquad \text{(Eq. 66)}$$

where

  WC_{ft} = water cushion, ft

  P_d = pressure differential, psi

  F_g = fluid gradient, psi/ft.

WELLBORE FORCE

$$WBF = (OD_{tj})^2 \times 0.7854 \times SICP + F \qquad \text{(Eq. 60)}$$

where

  WBF = wellbore force, lb

  OD_{tj} = outside diameter of tool joint, in.

  $SICP$ = shut-in casing pressure, psi

  F = friction factor, 1,000 lb.

$$L_{dp} = WBF \div (W_{dp} \times BF) \qquad \text{(Eq. 62)}$$

where

  L_{dp} = length of drill pipe string, ft

  WBF = wellbore force, lb

  W_{dp} = weight of drill pipe, pounds per foot (ppf)

  BF = buoyancy factor, dimensionless.

$$WBF = (OD_{dp})^2 \times 0.7854 \times SICP + F \qquad \text{(Eq. 65)}$$

where

  WBF = wellbore force

  OD_{dp} = OD of drill pipe, in.

  $SICP$ = shut-in casing pressure, psi

  F = friction factor, 1,000 lb.

Cross-Reference Index to Subpart O 30 CFR, Subparts D, G, C, and O

An important factor in ensuring that Outer Continental Shelf (OCS) oil and gas operations are carried out in a manner that emphasizes safe operations and minimizes the risk of environmental damage is the employment of qualified personnel to perform such operations. To ensure that lessees' or contractors' key operating personnel are properly trained, the Minerals Management Service (MMS) of the Department of the Interior has issued 30 Code of Federal Regulations (CFR), Part 250, Subpart O, Oil and Gas and Sulphur Operations in the Outer Continental Shelf Training.

Pursuant to the MMS training requirements and other federal regulations relevant to well control, this appendix has been prepared. Included here as a convenient reference for training students in well control are a cross-reference index to PETEX training material and 30 CFR, Subparts D, G, C, and O.

For more information, contact the U.S. Department of the Interior, Minerals Management Service, Mail Stop 4760, 381 Elden Street, Herndon, VA 22070-4817; (703) 787-1600.

U.S. Department of the Interior
Minerals Management Service
Effective July 1, 1995

CROSS-REFERENCE RELATING ELEMENTS IN *PRACTICAL WELL CONTROL* TO REQUIREMENTS IN 30 CFR, PART 250, SUBPART O

The following Subpart O standards are covered by the PETEX training manual *Practical Well Control.*

§250.212   **Drilling Well-Control Training.**

  (c) *Basic well-control course for drilling supervisors.*

    (2) Candidates shall receive instructions on the care, handling, and characteristics of drilling and completion fluids including the following:

      (i) Density
        Practical Well Control, pp. 1-3, 1-4, 2-2, 2-4 – 2-5, 8-2, 8-4 – 8-5, B-13 – B-14

      (ii) Viscosity
        Practical Well Control, pp. 2-4 – 2-5, 2-14

(iii) Fluid loss

Practical Well Control, pp. 2-13 – 2-14

(iv) Salinity

Practical Well Control, p. 2-10

(v) Gas cutting

Practical Well Control, p. 2-11

(vi) Procedure for increasing density

Practical Well Control, pp. 3-7, 6-12 – 6-14

(3) Candidates shall receive instruction on the major causes of an uncontrolled flow from a well including the following:

(i) Failure to keep the hole full

Practical Well Control, pp. 2-5 – 2-7

(ii) Swabbing effect of pulling the pipe

Practical Well Control, pp. 2-7 – 2-8

(iii) Loss of circulation

Practical Well Control, pp. 2-9, 7-14

(iv) Insufficient density of drilling fluid

Practical Well Control, p. 2-4

(v) Abnormally pressured formations

Practical Well Control, p. 2-9

(vi) Effect of too rapid lowering of pipe in the hole

Practical Well Control, p. 2-5

(4) Candidates shall receive instructions on the importance of measuring the volume of fluid required to fill the hole during the trips and methods for measuring and recording hole-fill volumes. These instructions shall include the importance of filling the hole as it relates to shallow-gas conditions.

Practical Well Control, pp. 2-2 – 2-4, 9-1

(5) Candidates shall receive instruction on the warning signals that indicate a kick is occurring or about to occur and on conditions that can lead to a kick including the following:

(i) Gain in pit volume

Practical Well Control, p. 2-13

(ii) Increase in return fluid-flow rate

Practical Well Control, p. 2-12

(iii) Hole not taking proper amount of fluid during trips

Practical Well Control, pp. 2-2 – 2-6

(iv) Drilling rate change

Practical Well Control, pp. 2-7, 2-12 – 2-13

(v) Decrease in circulating pressure or increase in pump strokes

Practical Well Control, p. 2-12

(vi) Trip, connection, and background gas change

Practical Well Control, p. 2-15

(vii) Gas-cut mud

Practical Well Control, pp. 2-10, 2-16 – 2-17

(viii) Water-cut mud or chloride concentration change

Practical Well Control, pp. 2-10, 2-15 – 2-16

(ix) Well flowing with pump shut down

Practical Well Control, p. 2-10

(6) Candidates shall receive instructions on the correct procedures for shutting in a well for well-control purposes including use of the BOP, choke manifold, and/or diverter system for well control. These instructions shall include the sequential steps to be followed.
Practical Well Control, pp. 3-1 – 3-5, 9-2 – 9-4

(7) Candidates shall receive instructions on one of the following constant bottomhole pressure methods of well control, including those conditions that may be unique to either a surface or a subsea stack:

(i) Driller's method
Practical Well Control, pp. 6-1 – 6-4

(ii) Wait-and-weight method
Practical Well Control, pp. 6-4 – 6-11

(iii) Concurrent (circulate and weight) method
Practical Well Control, pp. 6-12 – 6-14

(iv) Other applicable constant bottomhole pressure methods
Practical Well Control, pp. 6-15 – 6-17

(8) Candidates shall be instructed on calculations used in well control and the basis for their use including the following:

(i) Fluid-density increase required to control fluid flow into wellbore
Practical Well Control, pp. 3-4 – 3-7

(ii) Conversion between fluid density and pressure and the importance of that conversion in understanding danger of formation breakdown under the pressure caused by the fluid column particularly when setting casing in shallow formation
Practical Well Control, pp. 5-1 – 5-7

(iii) Calculation of equivalent pressures at the casing seat with emphasis on the importance of casing seat depth
Practical Well Control, pp. 5-1 – 5-7

(iv) Drop in pump pressure as fluid density increases during well-control operations; relationships between pump pressure, pump rate, and fluid density, and
Practical Well Control, pp. 4-7, 4-8

(v) Pressure limitations on casings
Practical Well Control, pp. 3-8 – 3-9, 7-16

(9) Candidates shall receive instructions on unusual well-control situations, including the following:

(i) Drill pipe is off bottom
Practical Well Control, p. 7-4 – 7-5

(ii) Drill pipe is out of the hole
Practical Well Control, p. 7-5

(iii) Lost circulation occurs
Practical Well Control, p. 7-11

(iv) Drill pipe is plugged
Practical Well Control, pp. 7-1 – 7-2

(v) There is excessive casing pressure, and
Practical Well Control, p. 7-16

(vi) There is a hole in drill pipe
Practical Well Control, p. 7-1

(10) Candidates shall receive instructions on the following:
  (i) Controlling shallow-gas kicks, and
      Practical Well Control, pp. 9-1 – 9-4
  (ii) Use of diverters
      Practical Well Control, pp. 9-2 – 9-4

(11) Candidates intending to receive subsea well-control qualification shall receive instructions on the special problems in well control when drilling with a subsea stack including:
  (i) Choke line friction determinations
      Practical Well Control, pp. 9-4 – 9-6
  (ii) Use of marine risers
      Practical Well Control, pp. 9-9 – 9-10
  (iii) Riser collapse
      Practical Well Control, p. 9-9 – 9-10
  (iv) Removal of trapped gas from the BOP stack after controlling a well kick
      Practical Well Control, p. 9-9
  (v) "U" tube effect as gas hits the choke line
      Practical Well Control, pp. 9-5 – 9-6

(12) Candidates shall receive instructions on the installation, operation, maintenance, and testing of BOP and diverter system.
   Practical Well Control, pp. 9-2 – 9-4, 9-6 – 9-8, 10-1 – 10-11

(13) Candidates shall receive instructions on the purpose, installation, operation, and general maintenance of the following auxiliary equipment:
  (i) Fluid pit-level indicator
      Practical Well Control, p. 10-19
  (ii) Fluid-volume measuring device
      Practical Well Control, p. 10-17
  (iii) Fluid-return indicator
      Practical Well Control, p. 10-19 – 10-20
  (iv) Gas detector
      Practical Well Control, pp. 10-20
  (v) Trip tank
      Practical Well Control, p. 10-21
  (vi) Gas separator
      Practical Well Control, pp. 10-21 – 10-23
  (vii) Degasser
      Practical Well Control, pp. 10-22 – 10-24
  (viii) Adjustable choke
      Practical Well Control, pp. 10-11 – 10-13

(14) Candidates shall receive instructions on the various items of equipment that will be subjected to pressure and/or wear.
   Practical Well Control, pp. 10-1 – 10-22

(15) Candidates shall receive instructions on the mechanics involved in various well-control situations, including the following:
  (i) Gas-bubble migration and expansion
      Practical Well Control, pp. 3-11 – 3-13

(ii) Bleeding volume from a shut-in well during gas migration
Practical Well Control, pp. 6-15 – 6-16

(iii) Excessive annular surface pressure
Practical Well Control, pp. 3-8 – 3-9, 5-3 – 5-7, 7-17

(iv) Differences between a gas kick and a saltwater and/or oil kick
Practical Well Control, pp. 3-4 – 3-5

(v) Special well-control techniques such as, but not limited to, barite plugs and cement plugs
Practical Well Control, pp. 7-14 – 7-16

(vi) Procedures and problems involved when experiencing lost circulation in well-control operations
Practical Well Control, pp. 7-11 – 7-16

(vii) Procedures and problems involved when experiencing a kick while drilling in a hydrogen sulfide (H_2S) environment
Practical Well Control, pp. 11-2, A-1 – A-6

(viii) Procedures and problems involved when experiencing a kick during snubbing, coil-tubing, or small-tubing operations
Practical Well Control, pp. 7-17 – 7-18

(16) Candidates shall receive instructions on organizing and directing a well-killing operation and shall subsequently direct such an operation using a model well or simulation device.

Practical Well Control, pp. 11-1 – 11-3

(17) Candidates shall receive instruction on the purpose and usage of BOP closing units, including the following:

(i) Charging procedures that include precharge and operating pressure
Practical Well Control, p. 10-18

(ii) Fluid volumes (usable and required)
Practical Well Control, pp. 10-15 – 10-18

(iii) Fluid pumps
Practical Well Control, pp. 10-13 – 10-14

(iv) Maintenance that includes charging fluid and inspection procedures
Practical Well Control, pp. 10-18 – 10-19

(18) Candidates shall receive stripping and snubbing operations instructions on the use of the entire BOP system for working pipe in or out of a wellbore that is under pressure.

Practical Well Control, pp. 7-6 – 7-11

(19) Candidates shall receive instructions for detecting entry into abnormally pressured formations and the accompanying warning signals including the following:

(i) Penetration rate change
Practical Well Control, pp. 2-12 – 2-13

(ii) Shale-density change
Practical Well Control, p. 2-16

(iii) Mud-chloride content change
Practical Well Control, pp. 2-15 – 2-16

(iv) Shale-cutting characteristics
Practical Well Control, pp. 2-9, 2-14

(v) Trip, connection, and background gas changes
Practical Well Control, p. 2-15

(20) Candidates shall receive instructions on the various types of completion fluids utilized and potential problems caused by their use in well control, including the following:

    (i) Gases
      Practical Well Control, p. 8-4

    (ii) Water-base system
      Practical Well Control, p. 8-5 – 8-6

    (iii) Oil-base system
      Practical Well Control, p. 8-4

    (iv) Packer fluids
      Practical Well Control, p. 8-6 – 8-7

(21) Candidates shall receive instructions on well-completion/well-control problems, including the following:

    (i) Multiple completions
      Practical Well Control, p. 8-14 – 8-19

    (ii) Running a drill-stem test
      Practical Well Control, p. 8-10 – 8-11

    (iii) Perforating
      Practical Well Control, p. 8-9 – 8-10

    (iv) Other completion operations
      Practical Well Control, p. 8-8 – 8-14

30 CFR, SUBPARTS C, D, E, F, G, H, AND O
SUBPART C-POLLUTION PREVENTION AND CONTROL

§250.40   Pollution prevention.

(a) During the exploration, development, production, and transportation of oil and gas or sulphur, the lessee shall take measures to prevent unauthorized discharge of pollutants into the offshore waters. The lessee shall not create conditions that will pose unreasonable risk to public health, life, property, aquatic life, wildlife, recreation, navigation, commercial fishing, or other uses of the ocean.

    (1) When pollution occurs as a result of operations conducted by or on behalf of the lessee and the pollution damages or threatens to damage life (including fish and other aquatic life), property, any mineral deposits (in areas leased or not leased), or the marine, coastal, or human environment, the control and removal of the pollution to the satisfaction of the District Supervisor shall be at the expense of the lessee. Immediate corrective action shall be taken in all cases where pollution has occurred. Corrective action shall be subject to modification when directed by the District Supervisor.

    (2) If the lessee fails to control and remove the pollution, the Director, in cooperation with other appropriate Agencies of Federal, State, and local governments, or in cooperation with the lessee, or both, shall have the right to control and remove the pollution at the lessee's expense. Such action shall not relieve the lessee of any responsibility provided for by law.

(b) (1) The District Supervisor may restrict the rate of drilling fluid discharges or prescribe alternative discharge methods. The District Supervisor may also restrict the use of components which could cause unreasonable degradation to the marine environment. No petroleum-based substances, including diesel fuel, may be added to the drilling mud system without prior approval of the District Supervisor.

(2) Approval of the method of disposal of drill cuttings, sand, and other well solids shall be obtained from the District Supervisor.

(3) All hydrocarbon-handling equipment for testing and production such as separators, tanks, and treaters shall be designed, installed, and operated to prevent pollution. Maintenance or repairs which are necessary to prevent pollution of offshore waters shall be undertaken immediately.

(4) Curbs, gutters, drip pans, and drains shall be installed in deck areas in a manner necessary to collect all contaminants not authorized for discharge. Oil drainage shall be piped to a properly designed, operated, and maintained sump system which will automatically maintain the oil at a level sufficient to prevent discharge of oil into offshore waters. All gravity drains shall be equipped with a water trap or other means to prevent gas in the sump system from escaping through the drains. Sump piles shall not be used as processing devices to treat or skim liquids but may be used to collect treated-produced water, treated-produced sand, or liquids from drip pans and deck drains and as a final trap for hydrocarbon liquids in the event of equipment upsets. Improperly designed, operated, or maintained sump piles which do not prevent the discharge of oil into offshore waters shall be replaced or repaired.

(5) On artificial islands, all vessels containing hydrocarbons shall be placed inside an impervious berm or otherwise protected to contain spills. Drainage shall be directed away from the drilling rig to a sump. Drains and sumps shall be constructed to prevent seepage.

(6) Disposal of equipment, cables, chains, containers, or other materials into offshore waters is prohibited.

(c) Materials, equipment, tools, containers, and other items used in the Outer Continental Shelf (OCS) which are of such shape or configuration that they are likely to snag or damage fishing devices shall be handled and marked as follows:

(1) All loose material, small tools, and other small objects shall be kept in a suitable storage area or a marked container when not in use and in a marked container before transport over offshore waters;

(2) All cable, chain, or wire segments shall be recovered after use and securely stored until suitable disposal is accomplished;

(3) Skid-mounted equipment, portable containers, spools or reels, and drums shall be marked with the owner's name prior to use or transport over offshore waters; and

(4) All markings must clearly identify the owner and must be durable enough to resist the effects of the environmental conditions to which they may be exposed.

(d) Any of the items described in paragraph (c) of this section that are lost overboard shall be recorded on the facility's daily operations report, as appropriate, and reported to the District Supervisor.

[53 FR 10690, Apr. 1, 1988, as amended at 56 FR 32099, July 15, 1991]

§250.41   Inspection of facilities.

(a) Drilling and production facilities shall be inspected daily or at intervals approved or prescribed by the District Supervisor to determine if pollution is occurring. Necessary maintenance or repairs shall be made immediately. Records of such inspections and repairs shall be maintained at the facility or at a nearby manned facility for 2 years.

[53 FR 10690, Apr. 1, 1988, as amended at 62 FR 13996, Mar. 25, 1997]

§250.44   Definitions concerning air quality.

For purposes of §§250.45 and 250.46 of this part:

Air pollutant means any combination of agents for which the Environmental Protection Agency (EPA) has established, pursuant to section 109 of the Clean Air Act, national primary or secondary ambient air quality standards.

Attainment area means, for any air pollutant, an area which is shown by monitored data or which is calculated by air quality modeling (or other methods determined by the Administrator of EPA to be reliable) not to exceed any primary or secondary ambient air quality standards established by EPA.

Best available control technology (BACT) means an emission limitation based on the maximum degree of reduction for each air pollutant subject to regulation, taking into account energy, environmental and economic impacts, and other costs. The BACT shall be verified on a case-by-case basis by the Regional Supervisor and may include reductions achieved through the application of processes, systems, and techniques for the control of each air pollutant.

Emission offsets means emission reductions obtained from facilities, either onshore or offshore, other than the facility or facilities covered by the proposed Exploration Plan or Development and Production Plan.

Existing facility is an OCS facility described in an Exploration Plan or a Development and Production Plan submitted or approved prior to June 2, 1980.

Facility means any installation or device permanently or temporarily attached to the seabed which is used for exploration, development, and production activities for oil, gas, or sulphur and which emits or has the potential to emit any air pollutant from one or more sources. All equipment directly associated with the installation or device shall be considered part of a single facility if the equipment is dependent on, or affects the processes of, the installation or device. During production, multiple installations or devices will be considered to be a single facility if the installations or devices are directly related to the production of oil, gas, or sulphur at a single site. Any vessel used to transfer production from an offshore facility shall be considered part of the facility while physically attached to it.

Nonattainment area means, for any air pollutant, an area which is shown by monitored data or which is calculated by air quality modeling (or other methods determined by the Administrator of FPA to be reliable) to exceed any primary or secondary ambient air quality standard established by EPA.

Projected emissions means emissions either controlled or uncontrolled, from a source(s).

Source means an emission point. Several sources may be included within a single facility.

Temporary facility means activities associated with the construction of platforms offshore or with facilities related to exploration for or development of offshore oil and gas resources which are conducted in one location for less than 3 years.

Volatile organic compound (VOC) means any organic compound which is emitted to the atmosphere as a vapor. The unreactive compounds are exempt from the above definition.

[53 FR 10690, Apr. 1, 1988, was amended at 56 FR 32100, July 15, 1991]

§250.45   Facilities described in a new or revised Exploration Plan or Development and Production Plan.

(a) *New plans.* All Exploration Plans and Development and Production Plans shall include the information required to make the necessary findings under paragraphs (d) through (i) of this section, and the lessee shall comply with the requirements of this section as necessary.

(b) *Applicability of §250.45 to existing facilities.*

(1) The Regional Supervisor may review any Exploration Plan or Development and Production Plan to determine whether any facility described in the plan should be subject to review under this section and has the potential to significantly affect the air quality of an onshore area. To make these decisions the Regional Supervisor shall consider the distance of the facility from shore, the size of the facility, the number of sources planned for the facility and their operational status, and the air quality status of the onshore area.

(2) For a facility identified by the Regional Supervisor in paragraph (b)(l) of this section, the Regional Supervisor shall require the lessee to refer to the information required in §250.33(b)(19) or §250.34(b)(12) of this part and to submit only that information required to make the necessary findings under paragraphs (d) through (i) of this section. The lessee shall submit this information within 120 days of the Regional Supervisor's determination or within a longer period of time at the discretion of the Regional Supervisor. The lessee shall comply with the requirements of this section as necessary.

(c) *Revised facilities.* All revised Exploration Plans and Development and Production Plans shall include the information required to make the necessary findings under paragraphs (d) through (i) of this section. The lessee shall comply with the requirements of this section as necessary.

(d) *Exemption formulas.* To determine whether a facility described in a new, modified, or revised Exploration Plan or Development and Production Plan is exempt from further air quality review, the lessee shall use the highest annual total amount of emissions from the facility for each air pollutant calculated in §250.33(b)(19)(i)(A) or §250.34(b)(12)(i)(A) of this part and compare these emissions to the emission exemption amount "E" for each air pollutant calculated using the following formulas: $E = 3400D^{2/3}$ for carbon monoxide (CO) and $E = 33.3D$ for total suspended particulates (TSP), sulphur dioxide (SO_2), nitrogen oxides (NO_x), and VOC (where E is the emission exemption amount expressed in tons per year, and D is the distance of the proposed facility from the closest onshore area of a State expressed in statute miles). If the amount of these projected emissions is less than or equal to the emission exemption amount "E" for the air pollutant, the facility is exempt from further air quality review required under paragraphs (e) through (i) of this section.

(e) *Significance levels.* For a facility not exempt under paragraph (d) of this section for air pollutants other than VOC, the lessee shall use an approved air quality model to determine whether the projected emissions of those air pollutants from the facility result in an onshore ambient air concentration above the following significance levels:

SIGNIFICANCE LEVELS: AIR POLLUTANT CONCENTRATIONS ($\mu g/m^3$)

Air pollutant	Averaging time (hours)				
	Annual	**24**	**8**	**3**	**1**
SO_2	1	5		25	
TSP	1	5			
NO_2	1				
CO			500		2,000

(f) *Significance determinations.*

(1) The projected emissions of any air pollutant other than VOC from any facility which result in an onshore ambient air concentration above the significance level determined under paragraph (e) of this section for that air pollutant, shall be deemed to significantly affect the air quality of the onshore area for that air pollutant.

(2) The projected emissions of VOC from any facility which is not exempt under paragraph (d) of this section for that air pollutant shall be deemed to significantly affect the air quality of the onshore area for VOC.

(g) *Controls required.*

(1) The projected emissions of any air pollutant other than VOC from any facility, except a temporary facility, which significantly affect the quality of a nonattainment area, shall be fully reduced. This shall be done through the application of BACT and, if additional reductions are necessary, through the application of additional emission controls or through the acquisition of offshore or onshore offsets.

(2) The projected emissions of any air pollutant other than VOC from any facility which significantly affect the air quality of an attainment or unclassifiable area shall be reduced through the application of BACT.

(i) Except for temporary facilities, the lessee also shall use an approved air quality model to determine whether the emissions of TSP or SO_2 that remain after the application of BACT cause the following maximum allowable increases over the baseline concentrations established in 40 CFR 52.21 to be exceeded in the attainment or unclassifiable area:

MAXIMUM ALLOWABLE CONCENTRATION INCREASES ($\mu G/M^3$)

Air pollutant	Averaging times		
	Annual mean[1]	24-hour maximum	3-hour maximum
Class I:			
TSP	5	10	
SO_2	2	5	25
Class II:			
TSP	19	37	
SO_2	20	91	512
Class III:			
TSP	37	75	
SO_2	40	182	700

[1] For TSP—geometric; For SO_2 —arithmetic.

No concentration of an air pollutant shall exceed the concentration permitted under the national secondary ambient air quality standard or the concentration permitted under the national primary air quality standard, whichever concentration is lowest for the air pollutant for the period of exposure. For any period other than the annual period, the applicable maximum allowable increase may be exceeded during one such period per year at any one onshore location.

(ii) If the maximum allowable increases are exceeded, the lessee shall apply whatever additional emission controls are necessary to reduce or offset the remaining emissions of TSP or SO_2 so that concentrations in the onshore ambient air of an attainment or unclassifiable area do not exceed the maximum allowable increases.

(3) (i) The projected emissions of VOC from any facility, except a temporary facility, which significantly affect the onshore air quality of a nonattainment area shall be fully reduced. This shall be done through the application of BACT and, if additional reductions are necessary, through the application of additional emission controls or through the acquisition of offshore or onshore offsets.

(ii) The projected emissions of VOC from any facility which significantly affect the onshore air quality of an attainment area shall be reduced through the application of BACT.

(4) (i) If projected emissions from a facility significantly affect the onshore air quality of both a nonattainment and an attainment or unclassifiable area, the regulatory requirements applicable to projected emissions significantly affecting a nonattainment area shall apply.

(ii) If projected emissions from a facility significantly affect the onshore air quality of more than one class of attainment area, the lessee must reduce projected emissions to meet the maximum allowable increases specified for each class in paragraph (g)(2)(i) of this section.

(h) *Controls required on temporary facilities.* The lessee shall apply BACT to reduce projected emissions of any air pollutant from a temporary facility which significantly affect the air quality of an onshore area of a State.

(i) *Emission offsets.* When emission offsets are to be obtained, the lessee must demonstrate that the offsets are equivalent in nature and quantity to the projected emissions that must be reduced after the application of BACT; a binding commitment exists between the lessee and the owner or owners of the source or sources; the appropriate air quality control jurisdiction has been notified of the need to revise the State Implementation Plan to include the information regarding the offsets; and the required offsets come from sources which affect the air quality of the area significantly affected by the lessee's offshore operations.

(j) *Review of facilities with emissions below the exemption amount.* If, during the review of a new, modified or revised Exploration Plan or Development and Production Plan, the Regional Supervisor determines or an affected State submits information to the Regional Supervisor which demonstrates, in the judgment of the Regional Supervisor, that projected emissions from an otherwise exempt facility will, either individually or in combination with other facilities in the area, significantly affect the air quality of an onshore area, then the Regional Supervisor shall require the lessee to submit additional information to determine whether emission control measures are necessary. The lessee shall be given the opportunity to present information to the Regional Supervisor which demonstrates that the exempt facility is not significantly affecting the air quality of an onshore area of the State.

(k) *Emission monitoring requirements.* The lessee shall monitor, in a manner approved or prescribed by the Regional Supervisor, emissions from the facility. The lessee shall submit this information monthly in a manner and form approved or prescribed by the Regional Supervisor.

(l) *Collection of meteorological data.* The Regional Supervisor may require the lessee to collect, for a period of time and in a manner approved or prescribed by the Regional Supervisor, and submit meteorological data from a facility.

[53 FR 10690, Apr. 1, 1988; 83 FR 19856, May 31, 1988; 53 FR 26067, July 11, 1988]

§250.46   Existing facilities.

(a) *Process leading to review of an existing facility.*

(1) An affected State may request that the Regional Supervisor supply basic emission data from existing facilities when such data are needed for the updating of the State's emission inventory. In submitting the request, the State must demonstrate that similar offshore and onshore facilities in areas under the State's jurisdiction are also included in the emission inventory.

(2) The Regional Supervisor may require lessees of existing facilities to submit basic emission data to a State submitting a request under paragraph (a)(1) of this section.

(3) The State submitting a request under paragraph (a)(1) of this section may submit information from its emission inventory which indicates that emissions from existing facilities may be significantly affecting the air quality of the onshore area of the State. The lessee shall be given the opportunity to present information to the Regional Supervisor which demonstrates that the facility is not significantly affecting the air quality of the State.

(4) The Regional Supervisor shall evaluate the information submitted under paragraph (a)(3) of this section and shall determine, based on the basic emission data, available meteorological data, and the distance of the facility or facilities from the onshore area, whether any existing facility has the potential to significantly affect the air quality of the onshore area of the State.

(5) If the Regional Supervisor determines that no existing facility has the potential to significantly affect the air quality of the onshore area of the State submitting information under paragraph (a)(3) of this section, the Regional Supervisor shall notify the State of and explain the reasons for this finding.

(6) If the Regional Supervisor determines that an existing facility has the potential to significantly affect the air quality of an onshore area of the State submitting information under paragraph (a)(3) of this section, the Regional Supervisor shall require the lessee to refer to the information requirements under §250.33(b)(19) or 250.34(b)(12) of this part and submit only that information required to make the necessary findings under paragraphs (b) through (e) of this section. The lessee shall submit this information within 120 days of the Regional Supervisor's determination or within a longer period of time at the discretion of the Regional Supervisor. The lessee shall comply with the requirements of this section as necessary.

(b) *Exemption formulas.* To determine whether an existing facility is exempt from further air quality review, the lessee shall use the highest annual total amount of emissions from the facility for each air pollutant calculated in §250.33(b)(19)(i)(A) or 250.34(b)(12)(i)(A) of this part and compare these emissions to the emission exemption amount "E" for each air pollutant calculated using the following formulas: $E=3400D^{2/3}$ for CO; and $E=33.3D$ for TSP, SO_2, NO_x, and VOC (where E is the emission exemption amount expressed in tons per year, and D is the distance of the facility from the closest onshore area of the State expressed in statute miles). If the amount of projected emissions is less than or equal to the emission exemption amount "E" for the air pollutant, the facility is exempt for that air pollutant from further air quality review required under paragraphs (c) through (e) of this section.

(c) *Significance levels.* For a facility not exempt under paragraph (b) of this section for air pollutants other than VOC, the lessee shall use an approved air quality model to determine whether projected emissions of those air pollutants from the facility result in an onshore ambient air concentration above the following significance levels:

SIGNIFICANCE LEVELS: AIR POLLUTANT CONCENTRATIONS ($\mu G/M^3$)

Air pollutant	Averaging time (hours)				
	Annual	24	8	3	1
SO_2	1	5		25	
TSP	1	5			
NO_2	1				
CO			500		2,000

(d) *Significance determinations.*

(1) The projected emissions of any air pollutant other than VOC from any facility which result in an onshore ambient air concentration above the significance levels determined under paragraph (c) of this section for that air pollutant shall be deemed to significantly affect the air quality of the onshore area for that air pollutant.

(2) The projected emissions of VOC from any facility which is not exempt under paragraph (b) of this section for that air pollutant shall be deemed to significantly affect the air quality of the onshore area for VOC.

(e) *Controls required.*

(1) The projected emissions of any air pollutant which significantly affect the air quality of an onshore area shall be reduced through the application of BACT.

(2) The lessee shall submit a compliance schedule for the application of BACT. If it is necessary to cease operations to allow for the installation of emission controls, the lessee may apply for a suspension of operations under the provisions of §250.10 of this part.

(f) *Review of facilities with emissions below the exemption amount.* If, during the review of the information required under paragraph (a)(6) of this section, the Regional Supervisor determines or an affected

State submits information to the Regional Supervisor which demonstrates, in the judgment of the Regional Supervisor, that projected emissions from an otherwise exempt facility will, either individually or in combination with other facilities in the area, significantly affect the air quality of an onshore area, then the Regional Supervisor shall require the lessee to submit additional information to determine whether control measures are necessary. The lessee shall be given the opportunity to present information to the Regional Supervisor which demonstrates that the exempt facility is not significantly affecting the air quality of an onshore area of the State.

(g) *Emission monitoring requirements.* The lessee shall monitor, in a manner approved or prescribed by the Regional Supervisor, emissions from the facility following the installation of emission controls. The lessee shall submit this information monthly in a manner and form approved or prescribed by the Regional Supervisor.

(h) *Collection of meteorological data.* The Regional Supervisor may require the lessee to collect, for a period of time and in a manner approved or prescribed by the Regional Supervisor, and submit meteorological data from a facility.

[53 FR 10690, Apr. 1, 1988; 53 FR 26067, July 11, 1988]

SUBPART D-OIL AND GAS DRILLING OPERATIONS

§250.50   Control of wells.

The lessee shall take necessary precautions to keep its wells under control at all times. The lessee shall utilize the best available and safest drilling technology in order to enhance the evaluation of conditions of abnormal pressure and to minimize the potential for the well to flow or kick. The lessee shall utilize personnel who are trained and competent and shall utilize and maintain equipment and materials necessary to assure the safety and protection of personnel, equipment, natural resources, and the environment.

§250.51   General requirements.

(a) *Fitness of drilling unit.*

(1) Drilling units shall be capable of withstanding the oceanographic, meteorological, and ice conditions for the proposed season and location of operations.

(2) Prior to commencing operation, drilling units shall be available for complete inspection by the District Supervisor.

(3) The lessee shall provide information and data on the fitness of the drilling unit to perform the proposed drilling operation. The information shall be submitted with or prior to the submission of Form MMS-123, Application for Permit to Drill (APD), in accordance with §250.64. The District Supervisor may require the submission of a third-party review of the design of drilling units which are of a unique design and/or not proven for use in the proposed environment if the District Supervisor believes that the information submitted by the lessee is insufficient to demonstrate suitability of the unit for use at the proposed drill site. A design Certified Verification Agent approved in accordance with §250.133 of this part shall be used for any required third-party review.

(b) *Drilling unit safety devices.*

(1) No later than May 31, 1989, all drilling units shall be equipped with a safety device which is designed to prevent the traveling block from striking the crown block. The device shall be checked for proper operation weekly and after each drill-line slipping operation. The results of the operational check shall be entered in the driller's report.

(2) No later than May 31, 1989, diesel engine air intakes shall be equipped with a device to shut down the diesel engine in the event of runaway. Diesel engines which are continuously attended shall be equipped with either remote operated manual or automatic shutdown devices. Diesel engines which are not continuously attended shall be equipped with automatic shutdown devices.

(c) *Oceanographic, meteorological, and drilling unit performance data.* Where such information is not otherwise readily available, upon request of the District Supervisor, lessees shall collect and report oceanographic, meteorological, and drilling unit performance data, and monitor ice conditions, if applicable, during the period of operations. The type of information to be collected and reported will be determined by the District Supervisor in the interests of safe conduct of operations and the structural integrity of the drilling unit.

(d) *Foundation requirements.* When the lessee fails to provide sufficient information pursuant to §§250.33 and 250.34 of this part to support a determination that the seafloor is capable of supporting a specific bottom-founded drilling unit under the site-specific soil and oceanographic conditions, the District Supervisor may require that additional surveys and soil borings be performed and the results be submitted for review and evaluation by the District Supervisor before approval is granted for commencing drilling operations.

(e) *Tests, surveys, and samples.*

(1) The lessee shall conduct tests, obtain well and mud logs or surveys, and take samples to determine the reservoir energy; the presence, quantity, and quality of oil, gas, sulphur, and water; and the amount of pressure in the formations penetrated. The lessee shall take formation samples or cores to determine the identity, fluid content, and characteristics of any penetrated formation in accordance with requirements approved or prescribed by the District Supervisor.

(2) Inclinational surveys shall be obtained on all vertical wells at intervals not exceeding 1,000 feet during the normal course of drilling. Directional surveys giving both inclination and azimuth shall be obtained on all directional wells at intervals not exceeding 500 feet during the normal course of drilling and at intervals not exceeding 100 feet in all portions of the hole when angle-changes are planned.

(3) On both vertical and directionally drilled wells, directional surveys giving both inclination and azimuth shall be obtained at intervals not exceeding 500 feet prior to or upon setting surface or intermediate casing, liners, and at total depth. Composite directional surveys shall be prepared with the interval shown from the bottom of the conductor casing or, in the absence of conductor casing, from the bottom of the drive or structural casing to total depth. In calculating all surveys, a correction from the true north to Universal-Transverse-Mercator-Grid-north or Lambert-Grid-north shall be made after making the magnetic-to-true-north correction. A composite dipmeter directional survey or a composite measurement-while-drilling (MWD) directional survey including a listing of the directionally computed inclinations and azimuths on a well classified as vertical will be acceptable as fulfilling the applicable requirements of this paragraph. In the event a composite MWD survey is run, a multishot survey shall be obtained at each casing point in order to confirm the MWD results.

(4) Wells are classified as vertical if the calculated average of inclination readings weighted by the respective interval lengths between readings from surface to drilled depth does not exceed 3 degrees from the vertical. When the calculated average inclination readings weighted by the length of the respective interval between readings from the surface to drilled depth exceeds 3 degrees, the well is classified as directional.

(5) The Regional Supervisor at the request of a holder of an adjoining lease may, for the protection of correlative rights, furnish a copy of the directional survey for a well drilled within 500 feet of the adjacent lease to that leaseholder.

(f) *Fixed drilling platforms.* Applications for installation of fixed drilling platforms or structures, including artificial islands, shall be submitted in accordance with the provisions of subpart I, Platforms and Structures, of this part. Mobile drilling units which have their jacking equipment removed or have been otherwise immobilized are classified as fixed drilling platforms.

(g) *Equipment movement.* The movement of drilling rigs and related equipment on and off an offshore platform or from well to well on the same offshore platform, including rigging up and rigging down, shall be conducted in a safe manner. All wells in the same well-bay which are capable of producing hydrocarbons shall be shut in below the surface with a pump-through-type tubing plug and at the surface with a closed master valve prior to moving such rigs and related equipment, unless otherwise approved by the District Supervisor. A closed surface-controlled subsurface safety valve of the pump-through-type may be used in lieu of the pump-through-type tubing plug, provided that the surface control has been locked out.

(h) *Emergency shutdown system.* When drilling operations are conducted on a platform where there are other hydrocarbon-producing wells or other hydrocarbon flow, an Emergency Shutdown System (ESD) manually controlled station shall be installed near the driller's console.

[53 FR 10690, Apr. 1, 1988; 53 FR 12227, Apr. 13, 1988, as amended at 54 FR 50616, Dec. 8, 1989; 55 FR 47752, Nov. 15, 1990; 58 FR 49928, Sept. 24, 1993]

§250.52   **Welding and burning practices and procedures.**

(a) *General requirements.*

(1)   For the purpose of this rule, the terms welding and burning are defined to include arc or fuel-gas (acetylene or other gas) cutting and arc or fuel-gas welding.

(2)   All offshore welding and burning shall be minimized by onshore fabrication when feasible. The requirements set forth in paragraphs (b), (c), and (d) of this section shall be applicable to any welding or burning practice or procedure performed on the following:

(i)   An offshore mobile drilling unit during the drilling mode;

(ii)   A mobile workover unit during any drilling, completion, recompletion, remedial, repair, stimulation, or other workover activity;

(iii)   A platform, structure, artificial island, or other installation during any drilling, well-completion, well-workover, or production operation; and

(iv)   A platform, structure, artificial island, or other installation which contains a well open to a hydrocarbon-bearing zone.

(3)   All water-discharge-point sources from hydrocarbon-handling vessels shall be monitored in order to stop welding and burning operations in case flammable fluids are discharged as a result of equipment upset or malfunction.

(4)   Equipment containing hydrocarbons or other flammable substances shall be relocated at least 35 feet horizontally from the work site. Similar equipment located at a lower elevation where slag, sparks, or other burning materials could fall, shall be relocated at least 35 feet from the point of impact. If relocation is impractical, either the equipment shall be protected with flame-proofed covers or otherwise shielded with metal or fire-resistant guards or curtains, or the contents shall have been rendered inert.

(b) *Welding, burning, and hot tapping plan.* Each lessee shall submit for approval by the District Supervisor a "Welding, Burning, and Hot Tapping Safe Practices and Procedures Plan" prior to beginning the first drilling and/or production operations on a lease. The plan shall include the qualification standards or requirements for personnel who the lessee will authorize to conduct welding, burning, and hot tapping operations and the methods by which the lessee will assure that only trained personnel who meet such standards or requirements are utilized. A copy of this plan and approval letter shall be available on the facility where the welding is conducted. Any person designated as a welding supervisor shall be thoroughly familiar with this plan. An approved plan is required prior to conducting any welding, burning, or hot tapping operation. All welding and burning equipment shall be inspected by the welding supervisor or the lessee's designated person in charge prior to beginning any welding, burning, or hot tapping. All engine-driven welding machines shall be equipped with spark arrestors and drip pans. Welding leads shall be completely insulated and in good condition, oxygen and fuel gas bottles shall be secured in a safe place, and leak-free hoses shall be equipped with proper fittings, gauges, and regulators.

(c) *Designated safe-welding and burning areas.* The lessee may establish and designate areas determined to be safe-welding areas. These designated areas shall be identified in the plan, and a drawing showing the location of these areas shall be maintained on the facility.

(d) *Undesignated welding and burning areas.* All welding and burning, which cannot be done in an approved safe-welding area, shall be performed in compliance with the following:

    (1) Prior to the commencement of any of these operations, the lessee's designated person in charge at the installation shall inspect the qualifications of the welder(s) to assure that the welder(s) is properly qualified in accordance with the approved qualification standards or requirements for welders. The designated person in charge and the welder(s) shall inspect the work area and area(s) at elevations below the work area where slag, sparks, or other hot materials could fall for potential fire and explosion hazards. After it has been determined that it is safe to proceed with the welding and burning operation, the designated person-in-charge shall issue a written authorization for the work.

    (2) During these welding or burning operations, one or more persons shall be designated as a fire watch. The person(s) assigned as a fire watch shall have no other duties while actual welding or burning operations are in progress. If the operation is to be in an area which is not equipped with a gas detector, the fire watch shall also maintain a continuous surveillance with a portable gas detector during the welding and burning operation. The fire watch shall remain on duty for a period of 30 minutes after welding or burning operations have been completed.

    (3) Prior to any of these operations, the fire watch shall have in their possession firefighting equipment in a usable condition.

    (4) No welding or burning operation, other than approved hot tapping, shall be done on piping, containers, tanks, or other vessels which have contained a flammable substance unless the contents have been rendered inert and are determined to be safe for welding or burning by the designated person in charge.

    (5) If drilling, well-completion, well-workover, or wireline operations are in progress, welding operations in other than approved safe-welding areas shall not be conducted unless the well(s) in the area where drilling, well-completion, well-workover, or wireline operations are in progress contain noncombustible fluids and the entry of formation hydrocarbons into the wellbore is precluded.

    (6) If welding or burning operations are conducted in or within 10 feet of a well-bay or production area, all producing wells in the well-bay or production area shall be shut in at the surface safety valve.

§250.53   Electrical equipment.

The following requirements shall be applicable to all electrical equipment on all platforms, artificial islands, fixed structures, and their facilities:

(a) All engines with electrical ignition systems shall be equipped with a low-tension ignition system designed and maintained to minimize the release of sufficient electrical energy to cause ignition of an external, combustible mixture or substance.

(b) All areas shall be classified in accordance with API RP 500, Recommended Practice for Classification of Locations for Electrical Installations at Petroleum Facilities.

(c) All electrical installations shall be made in accordance with API RP 14F, Design and Installation of Electrical Systems for Offshore Production Platforms, except sections 7.4, Emergency Lighting and 9.4, Aids to Navigation Equipment.

(d) Maintenance of electrical systems shall be by personnel who are trained and experienced with the area classifications, distribution system, performance characteristics and operation of the equipment, and with the hazards involved.

[53 FR 10690, Apr. 1, 1988, as amended at 54 FR 50616, Dec. 8, 1989; 61 FR 60024, Nov. 26, 1996]

§250.54   **Well casing and cementing.**

(a) *General requirements.*

   (1) For the purpose of this subpart, the casing strings in order of normal installation are as follows:

      (i)   Drive or structural,

      (ii) Conductor,

      (iii) Surface,

      (iv) Intermediate, and

      (v)   Production casing.

   (2) The lessee shall case and cement all wells with a sufficient number of strings of casing and quantity and quality of cement in a manner necessary to prevent release of fluids from any stratum through the wellbore (directly or indirectly) into offshore waters, prevent communication between separate hydrocarbon-bearing strata, protect freshwater aquifers from contamination, support unconsolidated sediments, and otherwise provide a means of control of the formation pressures and fluids. Cement composition, placement techniques, and waiting time shall be designed and conducted so that the cement in place behind the bottom 500 feet of casing or total length of annular cement fill, if less, attains a minimum compressive strength of 500 pounds per square inch (psi). Cement placed across permafrost zones shall be designed to set before freezing and have a low heat of hydration.

   (3) The lessee shall install casing designed to withstand the anticipated stresses imposed by tensile, compressive, and buckling loads; burst and collapse pressures; thermal effects; and combinations thereof. Safety factors in the casing program design shall be of sufficient magnitude to provide well control during drilling and to assure safe operations for the life of the well. Any portion of an annulus opposite a permafrost zone which is not protected by cement shall be filled with a liquid which has a freezing point below the minimum permafrost temperature to prevent internal freezeback and which is treated to minimize corrosion.

   (4) In cases where cement has filled the annular space back to the mud line, the cement may be washed out or displaced to a depth not exceeding the depth of the structural casing shoe to facilitate casing removal upon well abandonment if the District Supervisor determines that subsurface protection against damage to freshwater aquifers and permafrost zones and against damage caused by adverse loads, pressures, and fluid flows is not jeopardized.

   (5) If there are indications of inadequate cementing (such as lost returns, cement channeling, or mechanical failure of equipment), the lessee shall evaluate the adequacy of the cementing operations by pressure testing the casing shoe, running a cement bond log, running a temperature survey, or a combination thereof before continuing operations. If the evaluation indicates inadequate cementing, the lessee shall recement or take other remedial actions as approved by the District Supervisor.

(6) A pressure-integrity test shall be run below the surface casing, the intermediate casing(s), and liner(s) used as intermediate casing(s). The District Supervisor may require a pressure-integrity test to be run at the conductor casing shoe due to local geologic conditions or planned casing setting depths. Pressure-integrity tests shall be made after drilling new hole below the casing shoe and before drilling more than 50 feet of new hole below a respective casing string. These tests shall be conducted either by testing to formation leak-off or by testing to a predetermined equivalent mud weight as specified in the approved APD. A safe margin, as approved by the District Supervisor, shall be maintained between the mud weight in use and the equivalent mud weight at the casing shoe as determined in the pressure-integrity test. Drilling operations shall be suspended when the safe margin is not maintained. Pressure-integrity and pore-pressure test results and related hole-behavior observations, such as gas-cut mud and well kicks made during the course of drilling, shall be used in adjusting the drilling mud program and the approved setting depth of the next casing string. The results of all tests and of hole-behavior observations made during the course of drilling related to formation integrity and pore pressure shall be recorded in the driller's report.

(b) *Drive or structural casing.* This casing shall be set by driving, jetting, or drilling to a minimum depth as may be prescribed or approved by the District Supervisor, in order to support unconsolidated deposits and to provide hole stability for initial drilling operations. If this portion of the hole is drilled, a quantity of cement sufficient to fill the annular space back to the mud line shall be used.

(c) *Conductor and surface casing requirements.*

(1) *Conductor and surface casing setting depths.* Conductor and surface casing design and setting depths shall be based upon relevant engineering and geologic factors including the presence or absence of hydrocarbons, potential hazards, and water depths. The approved casing setting depths may be adjusted when the change is approved by the District Supervisor to permit the casing shoe to be set in a competent formation or below formations which should be isolated from the wellbore by casing for safer drilling operations. However, the conductor casing shall be set immediately prior to drilling into formations known to contain oil or gas or, if the presence of oil or gas is unknown, upon encountering a formation containing oil or gas. Upon encountering unexpected formation pressures, the lessee shall submit a revised casing program to the District Supervisor for approval. The District Supervisor may permit a lessee to drill a well without setting conductor casing provided the information from approved logging and mud-monitoring programs for wells previously drilled in the immediate vicinity combined with other available geologic data are sufficient to demonstrate the absence of shallow hydrocarbons or hazards.

(2) *Conductor casing cementing requirements.* Conductor casing shall be cemented with a quantity of cement that fills the calculated annular space back to the mud line except as applicable to the bottom of an excavation (glory hole) or to the surface of an artificial island. Cement fill in annular spaces shall be verified by the observation of cement returns. In the event that observation of cement returns is not feasible, additional quantities of cement shall be used to assure fill to the mud line.

(3) *Surface casing cementing requirements.*

(i) Surface casing shall be cemented with a quantity of cement that fills the calculated annular space to at least 200 feet inside the conductor casing. When geologic conditions such as near-surface fractures and faulting exist, surface casing shall be cemented with a quantity of cement that fills the calculated annular space to the mud line, or as approved or prescribed by the District Supervisor.

(ii) For floating drilling operations, a lesser volume of cement may be used to prevent sealing the annular space between the conductor casing and surface casing if the District Supervisor determines that the uncemented space is necessary to provide protection from burst and collapse pressures which may be applied inadvertently to the annulus between casings during blowout preventer (BOP) testing operations. Any annular space open to the drilled hole shall be sealed in accordance with the requirements for abandonment in subpart G, Abandonment of Wells, of this part.

(d) *Intermediate casing requirements.*

(1) Intermediate casing string(s) shall be set for protection when geologic characteristics or wellbore conditions, as anticipated or as encountered, so indicate.

(2) Quantities of cement that cover and isolate all hydrocarbon-bearing zones in the well and isolate abnormal pressure intervals from normal pressure intervals shall be used. This requirement for isolation may be satisfied by squeeze cementing prior to completion, suspension of operations, or abandonment, whichever occurs first. Sufficient cement shall be used to provide annular fill-up to a minimum of 500 feet above the zones to be isolated or 500 feet above the casing shoe in wells where zonal coverage is not required.

(3) If a liner is to be used as an intermediate string below a surface casing string, it shall be lapped a minimum of 100 feet into the previous casing string and cemented as required for intermediate casing. When a liner is to be used as production casing below a surface casing string, it shall be extended to the surface and cemented to avoid surface casing being used as production casing.

(e) *Production casing requirements.*

(1) Production casing shall be cemented to cover or isolate all zones above the shoe which contain hydrocarbons; but in any case, a volume sufficient to fill the annular space at least 500 feet above the uppermost hydrocarbon-bearing zone shall be used.

(2) When a liner is to be used as production casing below intermediate casing, it shall be lapped a minimum of 100 feet into the previous casing string and cemented as required for the production casing.

§250.55 Pressure testing of casing.

(a) Prior to drilling the plug after cementing and in the cases of plugs in production casing strings and liners not planned to be subsequently drilled out, all casings, except the drive or structural casing, shall be pressure tested to 70 percent of the minimum internal-yield pressure of the casing or as otherwise approved or required by the District Supervisor. If the pressure declines more than 10 percent in 30 minutes or if there is another indication of a leak, the casing shall be recemented, repaired, or an additional casing string run and the casing pressure tested again. Additional remedial actions shall be taken until a satisfactory pressure test is obtained. The results of all casing pressure tests shall be recorded in the driller's report.

(b) Each production liner lap shall be tested to a minimum of 500 psi above formation fracture pressure at the shoe of the casing into which the liner is lapped, or as otherwise approved or required by the District Supervisor. The drilling liner-lap test pressure shall be equal to or exceed the pressure that will be encountered at the liner lap when conducting the planned pressure-integrity test below the liner shoe. The test results shall be recorded on the driller's report. If the test indicates an improper seal, remedial action shall be taken which provides a proper seal as demonstrated by a satisfactory pressure test.

(c) In the event of prolonged drill-pipe rotation within a casing string run to the surface or extended operations such as milling, fishing, jarring, washing over, and other operations which could damage the casing, the casing shall be pressure tested or evaluated by a logging technique such as a caliper log every 30 days. The evaluation results shall be submitted to the District Supervisor with a determination of effects of operations on the integrity of the casing for continued service during drilling operations and over the producing life of the well. If the integrity of the casing in the well has deteriorated to an unsafe level, remedial operations shall be conducted or additional casing set in accordance with a plan approved by the District Supervisor prior to continuing drilling operations.

(d) After cementing any string of casing other than the structural casing string, drilling shall not be resumed until there has been a time lapse of 8 hours under pressure for the conductor casing string and 12 hours under pressure for all other casing strings. Cement is considered under pressure if one or more float valves are shown to be holding the cement in place or when other means of holding pressure are used.

§250.56   **Blowout preventer systems and system components.**

(a) *General.* The BOP systems and system components shall be designed, installed, used, maintained, and tested to assure well control.

(b) *BOP stacks.* The BOP stacks shall consist of an annular preventer and the number of ram-type preventers as specified under paragraphs (e)(1), (f), and (g) of this section. The pipe rams shall be of a proper size(s) to fit the drill pipe in use.

(c) *Working pressure.* The working-pressure rating of any BOP component shall exceed the anticipated surface pressure to which it may be subjected. The District Supervisor may approve a lower working pressure rating for the annular preventer if the lessee demonstrates that the anticipated or actual well conditions will not place demands above its rated working pressure. (Refer to related requirements in §250.64(f)(3)(ii) of this part.)

(d) *BOP equipment.* All BOP systems shall be equipped and provided with the following:

(1) An accumulator system which shall provide sufficient capacity to supply 1.5 times the volume of fluid necessary to close and hold closed all BOP equipment units with a minimum pressure of 200 psi above the precharge pressure without assistance from a charging system. No later than December 1, 1988, accumulator regulators supplied by rig air and without a secondary source of pneumatic supply, shall be equipped with manual overrides or alternately, other devices provided to ensure capability of hydraulic operations if rig air is lost.

(2) A backup to the primary accumulator-charging system which shall be automatic, supplied by a power source independent from the power source to the primary accumulator-charging system, and possess sufficient capability to close all BOP components and hold them closed.

(3) At least one operable remote BOP control station in addition to the one on the drilling floor. This control station shall be in a readily accessible location away from the drilling floor.

(4) A drilling spool with side outlets if side outlets are not provided in the body of the BOP stack to provide for separate kill and choke lines.

(5) For surface BOP systems, a choke and a kill line each equipped with two full-opening valves. At least one of the valves on the choke line shall be remotely controlled. At least one of the valves on the kill line shall be remotely controlled except that a check valve may be installed on the kill line in lieu of the remotely controlled valve provided two readily accessible manual valves are in place and the check valve is placed between the manual valves and the pump. For subsea BOP systems, a choke and a kill line each equipped with two full-opening valves. At least one of the valves on the choke line and at least one of the valves on the kill line shall be remotely controlled.

(6) A fill-up line above the uppermost preventer.

(7) A choke manifold suitable for the anticipated pressures to which it may be subjected, method of well control to be employed, surrounding environment, and corrosiveness, volume, and abrasiveness of fluids. The choke manifold shall also meet the following requirements:

     (i) Manifold and choke equipment subject to well and/or pump pressure shall have a rated working pressure at least as great as the rated working pressure of the ram-type BOP's or as otherwise approved by the District Supervisor;

(ii) All components of the choke manifold system shall be protected from the danger, if any, of freezing by heating, draining, or filling with proper fluids; and

(iii) When buffer tanks are installed downstream of the choke assemblies for the purpose of manifolding the bleed lines together, isolation valves shall be installed on each line.

(8) Valves, pipes, flexible steel hoses, and other fittings upstream of, and including, the choke manifold with pressure ratings at least as great as the rated working pressure of the ram-type BOP's or as otherwise approved by the District Supervisor.

(9) A wellhead assembly with a rated working pressure that exceeds the anticipated surface pressure to which it may be subjected.

(10) The following system components:

(i) On a conventional drilling rig, a kelly cock installed below the swivel (upper kelly cock), essentially full opening, and a similar valve of such design that it can be run through the BOP stack (strippable) installed at the bottom of the kelly (lower kelly cock). With a mud motor in service and while using drill pipe in lieu of a kelly, one kelly cock located above and one strippable kelly cock located below the joint of drill pipe employed in lieu of a kelly. On a top-drive system equipped with a remote controlled valve, a second and lower strippable valve of a conventional kelly cock or comparable type either manually or remotely controlled. All required manual and remotely controlled valves of a kelly cock or comparable type in a top-drive system shall be essentially full-opening and tested according to the test pressure and test frequency as stated in §250.57(d) of this part. A wrench to fit each manually operable valve in a conventional drilling rig, mud motor, and top-drive system shall be stored in a location readily accessible to the drilling crew.

(ii) An inside BOP and an essentially full-opening drill-string safety valve in the open position on the rig floor at all times while drilling operations are being conducted. These valves shall be maintained on the rig floor to fit all connections that are in the drill string. A wrench to fit the drill-string safety valve shall be stored in a location readily accessible to the drilling crew.

(iii) A safety valve available on the rig floor assembled with the proper connection to fit the casing string being run in the hole.

(11) Locking devices installed on the ram-type preventers.

(e) *Subsea BOP requirements.*

(1) Prior to drilling below surface and intermediate casing, a BOP system shall be installed consisting of at least four remote-controlled, hydraulically-operated BOP's including at least two equipped with pipe rams, one with blind-shear rams, and one annular type. A subsea accumulator closing unit or a suitable alternate approved by the District Supervisor is required to provide fast closure of the BOP components and to operate all critical functions in case of a loss of the power fluid connection to the surface. When proposed casing setting depths or local geology indicate the need for a BOP to provide safety during the drilling of the surface hole, the District Supervisor may require that a subsea BOP system be installed prior to drilling below the conductor casing.

(2) The BOP system shall include operable dual-pod control systems necessary to ensure proper and independent operation of the BOP system functions when drilling below the surface casing.

(3) Prior to the removal of the marine riser, the riser shall be displaced with seawater. Sufficient hydrostatic pressure or other suitable precautions, such as mechanical or cement plugs or closing the BOP, shall be maintained within the wellbore to compensate for the reduction in pressure and to maintain a safe controlled well condition.

(4) Any necessary repair or replacement of the BOP system or a system component after installation shall be accomplished under safe controlled conditions, (e.g., after casing has been cemented but prior to drilling out the casing shoe or by setting a cement plug, bridge plug, or a packer).

(5) When a subsea BOP system is to be used in an area which is subject to ice scour, the BOP stack shall be placed in an excavation (glory hole) of sufficient depth to assure that the top of the stack is below the deepest probable ice-scour depth.

(f) *Surface BOP requirements.* Prior to drilling below surface or intermediate casing, a BOP system shall be installed consisting of at least four remote-controlled, hydraulically-operated BOP's including at least two equipped with pipe rams, one with blind rams, and one annular type.

(g) *Tapered drill-string operations.*

(1) Prior to commencing tapered drill-pipe operations, the BOP stack shall be equipped with conventional and/or variable-bore pipe rams installed in two or more ram cavities to provide the following:

(i) Two sets of pipe rams capable of sealing around the larger size drill string, and

(ii) One set of pipe rams capable of sealing around the smaller size drill string.

(2) Subsea BOP systems shall have blind-shear ram capability. Surface BOP systems shall have blind ram capability.

§250.57   Blowout preventer systems tests, actuations, inspections, and maintenance.

(a) Prior to conducting high-pressure tests, all BOP systems shall be tested to a low pressure of 200 to 300 psi.

(b) Surface ram-type BOP's and the choke manifold shall be pressure tested with water to rated working pressure or as otherwise approved by the District Supervisor. The annular-type BOP shall be pressure tested with water to 70 percent of its rated working pressure or as otherwise approved by the District Supervisor.

(c) Subsea BOP system components shall be stump pressure tested at the surface with water to their rated working pressure, except that the annular-type BOP shall not be pressure tested above 70 percent of its rated working pressure. After the installation of the BOP stack on the seafloor, the ram-type BOP's and choke manifold shall be pressure tested to rated working pressure or as otherwise approved by the District Supervisor. The annular preventer shall be pressure tested to 70 percent of its rated working pressure or as otherwise approved by the District Supervisor.

(d) In conjunction with the weekly pressure test of surface and subsea BOP systems required in paragraph (e) of this section, the choke manifold valves; upper and lower kelly cocks; top-drive, inside-BOP, and the drill string safety valves shall be pressure tested to pipe-ram test pressures or otherwise approved by the District Supervisor. Safety valves assembled with proper casing connections shall be actuated prior to running casing.

(e) Surface and subsea BOP systems shall be pressure tested as follows:

(1) When installed.

(2) Before drilling out each string of casing or before continuing operations in cases where the cement is not drilled out.

(3) At least once each week, but not exceeding 7 days between pressure tests, alternating between control stations and pods. If either control station or pod is not functional, further drilling operations shall be suspended until that system becomes operable. A period of more than 7 days between BOP tests is allowed when there is stuck drill pipe or pressure-control operations and remedial efforts are being performed, provided that the pressure tests are conducted as soon as possible and before normal operations resume. The reason for postponing pressure testing shall

be entered into the driller's report. Pressure testing shall be performed at intervals to allow each drilling crew to operate the equipment. The weekly pressure test is not required for blind and blind-shear rams.

(4) Blind and blind-shear rams shall be actuated at least once every 7 days. Closing pressure on the blind and blind-shear rams greater than that necessary to indicate proper operation of the rams is not required.

(5) Variable bore-pipe rams shall be pressure-tested against all sizes of pipe in use, excluding drill collars and bottom-hole tools.

(6) Following the disconnection or repair of any well-pressure containment seal in the wellhead/ BOP stack assembly but limited to the affected component.

(f) All BOP systems and marine risers shall be inspected and maintained to assure that the equipment will function properly. The BOP systems and marine risers shall be visually inspected at least once each day if the weather and sea conditions permit the inspection. Inspection of BOP systems and marine risers may be accomplished by the use of television equipment. The District Supervisor may approve alternate methods of inspection of marine risers on dynamic-positioned rigs. Casing risers on fixed structures and jackup rigs are not subject to the daily underwater inspection requirement.

(g) The lessee shall record pressure conditions during BOP tests on pressure charts, unless otherwise approved by the District Supervisor. The test interval for each BOP component tested shall be sufficient to demonstrate that the component is effectively holding pressure. The charts shall be certified as correct by the operator's representative at the facility.

(h) The time, date, and results of all pressure tests, actuations, and inspections of the BOP system, system components, and marine risers shall be recorded in the driller's report. The BOP tests shall be documented in accordance with the following:

(1) The documentation shall indicate the sequential order of BOP and auxiliary equipment testing and the pressure and duration of each test. As an alternate, the documentation in the driller's report may reference a BOP test plan that contains the required information and is retained on file at the facility.

(2) The control station used during the test shall be identified in the driller's report. For a subsea system the pod used during the test shall be identified in the driller's report.

(3) Any problems or irregularities observed during BOP and auxiliary equipment testing and any actions taken to remedy such problems or irregularities shall be noted in the driller's report.

(4) Documentation required to be entered in the driller's report may instead be referenced in the driller's report. All records including pressure charts, driller's report, and referenced documents pertaining to BOP tests, actuations, and inspections, shall be available for MMS review at the facility for the duration of the drilling activity. Following completion of the drilling activity, all such records shall be retained for a period of 2 years at the facility, at the lessee's field office nearest the OCS facility, or at another location conveniently available to the District Supervisor.

[53 FR 10690, Apr. 1, 1988, as amended at 56 FR 1914, Jan. 18, 1991]

§250.58   Well-control drills.

(a) Well-control drills shall be conducted for each drilling crew in accordance with the following requirements:

(1) Drills shall be designed to acquaint each crew member with each member's function at the particular test station so each member can perform their functions promptly and efficiently.

(2) A well-control drill plan, applicable to the particular site, shall be prepared for each crew member outlining the assignments each member is to fulfill during the drill and establishing

a prescribed time for the completion of each portion of the drill. A copy of the complete well-control drill plan shall be posted on the rig floor and/or bulletin board.

(3) The drill shall be carried out during periods of activity selected to minimize the risk of sticking the drill pipe or otherwise endangering the operation. In each of these drills, the reaction time of participants shall be measured up to the point when the designated person is prepared to activate the closing sequence of the BOP system. The total time for the crew to complete its entire drill assignment shall also be measured. This operation shall be recorded on the driller's report as "Well-Control Drill." All drills shall be initiated by the toolpusher through the raising of the float on the pit-level device, activating the mud-return indicator, or its equivalent. This operation shall be performed at least once each week (well conditions permitting) with each crew. The drills shall be timed so they will cover a range of different operations which include on-bottom drilling and tripping. A diverter drill shall be developed and conducted in a similar manner for shallow operations.

(4) *On-bottom drilling.* A drill conducted while on bottom shall include the following as practicable:

    (i) Detect kick and sound alarm;

    (ii) Position kelly and tool joints so connections are accessible from the floor, but tool joints are clear of sealig elements in BOP systems, stop pumps, check for flow, close in the well;

    (iii) Record time;

    (iv) Record drill-pipe pressure and casing pressure;

    (v) Measure pit gain and mark new level;

    (vi) Estimate volume of additional mud in pits;

    (vii) Weight sample of mud from suction pit;

    (viii) Check all valves on choke manifold and BOP system for correct position (open or closed);

    (ix) Check BOP system components and choke manifold for leaks;

    (x) Check flow line and choke exhaust lines for flow;

    (xi) Check accumulator pressure;

    (xii) Prepare to extinguish sources of ignition;

    (xiii) Alert standby boat or prepare safety capsule for launching;

    (xiv) Place crane operator on duty for possible personnel evacuation;

    (xv) Prepare to lower escape ladders and prepare other abandonment devices for possible use;

    (xvi) Determine materials needed to circulate out kick; and

    (xvii) Time drill and enter drill report on driller's report.

(5) *Tripping pipe.* A drill conducted during a trip shall include the following as practicable:

    (i) Detect kick and sound alarm;

    (ii) Install safety valve, close safety valve;

    (iii) Position pipe, prepare to close annular preventer;

    (iv) Install inside preventer, open safety valve;

    (v) Record time;

    (vi) Record casing pressure;

    (vii) Check all valves on choke manifold and BOP system for correct position (open or closed);

    (viii) Check for leaks on BOP system component and choke manifold;

    (ix) Check flow line and choke exhaust lines for flow;

    (x) Check accumulator pressure;

(xi) Prepare to extinguish sources of ignition;

(xii) Alert standby boat or prepare safety capsule for launching;

(xiii) Place crane operator on duty for possible personnel evacuation;

(xiv) Prepare to lower escape ladders and prepare other abandonment devices for possible use;

(xv) Prepare to strip back to bottom; and

(xvi) Time drill and enter drill report on driller's report.

(b) A well-control drill may be required by a Minerals Management Service (MMS) authorized representative after consulting with the lessee's senior representative present.

§250.59   Diverter systems.

(a) When drilling a conductor or surface hole, all drilling units shall be equipped with a diverter system consisting of a diverter sealing element, diverter lines, and control systems unless otherwise approved by the District Supervisor for floating drilling operations. The diverter system shall be designed, installed, and maintained so as to divert gases, water, mud, and other materials away from the facilities and personnel.

(b) No later than May 31, 1990, diverter systems shall be in compliance with the requirements of this section. The requirements applicable to diverters which were in effect April 1, 1988, shall remain in effect until May 31, 1990.

(c) The diverter system shall be equipped with remote-controlled valves in the flow and vent lines that can be operated from at least one remote-control station in addition to the one on the drilling floor. Any valve used in a diverter system shall be full-opening. No manual or butterfly valve shall be installed in any part of the diverter system. There shall be a minimum number of turns in the vent line(s) downstream of the spool outlet flange and the radius of curvature of turns shall be as large as practicable. All right-angle and sharp turns shall be targeted. Flexible hose may be used for diverter lines instead of rigid pipe if the flexible hose has integral end couplings. The entire diverter system shall be firmly anchored and supported to prevent whipping and vibration. All diverter control instruments and lines shall be protected from physical damage from thrown and falling objects.

(d) For drilling operations conducted with a surface wellhead configuration, the following shall apply:

(1) If the diverter system utilizes only one spool outlet, branch lines shall be installed to provide downwind diversion capability; and

(2) No spool outlet or diverter line internal diameter shall be less than 10 inches, except that dual spool outlets are acceptable provided that each outlet has a minimum internal diameter of 8 inches and that both outlets are piped to overboard lines and that each line downstream of the changeover nipple at the spool has a minimum internal diameter of 10 inches.

(e) For drilling operations conducted where a floating or semisubmersible type of drilling vessel is used and drilling fluids are circulated to the drilling vessel, the following shall apply:

(1) If the diverter system utilizes only one spool outlet, branch lines shall be installed to provide downwind diversion capability;

(2) No spool outlet or diverter line internal diameter shall be less than 12 inches; and

(3) Dynamically positioned drill ships may be equipped with a single vent line provided appropriate vessel heading is maintained to allow for downwind diversion.

(f) The diverter sealing element and diverter valves shall be pressure tested to a minimum of 200 psi when nippled up on conductor casing with a surface wellhead configuration. No more than 7 days shall elapse between subsequent similar pressure tests. For surface and subsea wellhead configurations, the diverter sealing element, diverter valves, and diverter-control systems, including the remote control system, shall be actuation-tested and the vent lines flow tested when first installed.

Subsequent actuation tests shall be conducted not less than once every 24-hour period thereafter alternating between control stations. All pressure test, flow test, and actuation results shall be recorded in the driller's report.

(g) Diverter systems and components for use in subfreezing conditions shall be suitable for use under these conditions.

§250.60 Mud program.

(a) *General requirements.* The quantities, characteristics, use, and testing of drilling mud and the related drilling procedures shall be designed and implemented to prevent the loss of well control.

(b) *Mud control.*

(1) Before starting out of the hole with drill pipe, the mud shall be properly conditioned by circulation with the drill pipe just off bottom to the extent that a volume of drilling mud equal to the annular volume is displaced. This procedure may be omitted if proper documentation in the driller's report shows the following:

(i) There is no indication of influx of formation fluids prior to starting to pull the drill pipe from the hole.

(ii) The weight of the returning mud is essentially the same as the weight of the mud entering the hole. In the event that the returning mud is lighter than the entering mud by a weight differential equal to or greater than 0.2 pounds per gallon (1.5 pounds per cubic foot), the mud shall be circulated until a volume of drilling mud equal to the annular volume is displaced, and the mud properties measured to assure that there has been no influx of gas or liquid.

(iii) Other mud properties recorded on the daily drilling log are within the limits established by the approved mud program.

(2) When mud in the hole is circulated, the driller's report shall be so noted.

(3) When coming out of the hole with drill pipe, the annulus shall be filled with mud before the change in mud level decreases the hydrostatic pressure by 75 psi, or every five stands of drill pipe, whichever gives a lower decrease in hydrostatic pressure. The number of stands of drill pipe and drill collars that may be pulled prior to filling the hole and the equivalent mud volume shall be calculated and posted near the driller's station. A mechanical, volumetric, or electronic device for measuring the amount of mud required to fill the hole shall be utilized.

(4) Drill pipe and downhole tool running and pulling speeds shall be at controlled rates so as not to induce an influx of formation fluids from the effects of swabbing nor cause a loss of drilling fluid and corresponding hydrostatic pressure decrease from the effects of surging.

(5) When there is an indication of swabbing or influx of formation fluids, the safety devices and measures necessary to control the well shall be employed. The mud shall be circulated and conditioned, on or near bottom, unless well or mud conditions prevent running the drill pipe back to the bottom.

(6) For each casing string, the maximum pressure to be contained under the BOP shall be posted near the driller's station.

(7) In areas where permafrost and/or hydrate zones may be present or are known to be present, drilling fluid temperatures shall be controlled or other measures taken to drill safely through those zones.

(8) An operable mud-gas separator and operable degasser shall be installed in the mud system prior to commencement of drilling operations and shall be maintained for use throughout the drilling of the well.

(9) The mud in the hole shall be circulated or reverse-circulated prior to pulling the drill-stem test tools from the hole. If circulating out test fluid is not feasible, test fluids may be bullheaded out of the drill-stem test string and tools with an appropriate kill fluid prior to pulling the test tools.

(c) *Mud-testing and monitoring equipment.*

(1) Mud-testing equipment shall be maintained on the drilling rig at all times, and mud tests shall be performed once each tour, or more frequently, as conditions warrant. Such tests shall be conducted in accordance with industry-accepted practices and shall include mud density, viscosity, and gel strength, hydrogen-ion concentration (pH), filtration, and other tests as may be deemed necessary by the District Supervisor in the interests of monitoring and maintaining mud quality for safe operations, prevention of downhole equipment problems, and for kick detection. The results of these tests shall be recorded in the driller's report.

(2) The following mud-system monitoring equipment shall be installed with derrick floor indicators and used when mud returns are established and throughout subsequent drilling operations:

(i) Recording mud-pit level indicator to determine mud-pit volume gains and losses. This indicator shall include both a visual and an audible warning device.

(ii) Mud-volume measuring device to accurately determine mud volumes required to fill the hole on trips.

(iii) Mud-return indicator devices which indicate the relationship between mud-return flow rate and pump discharge rate. This indicator shall include both a visual and an audible warning device.

(iv) Gas-detecting equipment to monitor the drilling mud returns with indicators located in the mud-logging compartment or on the rig floor. If the indicators are only in the mud-logging compartment, there shall be a means of immediate communication with the rig floor, and the gas-detecting equipment shall be continually manned. If the indicators are on the rig floor only, an audible alarm shall be installed.

(d) *Mud quantities.*

(1) Quantities of mud and mud materials at the drill site shall be utilized, maintained, and replenished as necessary to ensure well control. Those quantities shall be based on known or anticipated drilling conditions to be encountered, rig storage capacity, weather conditions, and estimated time for delivery.

(2) Daily inventories of mud and mud materials including weight materials and additives at the drill site shall be recorded and those records maintained at the well site.

(3) Drilling operations shall be suspended in the absence of sufficient quantities of mud and mud materials to maintain well control.

(e) *Safety precautions in mud-handling areas.* Mud-handling areas which are classified as per API RP 500B where dangerous concentrations of combustible gas may accumulate shall be equipped with ventilation systems and gas monitors as described below no later than May 31, 1989. Regulatory requirements in effect on April 1, 1988 are applicable until May 31, 1989.

(1) Be ventilated with high-capacity mechanical ventilation systems capable of replacing the air once every 5 minutes or 1.0 cubic feet of air-volume flow per minute per square foot of area, whichever is greater, unless such ventilation is provided by natural means. If not continuously activated, mechanical ventilation systems shall be activated on signal from gas detectors that are operational at all times indicating the presence of 1 percent or more of gas by volume.

(2) Be maintained at a negative pressure relative to an adjacent area if mechanical ventilation is installed to meet the requirements in paragraph (e)(1) of this section and discharges may be hazardous. The negative pressure areas shall be protected with at least one of the following:

(i) A pressure-sensitive alarm, (ii) open-door alarms on each access to the area, (iii) automatic door-closing devices, (iv) air locks, or (v) other devices as approved by the District Supervisor.

(3) Be fitted with gas detectors and alarms except in open areas where adequate ventilation is provided by natural means.

(4) Be equipped with either explosion-proof or pressurized electrical equipment to prevent the ignition of explosive gases. Where air is used for pressuring, the air intake shall be located outside of, and as far as practicable from, hazardous areas.

(5) Mechanical ventilation systems shall be fitted with alarms which are activated upon a failure of the system.

(6) Gas detection systems shall be tested for operation and recalibrated at a frequency such that no more than 90 days shall elapse between tests.

[53 FR 10690, Apr. 1, 1988, as amended at 55 FR 47752, Nov. 15, 1990]

§250.61　Securing of wells.

A downhole safety device such as a cement plug, bridge plug, or packer shall be timely installed when drilling operations are interrupted by events such as those which force evacuation of the drilling crew, prevent station keeping, or require repairs to major drilling or well-control equipment. In floating drilling operations, the use of blind-shear rams or pipe rams and an inside BOP may be approved by the District Supervisor in lieu of the above requirements if supported by evidence of special circumstances and/or the lack of sufficient time.

§250.62　Field drilling rules.

When geological and engineering information available in a field enables a District Supervisor to determine specific operating requirements appropriate to wells to be drilled in the field, field drilling rules may be established on the initiative of the District Supervisor, or in response to a request from a lessee. Such rules may modify the requirements of this subpart. After field drilling rules have been established, development wells to which such rules apply shall be drilled in accordance with such rules and other requirements of this subpart. Field drilling rules may be amended or canceled for cause at any time upon the initiative of the District Supervisor or upon the approval of a request by a lessee.

§250.63　Supervision, surveillance, and training.

(a) The lessee shall provide onsite supervision of drilling operations on a 24-hour per day basis.

(b) From the time drilling operations are initiated and until the well is completed or abandoned, a member of the drilling crew or the toolpusher shall maintain rig-floor surveillance continuously, unless the well is secured with BOP's, bridge plugs, packers, or cement plugs.

(c) Lessee and drilling contractor personnel shall be trained and qualified in accordance with the provisions of Subpart O of this part and MMS Training Standard MMS-OCS-T 1, Training and Qualifications of Personnel in Well-Control Equipment and Techniques for Drilling Offshore Locations (Second Edition). Records of specific training which lessee and drilling contractor personnel have successfully completed, the dates of completion, and the names and dates of the courses shall be maintained at the drill site.

§250.64　Applications for permit to drill.

(a) Prior to commencing the drilling of a well under an approved Exploration Plan, Development and Production Plan, or Development Operations Coordination Document, the lessee shall file a Form MMS-123, APD, with the District Supervisor for approval. Prior to commencing operations, written approval from the District Supervisor must be received by the lessee unless oral approval has been given pursuant to §250.6(a).

(b) The APD's for wells to be drilled from mobile drilling units shall include the following:

(1) An identification of the maximum environmental and operational conditions the rig is designed to withstand.

(2) Applicable current documentation of operational limitations imposed by the American Bureau of Shipping classification or other appropriate classification society and either a U.S. Coast Guard Certificate of Inspection or a U.S. Coast Guard Letter of Compliance.

(3) For frontier areas, the design and operating limitations beyond which suspension, curtailment, or modification of drilling or rig operations are required (e.g., vessel motion, offset, riser angle, anchor tensions, wind speed, wave height, currents, icing or iceloading, settling, tilt or lateral movement, resupply capability) and the contingency plans which identify actions to be taken prior to exceeding the design or operating limitations of the rig.

(4) A program which provides for safety in drilling operations where a floating or semisubmersible type of drilling vessel is used and formation competency at the structural and/or conductor casing setting depth(s) is (are) not adequate to permit circulation of drilling fluids to the vessel while drilling the conductor and/or surface hole. This program shall include all known pertinent information including seismic and geologic data, water depth, drilling-fluid hydrostatic pressure, a schematic diagram indicating the equipment to be installed from the rotary table to the proposed conductor and/or surface casing seat(s), and the contingency plan for moving off location.

(c) The APD's shall include rated capacities of the proposed drilling unit and of major drilling equipment.

(d) In those areas which are subject to subfreezing conditions, the lessee shall furnish evidence that the drilling equipment, BOP system and components, drilling safety systems, diverter systems, and other associated equipment and materials are suitable for drilling operations under subfreezing conditions.

(e) After a drilling unit has been approved for use in an MMS District, the information listed in paragraphs (b) (1), (2), and (3), (c), and (d) of this section need not be resubmitted unless required by the District Supervisor or there are changes in equipment that affect the rated capacity of the unit.

(f) An APD shall include the following in addition to a fully completed Form MMS-123:

(1) A plat, drawn to a scale of 2,000 feet to the inch, showing the surface and subsurface location of the well to be drilled and of all the wells previously drilled in the vicinity from which information is available. Locations shall be indicated in feet from the block line.

(2) The design criteria considered for the well and for well control, including the following:

(i) Pore pressures.

(ii) Formation fracture gradients.

(iii) Potential lost circulation zones.

(iv) Mud weights.

(v) Casing setting depths.

(vi) Anticipated surface pressures (which for purposes of this section are defined as the pressure which can reasonably be expected to be exerted upon a casing string and its related wellhead equipment). In the calculation of an anticipated surface pressure, the lessee shall take into account the drilling, completion, and producing conditions. The lessee shall consider mud densities to be used below various casing strings, fracture gradients of the exposed formations, casing setting depths, total well depth, formation fluid type, and other pertinent conditions. Considerations for calculating anticipated surface pressure may vary for each segment of the well. The lessee shall include as a part of the statement of anticipated surface pressures the calculations used to determine these pressures during the drilling phase and the completion phase, including the anticipated surface pressure used for production string design.

    (vii) If a shallow hazards site survey is conducted, the lessee shall submit with or prior to the submittal of the APD, two copies of a summary report describing the geological and manmade conditions present. The lessee shall also submit two copies of the site maps and data records identified in the survey strategy.

    (viii) Permafrost zones, if applicable.

 (3) A BOP equipment program including the following:

    (i) The pressure rating of BOP equipment.

    (ii) A well-control procedure for use of the annular preventer for those wells where the anticipated surface pressure exceeds the rated working pressure of the annular preventer.

    (iii) A description of subsea BOP accumulator system or other type of closing system proposed for use.

    (iv) A schematic drawing of the diverter system to be used (plan and elevation views) showing spool outlet internal diameter(s); diverter-line lengths and diameters, burst strengths, and radius of curvature at each turn; valve type, size, working pressure rating, and location; the control instrumentation logic; and the operating procedure to be used by lessee or contractor personnel.

    (v) A schematic drawing of the BOP stack showing the inside diameter of the BOP stack, and the number of annular, pipe ram, variable-bore pipe ram, blind ram, and blind-shear ram preventers.

 (4) A casing program including the following:

    (i) Casing size, weight, grade, type of connection, and setting depth;

    (ii) Casing design safety factors for tension, collapse, and burst with the assumptions made to arrive at these values; and

    (iii) In areas containing permafrost, casing programs that incorporate setting depths for conductor and surface casing based on the anticipated depth of the permafrost at the proposed well location and which utilize the current state-of-the-art methods to safely drill and set casing. The casing program shall provide protection from thaw subsidence and freezeback effect, proper anchorage, and well control.

 (5) The drilling prognosis including the following:

    (i) Projected plans for coring at specified depths;

    (ii) Projected plans for logging;

    (iii) Estimated depths to the top of significant marker formations; and

    (iv) Estimated depths at which encounters with significant porous and permeable zones containing fresh water, oil, gas, or abnormally pressured water are expected.

 (6) A cementing program including type and amount of cement in cubic feet to be used for each casing string.

 (7) A mud program including the minimum quantities of mud and mud materials, including weight materials, to be kept at the site.

 (8) A directional survey program for directionally drilled wells.

 (9) A plot of the estimated pore pressures and formation fracture gradients and the proposed mud weights and casing setting depths on the same sheet.

 (10) A H_2S Contingency Plan, if applicable, and not submitted previously.

 (11) Such other information as may be required by the District Supervisor.

 (g) Public information copies of the APD shall be submitted in accordance with §250.17 of this part.

[53 FR 10690, Apr. 1, 1988, as amended at 58 FR 49928, Sept. 24, 1993]

§250.65   **Sundry notices and reports on wells.**

(a) Notices of the lessee's intention to change plans, make changes in major drilling equipment, deepen or plug back a well, or engage in similar activities and subsequent reports pertaining to such operations shall be submitted to the District Supervisor on Form MMS-124, Sundry Notices and Reports on Wells. Prior to commencing operations, written approval must be received from the District Supervisor unless oral approval is obtained.

(b) The Form MMS-124 submitted shall contain a detailed statement of the proposed work that will materially change from the approved work described in the APD. Information submitted shall include the present status of the well, including the production string or last string of casing, the well depth, the present production zones and productive capability, and all other information specified on Form MMS-124. Within 30 days after completion of the work, a subsequent detailed report of all the work done and the results obtained shall be submitted.

(c) A Form MMS-124 with a plat, certified by a registered land surveyor, shall be filed as soon as the well's final surveyed surface location, water depth, and the rotary kelly bushing elevation have been determined.

(d) Public information copies of Sundry Notices and Reports on Wells shall be submitted in accordance with §250.17 of this part.

[53 FR 10690, Apr. 1, 1988, as amended at 58 FR 49928, Sept. 24, 1993]

§250.66   **Well records.**

(a) Complete and accurate records for each well and of all well operations shall be retained for a period of 2 years at the lessee's field office nearest the OCS facility or at another location conveniently available to the District Supervisor. The records shall contain a description of any significant malfunction or problem; all the formations penetrated; the content and character of oil, gas, and other mineral deposits and water in each formation; the kind, weight, size, grade, and setting depth of casing; all well logs and surveys run in the wellbore; and all other information required by the District Supervisor in the interests of resource evaluation, waste prevention, conservation of natural resources, protection of correlative rights, safety, and environment.

(b) When drilling operations are suspended, or temporarily prohibited under the provisions of §250.10 of this part, the lessee shall, within 30 days after termination of the suspension or temporary prohibition or within 30 days after the completion of any activities related to the suspension or prohibition, transmit to the District Supervisor duplicate copies of the records of all activities related to and conducted during the suspension or temporary prohibition on, or attached to, Form MMS-125, Well Summary Report, or Form MMS-124, as appropriate.

(c) Upon request by the Regional or District Supervisor, the lessee shall furnish the following:

(1) Copies of the records of any of the well operations specified in paragraph (a) of this section;

(2) Paleontological reports identifying microscopic fossils by depth and/or washed samples of drill cuttings normally maintained by the lessee for paleontological determinations;

(3) Copies of the daily driller's report at a frequency as determined by the District Supervisor. Items to be reported include spud dates, casing setting depths, cement quantities, casing characteristics, pressure integrity tests, mud weights, kicks, lost returns, and any unusual activities; and

(4) Legible, exact copies of service company reports on cementing, perforating, acidizing, analyses of cores, testing, or other similar services.

(d) As soon as available, the lessee shall transmit copies (field or final prints of individual runs) of logs or charts of electrical, radioactive, sonic, and other well-logging operations, directional-well surveys, and analyses of cores. Composite logs of multiple runs and directional-well surveys shall be transmitted to the District Supervisor in duplicate as soon as available but not later than 30 days after completion of each well.

(e) If the drilling unit moves from the wellbore prior to completing the well, the lessee shall submit to the District Supervisor copies of the well records with completed Form MMS-124, within 30 days after moving from the wellbore.

(f) If the Regional or District Supervisor determines that circumstances warrant, the lessee shall submit any other reports and records of operations, including paleontological interpretations based upon identification of microscopic fossils, in the manner and form prescribed by the Regional or District Supervisor.

(g) Records relating to the drilling of a well shall be retained for a period of 90 days after drilling operations are completed. Records relating to the completion of a well or of any workover activity which materially alters the completion configuration or materially affects or alters a hydrocarbon-bearing zone shall be kept until the well is permanently plugged and abandoned.

[53 FR 10690, Apr. 1, 1988, as amended at 58 FR 49928, Sept. 24, 1993]

§250.67　**Hydrogen sulfide.**

(a) *What precautions must I take when operating in an H_2S area?* You must:

    (1) Take all necessary and feasible precautions and measures to protect personnel from the toxic effects of H_2S and to mitigate damage to property and the environment caused by H_2S. You must follow the requirements of this section when conducting drilling, well-completion/well-workover, and production operations in zones with H_2S present and when conducting operations in zones where the presence of H_2S is unknown. You do not need to follow these requirements when operating in zones where the absence of H_2S has been confirmed; and

    (2) Follow your approved contingency plan.

(b) *Definitions.* Terms used in this section have the following meanings:

Facility means a vessel, a structure or an artificial island used for drilling well-completion, well-workover, and/or production operations.

H_2S absent means:

    (1) Drilling, logging, coring, testing or producing operations have confirmed the absence of H_2S in concentrations that could potentially result in atmospheric concentrations of 20 ppm or more of H_2S; or

    (2) Drilling in the surrounding areas and correlation of geological and seismic data with equivalent stratigraphic units have confirmed an absence of H_2S throughout the area to be drilled.

H_2S present means that drilling, logging, coring, testing, or producing operations have confirmed the presence of H_2S in concentrations and volumes that could potentially result in atmospheric concentrations of 20 ppm or more of H_2S.

H_2S unknown means the designation of a zone or geologic formation where neither the presence nor absence of H_2S has been confirmed.

Well-control fluid means drilling mud and completion or workover fluid as appropriate to the particular operation being conducted.

(c) *Classifying an area for the presence of H_2S.* You must:

    (1) Request and obtain an approved classification for the area from the Regional Supervisor before you begin operations. Classifications are "H_2S absent," H_2S present," or "H_2S unknown;"

    (2) Submit your request with your application for permit to drill;

    (3) Support your request with available information such as geologic and geophysical data and correlations, well logs, formation tests, cores and analysis of formation fluids; and

    (4) Submit a request for reclassification of a zone when additional data indicate a different classification is needed.

(d) *What do I do if conditions change?* If you encounter H$_2$S that could potentially result in atmospheric concentrations of 20 ppm or more in areas not previously classified as having H$_2$S present, you must immediately notify MMS and begin to follow requirements for areas with H$_2$S present.

(e) *What are the requirements for conducting simultaneous operations?* When conducting any combination of drilling, well-completion, well-workover, and production operations simultaneously, you must follow the requirements in the section applicable to each individual operation.

(f) *Requirements for submitting an H$_2$S Contingency Plan.* Before you begin operations, you must submit an H$_2$S Contingency Plan to the District Supervisor for approval. Do not begin operations before the District Supervisor approves your plan. You must keep a copy of the approved plan in the field, and you must follow the plan at all times. Your plan must include:

(1) Safety procedures and rules that you will follow concerning equipment, drills, and smoking;

(2) Training you provide for employees, contractors, and visitors;

(3) Job position and title of the person responsible for the overall safety of personnel;

(4) Other key positions, how these positions fit into your organization, and what the functions, duties, and responsibilities of those job positions are;

(5) Actions that you will take when the concentration of H$_2$S in the atmosphere reaches 20 ppm, who will be responsible for those actions, and a description of the audible and visual alarms to be activated;

(6) Briefing areas where personnel will assemble during an H$_2$S alert. You must have at least two briefing areas on each facility and use the briefing area that is upwind of the H$_2$S source at any given time;

(7) Criteria you will use to decide when to evacuate the facility and procedures you will use to safely evacuate all personnel from the facility by vessel, capsule, or lifeboat. If you use helicopters during H$_2$S alerts, describe the types of H$_2$S emergencies during which you consider the risk of helicopter activity to be acceptable and the precautions you will take during the nights;

(8) Procedures you will use to safely position all vessels attendant to the facility. Indicate where you will locate the vessels with respect to wind direction. Include the distance from the facility and what procedures you will use to safely relocate the vessels in an emergency;

(9) How you will provide protective-breathing equipment for all personnel, including contractors and visitors;

(10) The agencies and facilities you will notify in case of a release of H$_2$S (that constitutes an emergency), how you will notify them, and their telephone numbers. Include all facilities that might be exposed to atmospheric concentrations of 20 ppm or more of H$_2$S;

(11) The medical personnel and facilities you will use if needed, their addresses, and telephone numbers;

(12) H$_2$S detector locations in production facilities producing gas containing 20 ppm or more of H$_2$S. Include an "H$_2$S Detector Location Drawing" showing:

    (i) All vessels, flare outlets, wellheads, and other equipment handling production containing H$_2$S;

    (ii) Approximate maximum concentration of H$_2$S in the gas stream; and

    (iii) Location of all H$_2$S sensors included in your contingency plan;

(13) Operational conditions when you expect to flare gas containing H$_2$S including the estimated maximum gas flow rate, H$_2$S concentration, and duration of flaring;

(14) Your assessment of the risks to personnel during flaring and what precautionary measures you will take;

(15) Primary and alternate methods to ignite the flare and procedures for sustaining ignition and monitoring the status of the flare (i.e., ignited or extinguished);

(16) Procedures to shut off the gas to the flare in the event the flare is extinguished;

(17) Portable or fixed sulphur dioxide (SO_2)-detection system(s) you will use to determine SO_2 concentration and exposure hazard when H_2S is burned;

(18) Increased monitoring and warning procedures you will take when the SO_2 concentration in the atmosphere reaches 2 ppm;

(19) Personnel protection measures or evacuation procedures you will initiate when the SO_2 concentration in the atmosphere reaches 5 ppm;

(20) Engineering controls to protect personnel from SO_2; and

(21) Any special equipment, procedures, or precautions you will use if you conduct any combination of drilling, well-completion, well-workover, and production operations simultaneously.

(g) *Training program.*

    (1) *When and how often do employees need to be trained?* All operators and contract personnel must complete an H_2S training program to meet the requirements of this section:

        (i) Before beginning work at the facility; and

        (ii) Each year, within 1 year after completion of the previous class.

    (2) *What training documentation do I need?* For each individual working on the platform, either:

        (i) You must have documentation of this training at the facility where the individual is employed; or

        (ii) The employee must carry a training completion card.

    (3) *What training do I need to give to visitors and employees previously trained on another facility?*

        (i) Trained employees or contractors transferred from another facility must attend a supplemental briefing on your H_2S equipment and procedures before beginning duty at your facility;

        (ii) Visitors who will remain on your facility more than 24 hours must receive the training required for employees by paragraph (g)(4) of this section; and

        (iii) Visitors who will depart before spending 24 hours on the facility are exempt from the training required for employees, but they must, upon arrival, complete a briefing that includes:

            (A) Information on the location and use of an assigned respirator; practice in donning and adjusting the assigned respirator; information on the safe briefing areas, alarm system, and hazards of H_2S and SO_2; and

            (B) Instructions on their responsibilities in the event of an H_2S release.

    (4) *What training must I provide to all other employees?* You must train all individuals on your facility on the:

        (i) Hazards of H_2S and of SO_2 and the provisions for personnel safety contained in the H_2S Contingency Plan;

        (ii) Proper use of safety equipment which the employee may be required to use;

        (iii) Location of protective breathing equipment, H_2S detectors and alarms, ventilation equipment, briefing areas, warning systems, evacuation procedures, and the direction of prevailing winds;

        (iv) Restrictions and corrective measures concerning beards, spectacles, and contact lenses in conformance with ANSI Z88.2;

(v) Basic first-aid procedures applicable to victims of H_2S exposure. During all drills and training sessions, you must address procedures for rescue and first aid for H_2S victims;

(vi) Location of:

(A) The first-aid kit on the facility;

(B) Resuscitators; and

(C) Litter or other device on the facility.

(vii) Meaning of all warning signals.

(5) *Do I need to post safety information?* You must prominently post safety information on the facility and on vessels serving the facility (i.e., basic first-aid, escape routes, instructions for use of life boats, etc.).

(h) *Drills.*

(1) *When and how often do I need to conduct drills on H_2S safety discussions on the facility?* You must:

(i) Conduct a drill for each person at the facility during normal duty hours at least once every 7-day period. The drills must consist of a dry-run performance of personnel activities related to assigned jobs.

(ii) At a safety meeting or other meetings of all personnel, discuss drill performance, new H_2S considerations at the facility, and other updated H_2S information at least monthly.

(2) *What documentation do I need?* You must keep records of attendance for:

(i) Drilling, well-completion, and well-workover operations at the facility until operations are completed; and

(ii) Production operations at the facility or at the nearest field office for 1 year.

(i) *Visual and audible warning systems—*

(1) *How must I install wind direction equipment?* You must install wind direction equipment in a location visible at all times to individuals on or in the immediate vicinity of the facility.

(2) *When do I need to display operational danger signs, display flags or activate visual or audible alarms?*

(i) You must display warning signs at all times on facilities with wells capable of producing H_2S and on facilities that process gas containing H_2S in concentrations of 20 ppm or more.

(ii) In addition to the signs, you must activate audible alarms and display flags or activate flashing red lights when atmospheric concentration of H_2S reaches 20 ppm.

(3) *What are the requirements for signs?* Each sign must be a high-visibility yellow color with black lettering as follows:

Letter height	Wording
12 inches	Danger.
	Poisonous Gas.
	Hydrogen Sulfide.
7 inches	Do not approach if red flag is flying.
(Use appropiate wording at right)	Do not approach if red lights are flashing.

(4) *May I use existing signs?* You may use existing signs containing the words "Danger-Hydrogen Sulfide-H₂S," provided the words "Poisonous Gas. Do Not Approach if Red Flag is Flying" or "Red Lights are Flashing" in lettering of a minimum of 7 inches in height are displayed on a sign immediately adjacent to the existing sign.

(5) *What are the requirements for flashing lights or flags?* You must activate a sufficient number of lights or hoist a sufficient number of flags to be visible to vessels and aircraft. Each light must be of sufficient intensity to be seen by approaching vessels or aircraft any time it is activated (day or night). Each flag must be red, rectangular, a minimum width of 3 feet, and a minimum height of 2 feet.

(6) *What is an audible warning system?* An audible warning system is a public address system or siren, horn, or other similar warning device with a unique sound used only for H₂S.

(7) *Are there any other requirements for visual or audible warning devices?* Yes, you must:

    (i) Illuminate all signs and flags at night and under conditions of poor visibility; and

    (ii) Use warning devices that are suitable for the electrical classification of the area.

(8) *What actions must I take when the alarms are activated?* When the warning devices are activated, the designated responsible persons must inform personnel of the level of danger and issue instructions on the initiation of appropriate protective measures.

(j) *H₂S-detection and H₂S monitoring equipment.*

    (1) *What are the requirements for an H₂S detection system?* An H₂S detection system must:

        (i) Be capable of sensing a minimum of 10 ppm of H₂S in the atmosphere; and

        (ii) Activate audible and visual alarms when the concentration of H₂S in the atmosphere reaches 20 ppm.

    (2) *Where must I have sensors for drilling, well-completion, and well-workover operations?* You must locate sensors at the:

        (i) Bell nipple;

        (ii) Mud-return line receiver tank (possum belly);

        (iii) Pipe-trip tank;

        (iv) Shale shaker;

        (v) Well-control fluid pit area;

        (vi) Driller's station;

        (vii) Living quarters; and

        (viii) All other areas where H₂S may accumulate.

    (3) *Do I need mud sensors?* The District Supervisor may require mud sensors in the possum belly in cases where the ambient air sensors in the mud-return system do not consistently detect the presence of H₂S.

    (4) *How often must I observe the sensors?* During drilling, well-completion and well-workover operations, you must continuously observe the H₂S levels indicated by the monitors in the work areas during the following operations:

        (i) When you pull a wet string of drill pipe or workover string;

        (ii) When circulating bottoms-up after a drilling break;

        (iii) During cementing operations;

        (iv) During logging operations; and

        (v) When circulating to condition mud or other well-control fluid.

(5) *Where must I have sensors for production operations?* On a platform where gas containing H₂S of 20 ppm or greater is produced, processed, or otherwise handled:

    (i) You must have a sensor in rooms, buildings, deck areas, or low-laying deck areas not otherwise covered by paragraph (j)(2) of this section, where atmospheric concentrations of H₂S could reach 20 ppm or more. You must have at least one sensor per 400 square feet of deck area or fractional part of 400 square feet;

    (ii) You must have a sensor in buildings where personnel have their living quarters;

    (iii) You must have a sensor within 10 feet of each vessel, compressor, wellhead, manifold, or pump, which could release enough H₂S to result in atmospheric concentrations of 20 ppm at a distance of 10 feet from the component;

    (iv) You may use one sensor to detect H₂S around multiple pieces of equipment, provided the sensor is located no more than 10 feet from each piece, except that you need to use at least two sensors to monitor compressors exceeding 50 horsepower;

    (v) You do not need to have sensors near wells that are shut in at the master valve and sealed closed;

    (vi) When you determine where to place sensors, you must consider:

        (A) The location of system fittings, flanges, valves, and other devices subject to leaks to the atmosphere; and

        (B) Design factors, such as the type of decking and the location of fire walls; and

    (vii) The District Supervisor may require additional sensors or other monitoring capabilities, if warranted by site specific conditions.

(6) *How must I functionally test the H₂S Detectors?*

    (i) Personnel trained to calibrate the particular H₂S detector equipment being used must test detectors by exposing them to a known concentration in the range of 10 to 30 ppm of H₂S.

    (ii) If the results of any functional test are not within 2 ppm or 10 percent, whichever is greater, of the applied concentration, recalibrate the instrument.

(7) *How often must I test my detectors?*

    (i) When conducting drilling, drill stem testing, well-completion, or well-workover operations in areas classified as H₂S present or H₂S unknown, test all detectors at least once every 24 hours. When drilling, begin functional testing before the bit is 1,500 feet (vertically) above the potential H₂S zone.

    (ii) When conducting production operations, test all detectors at least every 14 days between tests.

    (iii) If equipment requires calibration as a result of two consecutive functional tests, the District Supervisor may require that H₂S-detection and H₂S-monitoring equipment be functionally tested and calibrated more frequently.

(8) *What documentation must I keep?*

    (i) You must maintain records of testing and calibrations (in the drilling or production operations report, as applicable) at the facility to show the present status and history of each device, including dates and details concerning:

        (A) Installation;

        (B) Removal;

        (C) Inspection;

        (D) Repairs;

        (E) Adjustments; and

        (F) Reinstallation.

(ii) Records must be available for inspection by MMS personnel.

(9) *What are the requirements for nearby vessels?* If vessels are stationed overnight alongside facilities in areas of H_2S present or H_2S unknown, you must equip vessels with an H_2S-detection system that activates audible and visual alarms when the concentration of H_2S in the atmosphere reaches 20 ppm. This requirement does not apply to vessels positioned upwind and at a safe distance from the facility in accordance with the positioning procedure described in the approved H_2S Contingency Plan.

(10) *What are the requirements for nearby facilities?* The District Supervisor may require you to equip nearby facilities with portable or fixed H_2S detector(s) and to test and calibrate those detectors. To invoke this requirement the District Supervisor will consider dispersion modeling results from a possible release to determine if 20 ppm H_2S concentration levels could be exceeded at nearby facilities.

(11) *What must I do to protect against SO_2 if I burn gas containing H_2S?* You must:

(i) Monitor the SO_2 concentration in the air with portable or strategically placed fixed devices capable of detecting a minimum of 2 ppm of SO_2;

(ii) Take readings at least hourly and at any time personnel detect SO_2 odor or nasal irritation;

(iii) Implement the personnel protective measures specified in the H_2S Contingency Plan if the SO_2 concentration in the work area reaches 2 ppm; and

(iv) Calibrate devices every 3 months if you use fixed or portable electronic sensing devices to detect SO_2.

(12) *May I use alternative measures?* You may follow alternative measures instead of those in paragraph (j)(11) of this section if you propose and the Regional Supervisor approves the alternative measures.

(13) *What are the requirements for protective-breathing equipment?* In an area classified as H_2S present or H_2S unknown, you must:

(i) Provide all personnel, including contractors and visitors on a facility, with immediate access to self-contained pressure-demand-type respirators with hoseline capability and breathing time of at least 15 minutes.

(ii) Design, select, use, and maintain respirators to conform to ANSI Z88.2 American National Standard for Respiratory Protection.

(iii) Make available at least two voice-transmission devices, which can be used while wearing a respirator, for use by designated personnel.

(iv) Make spectacle kits available as needed.

(v) Store protective-breathing equipment in a location that is quickly and easily accessible to all personnel.

(vi) Label all breathing-air bottles as containing breathing-quality air for human use.

(vii) Ensure that vessels attendant to facilities carry appropriate protective-breathing equipment for each crew member. The District Supervisor may require additional protective-breathing equipment on certain vessels attendant to the facility.

(viii) During H_2S alerts, limit helicopter flights to and from facilities to the conditions specified in the H_2S Contingency Plan. During authorized flights, the flight crew and passengers must use pressure-demand-type respirators. You must train all members of night crews in the use of the particular type(s) of respirator equipment made available.

(ix) As appropriate to the particular operation(s), (production, drilling, well-completion or well-workover operations, or any combination of them), provide a system of breathing-air manifolds, hoses, and masks at the facility and the briefing areas. You must provide a cascade air-bottle system for the breathing-air manifolds to refill individual protective-breathing apparatus bottles. The cascade air-bottle system may be recharged by a high-pressure compressor suitable for providing breathing-quality air, provided the compressor suction is located in an uncontaminated atmosphere.

(k) *Personnel safety equipment.*

(1) *What additional personnel-safety equipment do I need?* You must ensure that your facility has:

(i) Portable H_2S detectors capable of detecting a 10 ppm concentration of H_2S in the air available for use by all personnel;

(ii) Retrieval ropes with safety harnesses to retrieve incapacitated personnel from contaminated areas;

(iii) Chalkboards and/or note pads for communication purposes located on the rig floor, shale-shaker area, the cement-pump rooms, well-bay areas, production processing equipment area, gas compressor area, and pipeline-pump area;

(iv) Bull horns and flashing lights; and

(v) At least three resuscitators on manned facilities, and a number equal to the personnel on board, not to exceed three, on normally unmanned facilities, complete with face masks, oxygen bottles, and spare oxygen bottles.

(2) *What are the requirements for ventilation equipment?* You must:

(i) Use only explosion-proof ventilation devices;

(ii) Install ventilation devices in areas where H_2S or SO_2 may accumulate; and

(iii) Provide movable ventilation devices in work areas. The movable ventilation devices must be multidirectional and capable of dispersing H_2S or SO_2 vapors away from working personnel.

(3) *What other personnel safety equipment do I need?* You must have the following equipment readily available on each facility:

(i) A first-aid kit of appropriate size and content for the number of personnel on the facility; and

(ii) At least one litter or an equivalent device.

(l) *Do I need to notify MMS in the event of an H_2S release?* You must notify MMS without delay in the event of a gas release which results in a 15 minute time weighted average atmospheric concentration of H_2S of 20 ppm or more anywhere on the facility.

(m) *Do I need to use special drilling, completion and workover fluids or procedures?* When working in an area classified as H_2S present or H_2S unknown:

(1) You may use either water- or oil-base muds in accordance with §250.40(b)(1).

(2) If you use water-base well-control fluids, and if ambient air sensors detect H_2S, you must immediately conduct either the Garrett-Gas-Train test or a comparable test for soluble sulfides to confirm the presence of H_2S.

(3) If the concentration detected by air sensors in over 20 ppm, personnel conducting the tests must don protective-breathing equipment conforming to paragraph (j)(13) of this section.

(4) You must maintain on the facility sufficient quantities of additives for the control of H_2S, well-control fluid pH, and corrosion equipment.

(i) *Scavengers.* You must have scavengers for control of H_2S available on the facility. When H_2S is detected, you must add scavengers as needed. You must suspend drilling until the scavenger is circulated throughout the system.

(ii) *Control pH.* You must add additives for the control of pH to water-base well-control fluids in sufficient quantities to maintain pH of at least 10.0.

(iii) *Corrosion inhibitors.* You must add additives to the well-control fluid system as needed for the control of corrosion.

(5) You must degas well-control fluids containing H_2S at the optimum location for the particular facility. You must collect the gases removed and burn them in a closed flare system conforming to paragraph (q)(6) of this section.

(n) *What must I do in the event of a kick?* In the event of a kick, you must use one of the following alternatives to dispose of the well-influx fluids giving consideration to personnel safety, possible environmental damage, and possible facility well-equipment damage:

(1) Contain the well-fluid influx by shutting in the well and pumping the fluids back into the formation.

(2) Control the kick by using appropriate well-control techniques to prevent formation fracturing in an open hole within the pressure limits of the well equipment (drill pipe, work string, casing, wellhead, BOP system, and related equipment). The disposal of H_2S and other gases must be through pressurized or atmospheric mud-separator equipment depending on volume, pressure and concentration of H_2S. The equipment must be designed to recover well-control fluids and burn the gases separated from the well-control fluid. The well-control fluid must be treated to neutralize H_2S and restore and maintain the proper quality.

(o) *Well testing in a zone known to contain H_2S.* When testing a well in a zone with H_2S present, you must do all of the following:

(1) Before starting a well test, conduct safety meetings for all personnel who will be on the facility during the test. At the meetings, emphasize the use of protective-breathing equipment, first-aid procedures, and the Contingency Plan. Only competent personnel who are trained and are knowledgeable of the hazardous effects of H_2S must be engaged in these tests.

(2) Perform well testing with the minimum number of personnel in the immediate vicinity of the rig floor and with the appropriate test equipment to safely and adequately perform the test. During the test, you must continuously monitor H_2S levels.

(3) Not burn produced gases except through a flare which meets the requirements of paragraph (q)(6) of this section. Before flaring gas containing H_2S, you must activate SO_2 monitoring equipment in accordance with paragraph (j)(11) of this section. If you detect SO_2 in excess of 2 ppm, you must implement the personnel protective measures in your H_2S Contingency Plan, required by paragraph (f)(13)(iv) of this section. You must also follow the requirements of §250.175. You must pipe gases from stored test fluids into the flare outlet and burn them.

(4) Use downhole test tools and wellhead equipment suitable for H_2S service.

(5) Use tubulars suitable for H_2S service. You must not use drill pipe for well testing without the prior approval of the District Supervisor. Water cushions must be thoroughly inhibited in order to prevent H_2S attack on metals. You must flush the test string fluid treated for this purpose after completion of the test.

(6) Use surface test units and related equipment that is designed for H_2S service.

(p) *Metallurgical properties of equipment.* When operating in a zone with H_2S present, you must use equipment that is constructed of materials with metallurgical properties that resist or prevent sulfide stress cracking (also known as hydrogen embrittlement, stress corrosion cracking, or H_2S embrittlement), chloride-stress cracking, hydrogen-induced cracking, and other failure modes. You must do all of the following:

(1) Use tubulars and other equipment, casing, tubing, drill pipe, couplings, flanges, and related equipment that is designed for H_2S service.

(2) Use BOP system components, wellhead, pressure-control equipment, and related equipment exposed to H_2S bearing fluids that conform to NACE Standard MR.01-75-96.

(3) Use temporary downhole well-security devices such as retrievable packers and bridge plugs that are designed for H_2S service.

(4) When producing in zones bearing H_2S, use equipment constructed of materials capable of resisting or preventing sulfide stress cracking.

(5) Keep the use of welding to a minimum during the installation or modification of a production facility. Welding must be done in a manner that ensures resistance to sulfide stress cracking.

(q) *General requirements when operating in an H_2S zone—*

(1) *Coring operations.* When you conduct coring operations in H_2S-bearing zones, all personnel in the working area must wear protective-breathing equipment at least 10 stands in advance of retrieving the core barrel. Cores to be transported must be sealed and marked for the presence of H_2S.

(2) *Logging operations.* You must treat and condition well-control fluid in use for logging operations to minimize the effects of H_2S on the logging equipment.

(3) *Stripping operations.* Personnel must monitor displaced well-control fluid returns and wear protective-breathing equipment in the working area when the atmospheric concentration of H_2S reaches 20 ppm or if the well is under pressure.

(4) *Gas-cut well-control fluid or well kick from H_2S-bearing zone.* If you decide to circulate out a kick, personnel in the working area during bottoms-up and extended-kill operations must wear protective-breathing equipment.

(5) *Drill- and workover-string design and precautions.* Drill- and workover-strings must be designed consistent with the anticipated depth, conditions of the hole, and reservoir environment to be encountered. You must minimize exposure of the drill- or workover-string to high stresses as much as practical and consistent with well conditions. Proper handling techniques must be taken to minimize notching and stress concentrations. Precautions must be taken to minimize stresses caused by doglegs, improper stiffness ratios, improper torque, whip, abrasive wear on tool joints, and joint imbalance.

(6) *Flare system.* The flare outlet must be of a diameter that allows easy nonrestricted flow of gas. You must locate flare line outlets on the downside of the facility and as far from the facility as is feasible, taking into account the prevailing wind directions, the wake effects caused by the facility and adjacent structure(s), and the height of all such facilities and structures. You must equip the flare outlet with an automatic ignition system including a pilot-light gas source or an equivalent system. You must have alternate methods for igniting the flare. You must pipe to the flare system used for H_2S all vents from production process equipment, tanks, relief valves, burst plates, and similar devices.

(7) *Corrosion mitigation.* You must use effective means of monitoring and controlling corrosion caused by acid gases (H_2S and CO_2) in both the downhole and surface portions of a production system. You must take specific corrosion monitoring and mitigating measures in areas of unusually severe corrosion where accumulation of water and/or higher concentration of H_2S exists.

(8) *Wireline lubricators.* Lubricators which may be exposed to fluids containing H_2S must be of H_2S-resistant materials.

(9) *Fuel and/or instrument gas.* You must not use gas containing H_2S for instrument gas. You must not use gas containing H_2S for fuel gas without the prior approval of the District Supervisor.

(10) *Sensing lines and devices.* Metals used for sensing line and safety-control devices which are necessarily exposed to H$_2$S-bearing fluids must be constructed of H$_2$S-corrosion resistant materials or coated so as to resist H$_2$S corrosion.

(11) *Elastomer seals.* You must use H$_2$S-resistant materials for all seals which may be exposed to fluids containing H$_2$S.

(12) *Water disposal.* If you dispose of produced water by means other than subsurface injection, you must submit to the District Supervisor an analysis of the anticipated H$_2$S content of the water at the final treatment vessel and at the discharge point. The District Supervisor may require that the water be treated for removal of H$_2$S. The District Supervisor may require the submittal of an updated analysis if the water disposal rate or the potential H$_2$S content increases.

(13) *Deck drains.* You must equip open deck drains with traps or similar devices to prevent the escape of H$_2$S gas into the atmosphere.

(14) *Sealed voids.* You must take precautions to eliminate sealed spaces in piping designs (e.g., slip-on flanges, reinforcing pads) which can be invaded by atomic hydrogen when H$_2$S is present.

[62 FR 3795, Jan. 27, 1997]

SUBPART E—OIL AND GAS WELL-COMPLETION OPERATIONS

§250.70   General requirements.

Well-completion operations shall be conducted in a manner to protect against harm or damage to life (including fish and other aquatic life), property, natural resources of the OCS including any mineral deposits (in areas leased and not leased), the national security or defense, or the marine, coastal, or human environment.

§250.71   Definition.

When used in this subpart, the following term shall have the meaning given below:

Well-completion operations means the work conducted to establish the production of a well after the production casing string has been set, cemented, and pressure-tested.

§250.72   Equipment movement.

The movement of well-completion rigs and related equipment on and off a platform or from well to well on the same platform, including rigging up and rigging down, shall be conducted in a safe manner. All wells in the same well-bay which are capable of producing hydrocarbons shall be shut in below the surface with a pump-through-type tubing plug and at the surface with a closed master valve prior to moving well-completion rigs and related equipment, unless otherwise approved by the District Supervisor. A closed surface-controlled subsurface safety valve of the pump-through type may be used in lieu of the pump-through-type tubing plug, provided that the surface control has been locked out of operation. The well from which the rig or related equipment is to be moved shall also be equipped with a back-pressure valve prior to removing the blowout preventer (BOP) system and installing the tree.

[53 FR 10690, Apr. 1, 1988, as amended at 55 FR 47752, Nov. 15, 1990]

§250.73   Emergency shutdown system.

When well-completion operations are conducted on a platform where there are other hydrocarbon-producing wells or other hydrocarbon flow, an emergency shutdown system (ESD) manually controlled station shall be installed near the driller's console or well-servicing unit operator's work station.

§250.74   Hydrogen sulfide.

When a well-completion operation is conducted in zones known to contain hydrogen sulfide (H_2S) or in zones where the presence of H_2S is unknown (as defined in §250.67 of this part), the lessee shall take appropriate precautions to protect life and property on the platform or completion unit, including, but not limited to operations such as blowing the well down, dismantling wellhead equipment and flow lines, circulating the well, swabbing, and pulling tubing, pumps, and packers. The lessee shall comply with the requirements in §250.67 of this part as well as the appropriate requirements of this subpart.

§250.75   Subsea completions.

No subsea well completion shall be commenced until the lessee obtains written approval from the District Supervisor in accordance with §250.83 of this part. That approval shall be based upon a case-by-case determination that the proposed equipment and procedures will adequately control the well and permit safe production operations.

§250.76   Crew instructions.

Prior to engaging in well-completion operations, crew members shall be instructed in the safety requirements of the operations to be performed, possible hazards to be encountered, and general safety considerations to protect personnel, equipment, and the environment. Date and time of safety meetings shall be recorded and available at the facility for review by MMS representatives.

§250.77   Welding and burning practices and procedures.

All welding, burning, and hot tapping activities involved in well-completion operations shall be conducted in accordance with the requirements in §250.52 of this part.

§250.78   Electrical requirements.

All electrical equipment and systems involved in well-completion operations shall be designed, installed, equipped, protected, operated, and maintained in accordance with the requirements in §250.53 of this part.

§250.79   Well-completion structures on fixed platforms.

Derricks, masts, substructures, and related equipment shall be selected, designed, installed, used, and maintained so as to be adequate for the potential loads and conditions of loading that may be encountered during the proposed operations. Prior to moving a well-completion rig or equipment onto a platform, the lessee shall determine the structural capability of the platform to safely support the equipment and proposed operations, taking into consideration the corrosion protection, age of platform, and previous stresses to the platform.

[53 FR 10690, Apr. 1, 1988, as amended at 54 FR 50616, Dec. 8, 1989]

§250.80   Diesel engine air intakes.

No later than May 31, 1989, diesel engine air intakes shall be equipped with a device to shut down the diesel engine in the event of runaway. Diesel engines which are continuously attended shall be equipped with either remote operated manual or automatic-shutdown devices. Diesel engines which are not continuously attended shall be equipped with automatic-shutdown devices.

§250.81 Traveling-block safety device.

After May 31, 1989, all units being used for well-completion operations which have both a traveling block and a crown block shall be equipped with a safety device which is designed to prevent the traveling block from striking the crown block. The device shall be checked for proper operation weekly and after each drill-line slipping operation. The results of the operational check shall be entered in the operations log.

§250.82 Field well-completion rules.

When geological and engineering information available in a field enables the District Supervisor to determine specific operating requirements, field well-completion rules may be established on the District Supervisor's initiative or in response to a request from a lessee. Such rules may modify the specific requirements of this subpart. After field well-completion rules have been established, well-completion operations in the field shall be conducted in accordance with such rules and other requirements of this subpart. Field well-completion rules may be amended or canceled for cause at any time upon the initiative of the District Supervisor or upon the request of a lessee.

§250.83 Approval and reporting of well-completion operations.

(a) No well-completion operation shall begin until the lessee receives written approval from the District Supervisor. If completion is planned and the data are available at the time the Application for Permit to Drill, Form MMS-123 (see §250.64 of this part), is submitted, approval for a well completion may be requested on that form. If the completion has not been approved or if the completion objective or plans have significantly changed, approval for such operations shall be requested on Form MMS-124, Sundry Notices and Reports on Wells.

(b) The following information shall be submitted with Form MMS-124 (or with Form MMS-123):

(1) A brief description of the well completion procedures to be followed, a statement of the expected surface pressure, and type and weight of completion fluids;

(2) A schematic drawing of the well showing the proposed producing zone(s) and the subsurface well-completion equipment to be used;

(3) For multiple completions, a partial electric log showing the zones proposed for completion, if logs have not been previously submitted; and

(4) When the well-completion is a zone known to contain H_2S or a zone where the presence of H_2S is unknown, information pursuant to §250.67 of this part.

(c) Within 30 days after completion. Form MMS-125, Well Summary Report, including a schematic of the tubing and subsurface equipment, shall be submitted to the District Supervisor.

(d) Public information copies of Form MMS-125 shall be submitted in accordance with §250.17.

[53 FR 10690, Apr. 1, 1988, as amended at 58 FR 49928, Sept. 24, 1993]

§250.84 Well-control fluids, equipment, and operations.

(a) Well-control fluids, equipment, and operations shall be designed, utilized, maintained, and/or tested as necessary to control the well in foreseeable conditions and circumstances, including subfreezing conditions. The well shall be continuously monitored during well-completion operations and shall not be left unattended at any time unless the well is shut in and secured.

(b) The following well-control-fluid equipment shall be installed, maintained, and utilized:

(1) A fill-up line above the uppermost BOP;

(2) A well-control, fluid-volume measuring device for determining fluid volumes when filling the hole on trips; and

(3) A recording mud-pit-level indicator to determine mud-pit-volume gains and losses. This indicator shall include both a visual and an audible warning device.

(c) When coming out of the hole with drill pipe, the annulus shall be filled with well-control fluid before the change in such fluid level decreases the hydrostatic pressure 75 pounds per square inch (psi) or every five stands of drill pipe, whichever gives a lower decrease in hydrostatic pressure. The number of stands of drill pipe and drill collars that may be pulled prior to filling the hole and the equivalent well-control fluid volume shall be calculated and posted near the operator's station. A mechanical, volumetric, or electronic device for measuring the amount of well-control fluid required to fill the hole shall be utilized.

§250.85   **Blowout prevention equipment.**

(a) The BOP system and system components and related well-control equipment shall be designed, used, maintained, and tested in a manner necessary to assure well control in foreseeable conditions and circumstances, including subfreezing conditions. The working pressure rating of the BOP system and BOP system components shall exceed the expected surface pressure to which they may be subjected. If the expected surface pressure exceeds the rated working pressure of the annular preventer, the lessee shall submit with Form MMS-124 or Form MMS-123, as appropriate, a well-control procedure that indicates how the annular preventer will be utilized, and the pressure limitations that will be applied during each mode of pressure control.

(b) The minimum BOP system for well-completion operations shall include the following:

(1) Three preventers, when the expected surface pressure is less than 5,000 psi, consisting of an annular preventer, one preventer equipped with pipe rams, and one preventer equipped with blind or blind-shear rams.

(2) Four preventers, when the expected surface pressure is 5,000 psi or greater, or for multiple tubing strings consisting of an annular preventer, two preventers equipped with pipe rams, and one preventer equipped with blind or blind-shear rams. When dual tubing strings are being handled simultaneously, dual pipe rams shall be installed on one of the pipe-ram preventers.

(3) When tapered drill string is used, the minimum BOP system shall include either of the following:

(i) Four preventers, when the expected surface pressure is less than 5,000 psi, consisting of an annular preventer, two sets of pipe rams, one capable of sealing around the larger size drill string and one capable of sealing around the smaller size drill string (one set of variable bore rams may be substituted for the two sets of pipe rams), and one preventer equipped with blind or blind shear rams; or

(ii) Five preventers, when the expected surface pressure is 5,000 psi or greater, consisting of an annular preventer, two sets of pipe rams, capable of sealing around the larger size drill string, one set of pipe rams capable of sealing around the smaller size drill string (one set of variable bore rams may be substituted for one set of pipe rams capable of sealing around the larger size drill string and the set of pipe rams capable of sealing around the smaller size drill string), and a preventer equipped with blind or blind-shear rams.

(c) The BOP systems for well completions shall be equipped with the following:

(1) A hydraulic-actuating system that provides sufficient accumulator capacity to supply 1.5 times the volume necessary to close all BOP equipment units with a minimum pressure of 200 psi above the precharge pressure without assistance from a charging system. No later than December 1, 1988, accumulator regulators supplied by rig air and without a secondary source of pneumatic supply, shall be equipped with manual overrides, or alternately, other devices provided to ensure capability of hydraulic operations if rig air is lost.

(2) A secondary power source, independent from the primary power source, with sufficient capacity to close all BOP system components and hold them closed.

(3) Locking devices for the pipe-ram preventers.

(4) At least one remote BOP-control station and one BOP-control station on the rig floor.

(5) A choke line and a kill line each equipped with two full opening valves and a choke manifold. At least one of the valves on the choke line shall be remotely controlled. At least one of the valves on the kill line shall be remotely controlled, except that a check valve on the kill line in lieu of the remotely controlled valve may be installed provided that two readily accessible manual valves are in place and the check valve is placed between the manual valves and the pump. This equipment shall have a pressure rating at least equivalent to the ram preventers.

(d) An inside BOP or a spring-loaded, back-pressure safety valve and an essentially full-opening, work-string safety valve in the open position shall be maintained on the rig floor at all times during well-completion operations. A wrench to fit the work-string safety valve shall be readily available. Proper connections shall be readily available for inserting valves in the work string.

[53 FR 10690, Apr. 1, 1988, as amended at 54 FR 50616, Dec. 8, 1989; 58 FR 49928, Sept. 24, 1993]

§250.86   Blowout preventer system testing, records, and drills.

(a) Prior to conducting high-pressure tests, all BOP system components shall be successfully tested to a low pressure of 200 psi to 300 psi. Ram-type BOP's, related control equipment, including the choke and kill manifolds, and safety valves shall be successfully tested to the rated working pressure of the BOP equipment or as otherwise approved by the District Supervisor. Variable bore rams shall be pressure-tested against all sizes of drill pipe in the well excluding drill collars. Surface BOP systems shall be pressure-tested with water. The annular BOP shall be successfully tested at 70 percent of its rated working pressure or as otherwise approved by the District Supervisor. Each valve in the choke and kill manifolds shall be successfully, sequentially pressure tested to the ram-type BOP test pressure.

(b) The BOP systems shall be tested at the following times:

(1) When installed.

(2) At least every 7 days, alternating between control stations and at staggered intervals to allow each crew to operate the equipment. If either control system is not functional, further operations shall be suspended until the nonfunctional system is operable. To test every 7 days is not required for blind or blind-shear rams. The blind or blind-shear rams shall be tested at least once every 30 days during operations. A longer period between blowout preventer tests is allowed when there is a stuck pipe or pressure-control operation and remedial efforts are being performed. The tests shall be conducted as soon as possible and before normal operations resume. The reason for postponing testing shall be entered into the operations log.

(3) Following repairs that require disconnecting a pressure seal in the assembly, the affected seal will be pressure tested.

(c) All personnel engaged in well completion operations shall participate in a weekly BOP drill to familiarize crew members with appropriate safety measures.

(d) The lessee shall record pressure conditions during BOP tests on pressure charts, unless otherwise approved by the District Supervisor. The test interval for each BOP component tested shall be sufficient to demonstrate that the component is effectively holding pressure. The charts shall be certified as correct by the operator's representative at the facility.

(e) The time, date, and results of all pressure tests, actuations, inspections, and crew drills of the BOP system, system components, and marine risers shall be recorded in the operations log. The BOP tests shall be documented in accordance with the following:

(1) The documentation shall indicate the sequential order of BOP and auxiliary equipment testing and the pressure and duration of each test. As an alternate, the documentation in the operations log may reference a BOP test plan that contains the required information and is retained on file at the facility.

(2) The control station used during the test shall be identified in the operations log. For a subsea system, the pod used during the test shall be identified in the operations log.

(3) Any problems or irregularities observed during BOP and auxiliary equipment testing and any actions taken to remedy such problems or irregularities shall be noted in the operations log.

(4) Documentation required to be entered in the operations log may instead be referenced in the operations, log. All records including pressure charts, operations log, and referenced documents pertaining to BOP tests, actuations, and inspections shall be available for MMS review at the facility for the duration of the well-completion activity. Following completion of the well-completion activity, all such records shall be retained for a period of 2 years at the facility, at the lessee's field office nearest the OCS facility, or at another location conveniently available to the District Supervisor.

[53 FR 10690, Apr. 1, 1988, as amended at 54 FR 50616, Dec. 8, 1989; 55 FR 47752, Nov. 15, 1990; 56 FR 1915, Jan. 18, 1991]

§250.87 Tubing and wellhead equipment.

(a) No tubing string shall be placed in service or continue to be used unless such tubing string has the necessary strength and pressure integrity and is otherwise suitable for its intended use.

(b) In the event of prolonged operations such as milling, fishing, jarring, or washing over that could damage the casing, the casing shall be pressure-tested, calipered, or otherwise evaluated every 30 days and the results submitted to the District Supervisor.

(c) When the tree is installed, the wellhead shall be equipped so that all annuli can be monitored for sustained pressure. If sustained casing pressure is observed on a well, the lessee shall immediately notify the District Supervisor.

(d) Wellhead, tree, and related equipment shall have a pressure rating greater than the shut-in tubing pressure and shall be designed, installed, used, maintained, and tested so as to achieve and maintain pressure control. New wells completed as flowing or gas-lift wells shall be equipped with a minimum of one master valve and one surface safety valve, installed above the master valve, in the vertical run of the tree.

(e) Subsurface safety equipment shall be installed, maintained, and tested in compliance with §250.121 of this part.

[53 FR 10690, Apr. 1, 1988, as amended at 55 FR 47753, Nov. 15, 1990]

SUBPART F—OIL AND GAS WELL-WORKOVER OPERATIONS

§250.90 General requirements.

Well-workover operations shall be conducted in a manner to protect against harm or damage to life (including fish and other aquatic life), property, natural resources of the Outer Continental Shelf (OCS) including any mineral deposits (in areas leased and not leased), the national security or defense, or the marine, coastal, or human environment.

§250.91   Definitions.

When used in this subpart, the following terms shall have the meanings given below:

Routine operations mean any of the following operations conducted on a well with the tree installed:

      (a)   Cutting paraffin;

      (b)   Removing and setting pump-through-type tubing plugs, gas-lift valves, and subsurface safety valves which can be removed by wireline operations;

      (c)   Bailing sand;

      (d)   Pressure surveys;

      (e)   Swabbing;

      (f)   Scale or corrosion treatment;

      (g)   Caliper and gauge surveys;

      (h)   Corrosion inhibitor treatment;

      (i)   Removing or replacing subsurface pumps;

      (j)   Through-tubing logging (diagnostics);

      (k)   Wireline fishing; and

      (l)   Setting and retrieving other subsurface flow-control devices.

Workover operations mean the work conducted on wells after the initial completion for the purpose of maintaining or restoring the productivity of a well.

§250.92   Equipment movement.

The movement of well-workover rigs and related equipment on and off a platform or from well to well on the same platform, including rigging up and rigging down, shall be conducted in a safe manner. All wells in the same well-bay which are capable of producing hydrocarbons shall be shut in below the surface with a pump-through-type tubing plug and at the surface with a closed master valve prior to moving well-workover rigs and related equipment unless otherwise approved by the District Supervisor. A closed surface-controlled subsurface safety valve of the pump-through-type may be used in lieu of the pump-through-type tubing plug provided that the surface control has been locked out of operation. The well to which a well-workover rig or related equipment is to be moved shall also be equipped with a back-pressure valve prior to removing the tree and installing and testing the blowout-preventer (BOP) system. The well from which a well-workover rig or related equipment is to be moved shall also be equipped with a back pressure valve prior to removing the BOP system and installing the tree. Coiled tubing units, snubbing units, or wireline units may be moved onto a platform without shutting in wells.

§250.93   Emergency shutdown system.

When well-workover operations are conducted on a well with the tree removed, an emergency shutdown system (ESD) manually controlled station shall be installed near the driller's console or well-servicing unit operator's work station, except when there is no other hydrocarbon-producing well or other hydrocarbon flow on the platform.

§250.94   Hydrogen sulfide.

When a well-workover operation is conducted in zones known to contain hydrogen sulfide (H_2S) or in zones where the presence of H_2S is unknown (as defined in §250.67 of this part), the lessee shall take appropriate precautions to protect life and property on the platform or rig, including but not limited to operations such as blowing the well down, dismantling wellhead equipment and flow lines, circulating the well, swabbing, and pulling tubing, pumps and packers. The lessee shall comply with the requirements in §250.67 of this part as well as the appropriate requirements of this subpart.

§250.95   Subsea workovers.

No subsea well-workover operation including routine operations shall be commenced until the lessee obtains written approval from the District Supervisor in accordance with §250.103 of this part. That approval shall be based upon a case-by-case determination that the proposed equipment and procedures will maintain adequate control of the well and permit continued safe production operations.

§250.96   Crew instructions.

Prior to engaging in well-workover operations, crew members shall be instructed in the safety requirements of the operations to be performed, possible hazards to be encountered, and general safety considerations to protect personnel, equipment, and the environment. Date and time of safety meetings shall be recorded and available at the facility for review by a Minerals Management Service representative.

§250.97   Welding and burning practices and procedures.

All welding, burning, and hot-tapping activities involved in well-workover operations shall be conducted in accordance with the requirements of §250.52 of this part.

§250.98   Electrical requirements.

All electrical equipment and systems involved in well-workover operations shall be designed, installed, equipped, protected, operated, and maintained in accordance with the requirements in §250.53 of this part.

§250.99   Well-workover structures on fixed platforms.

Derricks, masts, substructures, and related equipment shall be selected, designed, installed, used, and maintained so as to be adequate for the potential loads and conditions of loading that may be encountered during the operations proposed. Prior to moving a well-workover rig or well-servicing equipment onto a platform, the lessee shall determine the structural capability of the platform to safely support the equipment and proposed operations, taking into consideration the corrosion protection, age of the platform, and previous stresses to the platform.

§250.100   Diesel engine air intakes.

No later than May 31, 1989, diesel engine air intakes shall be equipped with a device to shut down the diesel engine in the event of runaway. Diesel engines which are continuously attended shall be equipped with either remote operated manual or automatic shutdown devices. Diesel engines which are not continuously attended shall be equipped with automatic shutdown devices.

[53 FR 10690, Apr. 1, 1988, as amended at 54 FR 50616, Dec. 8, 1989]

§250.101   Traveling-block safety device.

After May 31, 1989, all units being used for well-workover operations which have both a traveling block and a crown block shall be equipped with a safety device which is designed to prevent the traveling block from striking the crown block. The device shall be checked for proper operation weekly and after each drill-line slipping operation. The results of the operational check shall be entered in the operations log.

§250.102   Field well-workover rules.

When geological and engineering information available in a field enables the District Supervisor to determine specific operating requirements, field well-workover rules may be established on the District Supervisor's initiative or in response to a request from a lessee. Such rules may modify the specific requirements of this subpart. After field well-workover rules have been established, well-workover operations in the field shall be conducted in accordance with such rules and other requirements of this subpart. Field well-workover rules may be amended or canceled for cause at any time upon the initiative of the District Supervisor or upon the request of a lessee.

§250.103   Approval and reporting for well-workover operations.

(a) No well-workover operation except routine ones, as defined in §250.91 of this part, shall begin until the lessee receives written approval from the District Supervisor. Approval for such operations shall be requested on Form MMS-124, Sundry Notices and Reports on Wells.

(b) The following information shall be submitted with Form MMS-124:

    (1) A brief description of the well-workover procedures to be followed, a statement of the expected surface pressure, and type and weight of workover fluids;

    (2) When changes in existing subsurface equipment are proposed, a schematic drawing of the well showing the zone proposed for workover and the workover equipment to be used; and

    (3) Where the well-workover is in a zone known to contain H_2S or a zone where the presence of H_2S is unknown, information pursuant to §250.67 of this part.

(c) The following additional information shall be submitted with Form MMS-124 if completing to a new zone is proposed:

    (1) Reason for abandonment of present producing zone including supportive well test data, and

    (2) A statement of anticipated or known pressure data for the new zone.

(d) Within 30 days after completing the well-workover operation, except routine operations, Form MMS-124, Sundry Notices and Reports on Wells, shall be submitted to the District Supervisor, showing the work as performed. In the case of a well-workover operation resulting in the initial recompletion of a well into a new zone, a Form MMS-125, Well Summary Report, shall be submitted to the District Supervisor and shall include a new schematic of the tubing subsurface equipment if any subsurface equipment has been changed.

[53 FR 10690, Apr. 1, 1988, as amended at 58 FR 49928, Sept. 24, 1993]

§250.104   Well-control fluids, equipment, and operations.

The following requirements apply during all well-workover operations with the tree removed:

(a) Well-control fluids, equipment, and operations shall be designed, utilized, maintained, and/or tested as necessary to control the well in foreseeable conditions and circumstances, including subfreezing conditions. The well shall be continuously monitored during well-workover operations and shall not be left unattended at anytime unless the well is shut in and secured.

(b) When coming out of the hole with drill pipe or a workover string, the annulus shall be filled with well-control fluid before the change in such fluid level decreases the hydrostatic pressure 75 pounds per square inch (psi) or every five stands of drill pipe or workover string, whichever gives a lower decrease in hydrostatic pressure. The number of stands of drill pipe or workover string and drill collars that may be pulled prior to filling the hole and the equivalent well-control fluid volume shall be calculated and posted near the operator's station. A mechanical, volumetric, or electronic device for measuring the amount of well-control fluid required to fill the hold shall be utilized.

(c) The following well-control-fluid equipment shall be installed, maintained, and utilized:

(1) A fill-up line above the uppermost BOP;

(2) A well-control, fluid-volume measuring device for determining fluid volumes when filling the hole on trips; and

(3) A recording mud-pit-level indicator to determine mud-pit-volume gains and losses. This indicator shall include both a visual and an audible warning device.

§250.105   **Blowout prevention equipment.**

(a) The BOP system, system components and related well-control equipment shall be designed, used, maintained, and tested in a manner necessary to assure well control in foreseeable conditions and circumstances, including subfreezing conditions. The working pressure rating of the BOP system and system components shall exceed the expected surface pressure to which they may be subjected. If the expected surface pressure exceeds the rated working pressure of the annular preventer, the lessee shall submit with Form MMS-124, requesting approval of the well-workover operation, a well-control procedure that indicates how the annular preventer will be utilized, and the pressure limitations that will be applied during each mode of pressure control.

(b) The minimum BOP system for well-workover operations with the tree removed shall include one of the following:

(1) Three preventers, when the expected surface pressure is less than 5,000 psi, consisting of an annular preventer, one preventer equipped with pipe rams, and one preventer equipped with blind or blind-shear rams.

(2) Four preventers, when the expected surface pressure is 5,000 psi or greater, or for multiple tubing strings consisting of an annular preventer, two preventers equipped with pipe rams, and one preventer equipped with blind or blind-shear rams. When dual tubing strings are being handled simultaneously, dual pipe rams shall be installed on one of the pipe-ram preventers.

(3) When a tapered drill string is used, the minimum BOP system shall include either of the following:

(i) Four preventers, when the expected surface pressure is less than 5,000 psi, consisting of an annular preventer, two sets of pipe rams, one capable of sealing around the larger size drill string, and one capable of sealing around the smaller size drill string (one set of variable bore rams may be substituted for the two sets of pipe rams), and one preventer equipped with blind or blind-shear rams; or

(ii) Five preventers, when the expected surface pressure is 5,000 psi or greater, consisting of an annular preventer, two sets of pipe rams capable of sealing around the larger size drill string, one set of pipe rams capable of sealing around the smaller size drill string (one set of variable bore rams may be substituted for one set of pipe rams capable of sealing around the larger size drill string and the set of pipe rams capable of sealing around the smaller size drill string), and a preventer equipped with blind or blind-shear rams.

(c) The BOP systems for well-workover operations with the tree removed shall be equipped with the following:

(1) A hydraulic-actuating system that provides sufficient accumulator capacity to supply 1.5 times the volume necessary to close all BOP equipment units with a minimum pressure of 200 psi above the precharge pressure without assistance from a charging system. No later than December 1, 1988, accumulator regulators supplied by rig air and without a secondary source of pneumatic supply, shall be equipped with manual overrides, or alternately, other devices provided to ensure capability of hydraulic operations if rig air is lost;

(2) A secondary power source, independent from the primary power source, with sufficient capacity to close all BOP system components and hold them closed;

(3) Locking devices for the pipe-ram preventers;

(4) At least one remote BOP-control station and one BOP-control station on the rig floor; and

(5) A choke line and a kill line each equipped with two full opening valves and a choke manifold. At least one of the valves on the choke-line shall be remotely controlled. At least one of the valves on the kill line shall be remotely controlled, except that a check valve on the kill line in lieu of the remotely controlled valve may be installed provided two readily accessible manual valves are in place and the check valve is placed between the manual valves and the pump. This equipment shall have a pressure rating at least equivalent to the ram preventers.

(d) The minimum BOP-system components for well-workover operations with the tree in place and performed through the wellhead inside of conventional tubing using small-diameter jointed pipe (usually ¾-inch to 1¼-inch) as a work string, i.e., small-tubing operations, shall include the following:

(1) Two sets of pipe rams, and

(2) One set of blind rams.

(e) The minimum BOP-system components for well-workover operations with the tree in place and performed by manipulating spooled, nonjointed pipe through the wellhead, i.e., coiled-tubing operations, shall include the following:

(1) One set of pipe rams hydraulically operated,

(2) One two-way slip assembly hydraulically operated,

(3) One pipe-cutter assembly hydraulically operated,

(4) One set of blind rams hydraulically operated,

(5) One pipe-stripper assembly, and

(6) One spool with side outlets.

(f) The minimum BOP-system components for well-workover operations with the tree in place and performed by moving tubing or drill pipe in or out of a well under pressure utilizing equipment specifically designed for that purpose, i.e., snubbing operations, shall include the following:

(1) One set of pipe rams hydraulically operated, and

(2) Two sets of stripper-type pipe rams hydraulically operated with spacer spool.

(g) An inside BOP or a spring-loaded, back-pressure safety valve and an essentially full-opening, work-string safety valve in the open position shall be maintained on the rig floor at all times during well-workover operations when the tree is removed or during well-workover operations with the tree installed and using small tubing as the work string. A wrench to fit the work-string safety valve shall be readily available. Proper connections shall be readily available for inserting valves in the work string. The full-opening safety valve is not required for coiled tubing or snubbing operations.

[53 FR 10690, Apr. 1, 1988, as amended at 54 FR 50616, Dec. 8, 1989; 58 FR 49928, Sept. 24, 1993]

§250.106   Blowout preventer system testing, records, and drills.

(a) Prior to conducting high-pressure tests, all BOP system components shall be successfully tested to a low pressure of 200 to 300 psi. Ram-type BOP's, related control equipment, including the choke and kill manifolds, and safety valves shall be successfully tested to the rated working pressure of the BOP equipment or as otherwise approved by the District Supervisor. Variable bore rams shall be pressure-tested against all sizes of drill pipe in the well excluding drill collars. Surface BOP systems shall be pressure tested with water. The annular-type BOP shall be successfully tested at 70 percent of its rated working pressure or as otherwise approved by the District Supervisor. Each valve in the choke and kill manifolds shall be successfully, sequentially pressure tested to the ram-type BOP test pressure.

(b) The BOP systems shall be tested at the following times:

(1) When installed;

(2) At least every 7 days, alternating between control stations and at staggered intervals to allow each crew to operate the equipment. If either control system is not functional, further operations shall be suspended until the nonfunctional, system is operable. The test every 7 days is not required for blind or blind-shear rams. The blind or blind-shear rams shall be tested at least once every 30 days during operation. A longer period between blowout preventer tests is allowed when there is a stuck pipe or pressure-control operation and remedial efforts are being performed. The tests shall be conducted as soon as possible and before normal operations resume. The reason for postponing testing shall be entered into the operations log.

(3) Following repairs that require disconnecting a pressure seal in the assembly, the affected seal will be pressure tested.

(c) All personnel engaged in well-workover operations shall participate in a weekly BOP drill to familiarize crew members with appropriate safety measures.

(d) The lessee shall record pressure conditions during BOP tests on pressure charts, unless otherwise approved by the District Supervisor. The test interval for each BOP component tested shall be sufficient to demonstrate that the component is effectively holding pressure. The charts shall be certified as correct by the operator's representative at the facility.

(e) The time, date, and results of all pressure tests, actuations, inspections, and crew drills of the BOP system, system components, and marine risers shall be recorded in the operations log. The BOP tests shall be documented in accordance with the following:

(1) The documentation shall indicate the sequential order of BOP and auxiliary equipment testing and the pressure and duration of each test. As an alternate, the documentation in the operations log may reference a BOP test plan that contains the required information and is retained on file at the facility.

(2) The control station used during the test shall be identified in the operations log. For a subsea system, the pod used during the test shall be identified in the operations log.

(3) Any problems or irregularities observed during BOP and auxiliary equipment testing and any actions taken to remedy such problems or irregularities shall be noted in the operations log.

(4) Documentation required to be entered in the operation log may instead be referenced in the operations log. All records including pressure charts, operations log, and referenced documents pertaining to BOP tests, actuations, and inspections, shall be available for MMS review at the facility for the duration of well-workover activity. Following completion of the well-workover activity, all such records shall be retained for a period of 2 years at the facility, at the lessee's field office nearest the OCS facility, or at another location conveniently available to the District Supervisor.

[53 FR 10690, Apr. 1, 1988, as amended at 54 FR 50617, Dec. 8, 1989; 56 FR 1915, Jan. 18, 1991]

§250.107 Tubing and wellhead equipment.

The lessee shall comply with the following requirements during well-workover operations with the tree removed:

(a) No tubing string shall be placed in service or continue to be used unless such tubing string has the necessary strength and pressure integrity and is otherwise suitable for its intended use.

(b) In the event of prolonged operations such as milling, fishing, jarring, or washing over that could damage the casing, the casing shall be pressure tested, calipered, or otherwise evaluated every 30 days and the results submitted to the District Supervisor.

(c) When reinstalling the tree, the wellhead shall be equipped so that all annuli can be monitored for sustained pressure. If sustained casing pressure is observed on a well, the lessee shall immediately notify the District Supervisor.

(d) Wellhead, tree, and related equipment shall have a pressure rating greater than the shut-in tubing pressure and shall be designed, installed, used, maintained, and tested so as to achieve and maintain pressure control. The tree shall be equipped with a minimum of one master valve and one surface safety valve in the vertical run of the tree when it is reinstalled.

(e) Subsurface safety equipment shall be installed, maintained, and tested in compliance with §250.121 of this part.

[53 FR 10690, Apr. 1, 1988, as amended at 54 FR 50617, Dec. 8, 1989; 55 FR 47753, Nov. 15, 1990]

§250.108   Wireline operations.

The lessee shall comply with the following requirements during routine, as defined in §250.91 of this part, and nonroutine wireline workover operations:

(a) Wireline operations shall be conducted so as to minimize leakage of well fluids. Any leakage that does occur shall be contained to prevent pollution.

(b) All wireline perforating operations and all other wireline operations where communication exists between the completed hydrocarbon-bearing zone(s) and the wellbore shall use a lubricator assembly containing at least one wireline valve.

(c) When the lubricator is initially installed on the well, it shall be successfully pressure tested to the expected shut-in surface pressure.

SUBPART G—ABANDONMENT OF WELLS

§250.110   General requirements.

(a) The lessee shall abandon all wells in a manner to assure downhole isolation of hydrocarbon zones, protection of freshwater aquifers, clearance of sites so as to avoid conflict with other uses of the Outer Continental Shelf (OCS), and prevention of migration of formation fluids within the wellbore or to the seafloor. Any well which is no longer used or useful for lease operations shall be plugged and abandoned in accordance with the provisions of this subpart. However, no production well shall be abandoned until its lack of capacity for further profitable production of oil, gas, or sulphur has been demonstrated to the satisfaction of the District Supervisor. No well shall be plugged if the plugging operations would jeopardize safe and economic operations of nearby wells, unless the well poses a hazard to safety or the environment.

(b) Lessees must plug and abandon all well bores, remove all platforms or other facilities, and clear the ocean of all obstructions to other users. This obligation:

(1) Accrues to the lessee when the well is drilled, the platform or other facility is installed, or the obstruction is created; and

(2) Is the joint and several responsibility of all lessees and owners of operating rights under the lease at the time the obligation accrues, and of each future lessee or owner of operating rights, until the obligation is satisfied under the requirements of this part.

[53 FR 10690, Apr. 1, 1988, as amended at 62 FR 27955, May 22, 1997]

EFFECTIVE DATE NOTE: At 62 FR 27955, May 22, 1997, §250.110 was amended by designating the existing text as paragraph (a) and adding a new paragraph (b), effective Aug. 20, 1997.

§250.111   Approvals.

The lessee shall not commence abandonment operations without prior approval of the District Supervisor. The lessee shall submit a request on Form MMS-124, Sundry Notices and Reports on Wells, for approval to abandon a well and a subsequent report of abandonment within 30 days from completion of the work in accordance with the following:

(a) *Notice of Intent to Abandon Well.* A request for approval to abandon a well shall contain the reason for abandonment including supportive well logs and test data, a description and schematic of proposed work including depths, type, location, length of plugs, the plans for mudding, cementing, shooting, testing, casing removal, and other pertinent information.

(b) *Subsequent report of abandonment.* The subsequent report of abandonment shall include a description of the manner in which the abandonment or plugging work was accomplished, including the nature and quantities of materials used in the plugging, and all information listed in paragraph (a) of this section with a revised schematic. If an attempt was made to cut and pull any casing string, the subsequent report shall include a description of the methods used, size of casing removed, depth of the casing removal point, and the amount of the casing removed from the well.

[53 FR 10690, Apr. 1, 1988, as amended at 58 FR 49928, Sept. 24, 1993]

§250.112   Permanent abandonment.

(a) *Isolation of zones in open hole.* In uncased portions of wells, cement plugs shall be set to extend from a minimum of 100 feet below the bottom to 100 feet above the top of any oil, gas, or freshwater zones to isolate fluids in the strata in which they are found and to prevent them from escaping into other strata or to the seafloor. The placement of additional cement plugs to prevent the migration of formation fluids in the wellbore may be required by the District Supervisor.

(b) *Isolation of open hole.* Where there is an open hole below the casing, a cement plug shall be placed in the deepest casing by the displacement method and shall extend a minimum of 100 feet above and 100 feet below the casing shoe. In lieu of setting a cement plug across the casing shoe, the following methods are acceptable:

(1) A cement retainer and a cement plug shall be set. The cement retainer shall have effective back-pressure control and shall be set not less than 50 feet and not more than 100 feet above the casing shoe. The cement plug shall extend at least 100 feet below the casing shoe and at least 50 feet above the retainer.

(2) If lost circulation conditions have been experienced or are anticipated, a permanent-type bridge plug may be placed within the first 150 feet above the casing shoe with a minimum of 50 feet of cement on top of the bridge plug. This bridge plug shall be tested in accordance with paragraph (g) of this section.

(c) *Plugging or isolating perforated intervals.* A cement plug shall be set by the displacement method opposite all perforations which have not been squeezed with cement. The cement plug shall extend a minimum of 100 feet above the perforated interval and either 100 feet below the perforated interval or down to a casing plug, whichever is the lesser. In lieu of setting a cement plug by the displacement method, the following methods are acceptable, provided the perforations are isolated from the hole below:

(1) A cement retainer and a cement plug shall be set. The cement retainer shall have effective back-pressure control and shall be set not less than 50 feet and not more than 100 feet above the top of the perforated interval. The cement plug shall extend at least 100 feet below the bottom of the perforated interval with 50 feet placed above the retainer.

(2) A permanent-type bridge plug shall be set within the first 150 feet above the top of the perforated interval with at least 50 feet of cement on top of the bridge plug.

(3)   A cement plug which is at least 200 feet long shall be set by the displacement method with the bottom of the plug within the first 100 feet above the top of the perforated interval.

(d)   *Plugging of casing stubs.* If casing is cut and recovered leaving a stub, the stub shall be plugged in accordance with one of the following methods:

(1)   A stub terminating inside a casing string shall be plugged with a cement plug extending at least 100 feet above and 100 feet below the stub. In lieu of setting a cement plug across the stub, the following methods are acceptable:

(i)   A cement retainer or a permanent-type bridge plug shall be set not less than 50 feet above the stub and capped with at least 50 feet of cement, or

(ii)   A cement plug which is at least 200 feet long shall be set with the bottom of the plug within 100 feet above the stub.

(2)   If the stub is below the next larger string, plugging shall be accomplished as required to isolate zones or to isolate an open hole as described in paragraphs (a) and (b) of this section.

(e)   *Plugging of annular space.* Any annular space communicating with any open hole and extending to the mud line shall be plugged with at least 200 feet of cement.

(f)   *Surface plug.* A cement plug which is at least 150 feet in length shall be set with the top of the plug within the first 150 feet below the mud line. The plug shall be placed in the smallest string of casing which extends to the mud line.

(g)   *Testing of plugs.* The setting and location of the first plug below the surface plug shall be verified by one of the following methods:

(1)   The lessee shall place a minimum pipe weight of 15,000 pounds on the cement plug, cement retainer, or bridge plug. The cement placed above the bridge plug or retainer is not required to be tested.

(2)   The lessee shall test the plug with a minimum pump pressure of 1,000 pounds per square inch with a result of no more than a 10-percent pressure drop during a 15-minute period.

(h)   *Fluid left in hole.* Each of the respective intervals of the hole between the various plugs shall be filled with fluid of sufficient density to exert a hydrostatic pressure exceeding the greatest formation pressure in the intervals between the plugs at time of abandonment.

(i)   *Clearance of location.* All wellheads, casings, pilings, and other obstructions shall be removed to a depth of at least 15 feet below the mud line or to a depth approved by the District Supervisor. The lessee shall verify that the location has been cleared of all obstructions in accordance with §250.114 of this part. The requirement for removing subsea wellheads or other obstructions and for verifying location clearance may be reduced or eliminated when, in the opinion of the District Supervisor, the wellheads or other obstructions would not constitute a hazard to other users of the seafloor or other legitimate uses of the area.

(j)   *Requirements for permafrost areas.* The following requirements shall be implemented for permafrost areas:

(1)   Fluid left in the hole adjacent to permafrost zones shall have a freezing point below the temperature of the permafrost and shall be treated to inhibit corrosion.

(2)   The cement used for cement plugs placed across permafrost zones shall be designed to set before freezing and to have a low heat of hydration.

§250.113   Temporary abandonment.

(a)   Any drilling well which is to be temporarily abandoned shall meet the requirements for permanent abandonment (except for the provisions in §§250.112 (f) and (i), and 250.114 and the following:

(1) A bridge plug or a cement plug at least 100 feet in length shall be set at the base of the deepest casing string unless the casing string has been cemented and has not been drilled out. If a cement plug is set, it is not necessary for the cement plug to extend below the casing shoe into the open hole.

(2) A retrievable or a permanent-type bridge plug or a cement plug at least 100 feet in length, shall be set in the casing within the first 200 feet below the mud line.

(b) Subsea wellheads, casing stubs, or other obstructions above the seafloor remaining after temporary abandonment will be protected in such a manner as to allow commercial fisheries gear to pass over the structure without damage to the structure or fishing gear. Depending on water depth, nature and height of obstruction above the seafloor, and the types and periods of fishing activity in the area, the District Supervisor may waive this requirement.

(c) In order to maintain the temporarily abandoned status of a well, the lessee shall provide, within 1 year of the original temporary abandonment and at successive 1-year intervals thereafter, an annual report describing plans for reentry to complete or permanently abandon the well.

(d) Identification and reporting of subsea wellheads, casing stubs, or other obstructions extending above the mud line will be accomplished in accordance with the requirements of the U.S. Coast Guard.

§250.114   Site clearance verification.

(a) The lessees shall verify site clearance after abandonment by one or more of the following methods as approved by the District Supervisor:

(1) Drag a trawl in two directions across the location,

(2) Perform a diver search around the wellbore,

(3) Scan across the location with a side-scan or on-bottom scanning sonar, or

(4) Use other methods based on particular site conditions.

(b) Certification that the area was cleared of all obstructions, the date the work was performed, the extent of the area searched around the location, and the search method utilized shall be submitted on Form MMS-124.

[53 FR 10690, Apr. l, 1988, as amended at 58 FR 49928, Sept. 24,1993]

SUBPART H—OIL AND GAS PRODUCTION SAFETY SYSTEMS

§250.120   General requirements.

Production safety equipment shall be designed, installed, used, maintained, and tested in a manner to assure the safety and protection of the human, marine, and coastal environments. Production safety systems operated in subfreezing climates shall utilize equipment and procedures selected with consideration of floating ice, icing, and other extreme environmental conditions that may occur in the area. Production shall not commence until the production safety system has been approved and a preproduction inspection has been requested by the lessee.

§250.121   Subsurface safety devices.

(a) *General.* All tubing installations open to hydrocarbon-bearing zones shall be equipped with subsurface safety devices that will shut off the flow from the well in the event of an emergency unless, after application and justification, the well is determined by the District Supervisor to be incapable of natural flowing. These devices may consist of a surface-controlled subsurface safety valve (SSSV), a subsurface-controlled SSSV, an injection valve, a tubing plug, or a tubing/annular subsurface safety device, and any associated safety valve lock or landing nipple.

(b) *Specifications for SSSV's.* Surface-controlled and subsurface-controlled SSSV's and safety valve locks and landing nipples installed in the OCS shall conform to the requirements in §250.126 of this part.

(c) *Surface-controlled SSSV's.* All tubing installations open to a hydrocarbon-bearing zone which is capable of natural flow shall be equipped with a surface-controlled SSSV, except as specified in paragraphs (d), (f), and (g) of this section. The surface controls may be located on the site or a remote location. Wells not previously equipped with a surface-controlled SSSV and wells in which a surface-controlled SSSV has been replaced with a subsurface-controlled SSSV in accordance with paragraph (d)(2) of this section shall be equipped with a surface-controlled SSSV when the tubing is first removed and reinstalled.

(d) *Subsurface-controlled SSSV's.* Wells may be equipped with subsurface-controlled SSSV's in lieu of a surface-controlled SSSV provided the lessee demonstrates to the District Supervisor's satisfaction that one of the following criteria are met:

   (1) Wells not previously equipped with surface-controlled SSSV's shall be so equipped when the tubing is first removed and reinstalled,

   (2) The subsurface-controlled SSSV is installed in wells completed from a single-well or multiwell satellite caisson or seafloor completions, or

   (3) The subsurface-controlled SSSV is installed in wells with a surface-controlled SSSV that has become inoperable and cannot be repaired without removal and reinstallation of the tubing.

(e) *Design, installation, and operation of SSSV's.* The SSSV's shall be designed, installed, operated, and maintained to ensure reliable operation.

   (1) The device shall be installed at a depth of 100 feet or more below the seafloor within 2 days after production is established. When warranted by conditions such as permafrost, unstable bottom conditions, hydrate formation, or paraffins, an alternate setting depth of the subsurface safety device may be approved by the District Supervisor.

   (2) Until a subsurface safety device is installed, the well shall be attended in the immediate vicinity so that emergency actions may be taken while the well is open to flow. During testing and inspection procedures, the well shall not be left unattended while open to production unless a properly operating subsurface-safety device has been installed in the well.

   (3) The well shall not be open to flow while the subsurface safety device is removed, except when flowing of the well is necessary for a particular operation such as cutting paraffin, bailing sand, or similar operations.

   (4) All SSSV's shall be inspected, installed, maintained, and tested in accordance with American Petroleum Institute Recommended Practice 14B, Recommended Practice for Design, Installation, and Operation of Subsurface Safety Valve Systems.

(f) *Subsurface safety devices in shut-in wells.* New completions (perforated but not placed on production) and completions shut in for a period of 6 months shall be equipped with either (1) a pump-through-type tubing plug; (2) a surface-controlled SSSV, provided the surface control has been rendered inoperative; or (3) an injection valve capable of preventing backflow. The setting depth of the subsurface safety device shall be approved by the District Supervisor on a case-by-case basis, when warranted by conditions such as permafrost, unstable bottom conditions, hydrate formations, and paraffins.

(g) *Subsurface safety devices in injection wells.* A surface-controlled SSSV or an injection valve capable of preventing backflow shall be installed in all injection wells. This requirement is not applicable if the District Supervisor concurs that the well is incapable of flowing. The lessee shall verify the no-flow condition of the well annually.

(h) *Temporary removal for routine operations.*

    (1) Each wireline- or pumpdown-retrievable subsurface safety device may be removed, without further authorization or notice, for a routine operation which does not require the approval of a Form MMS-124, Sundry Notices and Reports on Wells, in §250.91 of this part for a period not to exceed 15 days.

    (2) The well shall be identified by a sign on the wellhead stating that the subsurface safety device has been removed. The removal of the subsurface safety device shall be noted in the records as required in §250-124(b) of this part. If the master valve is open, a trained person shall be in the immediate vicinity of the well to attend the well so that emergency actions may be taken, if necessary.

    (3) A platform well shall be monitored, but a person need not remain in the well-bay area continuously if the master valve is closed. If the well is on a satellite structure, it must be attended or a pump-through plug installed in the tubing at least 100 feet below the mud line and the master valve closed, unless otherwise approved by the District Supervisor.

    (4) The well shall not be allowed to flow while the subsurface safety device is removed, except when flowing the well is necessary for that particular operation. The provisions of this paragraph are not applicable to the testing and inspection procedures in §250.124 of this part.

(i) *Additional safety equipment.* All tubing installations in which a wireline- or pumpdown-retrievable subsurface safety device is installed after the effective date of this subpart shall be equipped with a landing nipple with flow couplings or other protective equipment above and below to provide for the setting of the SSSV. The control system for all surface-controlled SSSV's shall be an integral part of the platform Emergency Shutdown System (ESD). In addition to the activation of the ESD by manual action on the platform, the system may be activated by a signal from a remote location. Surface-controlled SSSV's shall close in response to shut-in signals from the ESD and in response to the fire loop or other fire detection devices.

(j) *Emergency action.* In the event of an emergency, such as an impending storm, any well not equipped with a subsurface safety device and which is capable of natural flow shall have the device properly installed as soon as possible with due consideration being given to personnel safety.

[53 FR 10690, Apr. 1, 1988, as amended at 54 FR 50617, Dec. 8, 1989; 58 FR 49928, Sept. 24, 1993]

§250.122   Design, installation, and operation of surface production-safety system.

(a) *General.* All production facilities, including separators, treaters, compressors, headers, and flowlines shall be designed, installed, and maintained in a manner which provides for efficiency, safety of operation, and protection of the environment.

(b) *Platforms.* All platform production facilities shall be protected with a basic and ancillary surface safety system designed, analyzed, installed, tested, and maintained in operating condition in accordance with the provisions of API RP 14C, Recommended Practice for Analysis, Design, Installation and Testing of Basic Surface Safety Systems for Offshore Production Platforms. If processing components are to be utilized, other than those for which Safety Analysis Checklists are included in API RP 14C, the analysis technique and documentation specified therein shall be utilized to determine the effects and requirements of these components upon the safety system. Safety device requirements for pipelines are contained in §250.154 of this part.

(c) *Specification for surface safety valves (SSV) and underwater safety valves (USV).* All wellhead SSV's, USV's and their actuators which are installed in the OCS shall conform to the requirements in §250.126 of this part.

(d) *Use of SSV's and USV's.* All SSV's and USV's shall be inspected, installed, maintained, and tested in accordance with API RP 14H, Recommended Practice for Use of Surface Safety Valves and

Underwater Safety Valves Offshore. If any SSV or USV does not operate properly or if any fluid flow is observed during the leakage test, the valve shall be repaired or replaced.

(e) *Approval of safety-systems design and installation features.* Prior to installation, the lessee shall submit, in duplicate for approval to the District Supervisor a production safety system application containing information relative to design and installation features. Information concerning approved design and installation features shall be maintained by the lessee at the lessee's offshore field office nearest the OCS facility or other location conveniently available to the District Supervisor. All approvals are subject to field verifications. The application shall include the following:

(1) A schematic flow diagram showing tubing pressure, size, capacity, design working pressure of separators, flare scrubbers, treaters, storage tanks, compressors, pipeline pumps, metering devices, and other hydrocarbon-handling vessels.

(2) A schematic flow diagram (API RP 14C, Figure E1) and the related Safety Analysis Function Evaluation chart (API RP 14C, subsection 4.3c).

(3) A schematic piping diagram showing the size and maximum allowable working pressures as determined in accordance with API RP 14E, Design and Installation of Offshore Production Platform Piping Systems.

(4) Electrical system information including the following:

(i) A plan for each platform deck outlining all hazardous areas classified in accordance with API RP 500, Recommended Practice for Classification of Locations for Electrical Installations at Petroleum Facilities, and outlining areas in which potential ignition sources, other than electrical, are to be installed. The area outlined shall include the following information:

(A) All major production equipment, wells, and other significant hydrocarbon sources and a description of the type of decking, ceiling, walls (e.g., grating or solid) and firewalls; and

(B) Location of generators, control rooms, panel boards, major cabling/conduit routes, and identification of the primary wiring method (e.g., type cable, conduit, or wire).

(ii) Elementary electrical schematic of any platform safety shut-down system with a functional legend.

(5) Certification that the design for the mechanical and electrical systems to be installed were approved by registered professional engineers. After these systems are installed, the lessee shall submit a statement to the District Supervisor certifying that new installations conform to the approved designs of this subpart.

(6) The design and schematics of the installation and maintenance of all fire- and gas-detection systems shall include the following:

(i) Type, location, and number of detection sensors;

(ii) Type and kind of alarms, including emergency equipment to be activated;

(iii) Method used for detection;

(iv) Method and frequency of calibration; and

(v) A functional block diagram of the detection system, including the electric power supply.

[53 FR 10690, Apr. 1, 1988, as amended at 61 FR 60024, Nov. 26, 1996]

§250.123   **Additional production system requirements.**

(a) *General.* Lessees shall comply with the following production safety system requirements (some of which are in addition to those contained in API RP 14C) incorporated by reference in §250.122 (b); of this part.

(b) *Design, installation, and operation of additional production systems.*

(1) *Pressure and fired vessels.* Pressure and fired vessels shall be designed, fabricated, code stamped, and maintained in accordance with applicable provisions of sections I, IV, and VIII of the ASME Boiler and Pressure Vessel Code. All existing uncoded vessels in use must be justified and approval for continued use obtained from the District Supervisor no later than August 29, 1988.

(i) Pressure relief valves shall be designed, installed, and maintained in accordance with applicable provisions of sections I, IV, and VIII of the ASME Boiler and Pressure Vessel Code. The relief valves shall conform to the valve-sizing and pressure-relieving requirements specified in these documents; however, the relief valves, except completely redundant relief valves, shall be set no higher than the maximum-allowable working pressure of the vessel. All relief valves and vents shall be piped in such a way as to prevent fluid from striking personnel or ignition sources.

(ii) Steam generators operating at less than 15 pounds per square inch gauge (psig) shall be equipped with a level safety low (LSL) sensor which will shut off the fuel supply when the water level drops below the minimum safe level. Steam generators operating at greater than 15 psi require, in addition to an LSL, a water-feeding device which will automatically control the water level.

(iii) The lessee shall use pressure recorders to establish the new operating pressure ranges of pressure vessels at any time when there is a change in operating pressures that requires new settings for the high-pressure shut-in sensor and/or the low-pressure shut-in sensor as provided herein. The pressure-recorder charts used to determine current operating pressure ranges shall be maintained at the lessee's field office nearest the OCS facility or at other locations conveniently available to the District Supervisor. The high-pressure shut-in sensor shall be set no higher than 15 percent or 5 psi, whichever is greater, above the highest operating pressure of the vessel. This setting shall also be set sufficiently below (5 percent or 5 psi, whichever is greater) the relief valve's set pressure to assure that the pressure source is shut in before the relief valve activates. The low-pressure shut-in sensor shall activate no lower than 15 percent or 5 psi, whichever is greater, below the lowest pressure in the operating range. The activation of low-pressure sensors on pressure vessels which operate at less than 5 psi shall be approved by the District Supervisor on a case-by-case basis.

(2) *Flowlines.*

(i) Flowlines from wells shall be equipped with high- and low-pressure shut-in sensors located in accordance with section A.1 and Figure A1 of API RP 14C. The lessee shall use pressure recorders to establish the new operating pressure ranges of flowlines at any time when there is a significant change in operating pressures. The most recent pressure-recorder charts used to determine operating pressure ranges shall be maintained at the lessee's field office nearest the OCS facility or at other locations conveniently available to the District Supervisor. The high-pressure shut-in sensor(s) shall be set no higher than 15 percent or 5 psi, whichever is greater, above the highest operating pressure of the line. But in all cases; it shall be set sufficiently below the maximum shut-in wellhead pressure or the gas-lift supply pressure to assure actuation of the SSV. The low-pressure shut-in sensor(s) shall be set no lower than 15 percent or 5 psi, whichever is greater, below the lowest operating pressure of the line in which it is installed.

(ii) If a well flows directly to the pipeline before separation, the flowline and valves from the well located upstream of and including the header inlet valve(s) shall have a working pressure equal to or greater than the maximum shut-in pressure of the well unless the flowline is protected by one of the following:

  (A) A relief valve which vents into the platform flare scrubber or some other location approved by the District Supervisor. The platform flare scrubber shall be designed to handle, without liquid-hydrocarbon carryover to the flare, the maximum-anticipated flow of liquid hydrocarbons which may be relieved to the vessel.

  (B) Two SSV's with independent high-pressure sensors installed with adequate volume upstream of any block valve to allow sufficient time for the valve(s) to close before exceeding the maximum allowable working pressure.

(3) *Safety sensors.* All shutdown devices, valves, and pressure sensors shall function in a manual reset mode. Sensors with integral automatic reset shall be equipped with an appropriate device to override the automatic reset mode. All pressure sensors shall be equipped to permit testing with an external pressure source.

(4) *ESD.* The ESD shall conform to the requirements of Appendix C, section C1, of API RP 14C, and the following:

  (i) The manually operated ESD valve(s) shall be quick-opening and nonrestricted to enable the rapid actuation of the shutdown system. Only ESD stations at the boat landing may utilize a loop of breakable synthetic tubing in lieu of a valve.

  (ii) Closure of the SSV shall not exceed 45 seconds after automatic detection of an abnormal condition or actuation of an ESD. The surface-controlled SSSV shall close in not more than 2 minutes after the shut-in signal has closed the SSV. Design-delayed closure time greater than 2 minutes shall be justified by the lessee based on the individual well's mechanical/production characteristics and be approved by the District Supervisor.

  (iii) A schematic of the ESD which indicates the control functions of all safety devices for the platforms shall be maintained by the lessee on the platform or at the lessee's field office nearest the OCS facility or other location conveniently available to the District Supervisor.

(5) *Engines.*

  (i) *Engine exhaust.* Engine exhausts shall be equipped to comply with the insulation and personnel protection requirements of API RP 14C, section 4.2c(4). Exhaust piping from diesel engines shall be equipped with spark arresters.

  (ii) *Diesel engine air intake.* No later than May 31, 1989, diesel engine air intakes shall be equipped with a device to shutdown the diesel engine in the event of runaway. Diesel engines which are continuously attended shall be equipped with either remote operated manual or automatic shutdown devices. Diesel engines which are not continuously attended shall be equipped with automatic shutdown devices.

(6) *Glycol dehydration units.* A pressure relief system or an adequate vent shall be installed on the glycol regenerator (reboiler) which will prevent overpressurization. The discharge of the relief valve shall be vented in a nonhazardous manner.

(7) *Gas compressors.* Compressor installations shall be equipped with the following protective equipment as required in API RP 14C, sections A4 and A8.

  (i) A Pressure Safety High (PSH), a Pressure Safety Low (PSL), a Pressure Safety Valve (PSV), and a Level Safety High (LSH), and an LSL to protect each interstage and suction scrubber.

  (ii) A Temperature Safety High (TSH) on each compressor discharge cylinder.

(iii) The PSH and PSL shut-in sensors and LSH shut-in controls protecting compressor suction and interstage scrubbers shall be designed to actuate automatic shutdown valves (SDV) located in each compressor suction and fuel gas line so that the compressor unit and the associated vessels can be isolated from all input sources. All automatic SDV's installed in compressor suction and fuel gas piping shall also be actuated by the shutdown of the prime mover. Unless otherwise approved by the District Supervisor, gas-well gas affected by the closure of the automatic SDV on a compressor suction shall be diverted to the pipeline or shut in at the wellhead.

(iv) A blowdown valve is required on the discharge line of all compressor installations of 1,000 horsepower (746 kilowatts) or greater.

(8) *Firefighting systems.* Firefighting systems for both open and totally enclosed platforms installed for extreme weather conditions or other reasons shall conform to subsection 5.2, Firewater systems, of API RP 14G, Fire Prevention and Control Open Type Offshore Production Platforms, and shall require approval of the District Supervisor. The following additional requirements shall apply for both open- and closed-production platforms:

(i) A firewater system consisting of rigid pipe with firehose stations or fixed firewater monitors shall be installed. The firewater system shall be installed to provide needed protection in all areas where production-handling equipment is located. A fixed waterspray system shall be installed in enclosed well-bay areas where hydrocarbon vapors may accumulate.

(ii) Fuel or power for firewater pump drivers shall be available for at least 30 minutes of run time during a platform shut-in. If necessary, an alternate fuel or power supply shall be installed to provide for this pump-operating time unless an alternate firefighting system has been approved by the District Supervisor.

(iii) A firefighting system using chemicals may be used in lieu of a water system if the District Supervisor determines that the use of a chemical system provides equivalent fire-protection control.

(iv) A diagram of the firefighting system showing the location of all firefighting equipment shall be posted in a prominent place on the facility or structure.

(v) For operations in subfreezing climates, the lessee shall furnish evidence to the District Supervisor that the firefighting system is suitable for the conditions.

(9) *Fire- and gas-detection system.*

(i) Fire (flame, heat, or smoke) sensors shall be installed in all enclosed classified areas. Gas sensors shall be installed in all inadequately ventilated enclosed classified areas. Adequate ventilation is defined as ventilation which is sufficient to prevent accumulation of significant quantities of vapor-air mixture in concentrations over 25 percent of the lower explosive limit (LEL). One approved method of providing adequate ventilation is a change of air volume each 5 minutes or 1 cubic foot of air-volume flow per minute per square foot of solid floor area, whichever is greater. Enclosed areas (e.g., buildings, living quarters, or doghouses) are defined as those areas confined on more than four of their six possible sides by walls, floors, or ceilings more restrictive to air flow than grating or fixed open louvers and of sufficient size to all entry of personnel. A classified area is any area classified Class I, Group D, Division 1 or 2, following the guidelines of API RP 500.

(ii) All detection systems shall be capable of continuous monitoring. Fire-detection systems and portions of combustible gas-detection systems related to the higher gas concentration levels shall be of the manual-reset type. Combustible gas-detection systems related to the lower gas-concentration level may be of the automatic-reset type.

(iii) A fuel-gas odorant or an automatic gas-detection and alarm system is required in enclosed, continuously manned areas of the facility which are provided with fuel gas. Living quarters and doghouses not containing a gas source and not located in a classified area do not require a gas detection system.

(iv) The District Supervisor may require the installation and maintenance of a gas detector or alarm in any potentially hazardous area.

(v) Fire- and gas-detection systems shall be an approved type, designed and installed in accordance with API RP 14C, API RP 14G, and API RP 14F, Design and Installation of Electrical Systems for Offshore Production Platforms.

(10) *Electrical equipment.* Electrical equipment and systems shall be designed, installed, and maintained in accordance with the requirements in §250.53 of this part.

(11) *Erosion.* A program of erosion control shall be in effect for wells or fields having a history of sand production. The erosion-control program may include sand probes, X-ray, ultrasonic, or other satisfactory monitoring methods. Records by lease, indicating the wells which have erosion-control programs in effect and the results of the programs, shall be maintained by the lessee for a period of 2 years and shall be made available to MMS upon request.

(c) *General platform operations.*

(1) Surface or subsurface safety devices shall not be bypassed or blocked out of service unless they are temporarily out of service for startup, maintenance, or testing procedures. Only the minimum number of safety devices shall be taken out of service. Personnel shall monitor the bypassed or blocked-out functions until the safety devices are placed back in service. Any surface or subsurface safety device which is temporarily out of service shall be flagged.

(2) When wells are disconnected from producing facilities and blind flanged, equipped with a tubing plug, or the master valves have been locked closed, compliance is not required with the provisions of API RP 14C or this regulation concerning the following:

(i) Automatic fail-close SSV's on wellhead assemblies, and

(ii) The PSH and PSL shut-in sensors in flowlines from wells.

(3) When pressure or atmospheric vessels are isolated from production facilities (e.g., inlet valve locked closed or inlet blind-flanged) and are to remain isolated for an extended period of time, safety device compliance with API RP 14C or this subpart is not required.

(4) All open-ended lines connected to producing facilities and wells shall be plugged or blind-flanged, except those lines designed to be open-ended such as flare or vent lines.

(d) *Welding and burning practices and procedures.* All welding, burning, and hot-tapping activities shall be conducted according to the specific requirements in §250.52 of this part.

[53 FR 10690, Apr. 1, 1988, 53 FR 12227, Apr. 13, 1988, as amended at 55 FR 47753, Nov. 15, 1990, 61 FR 60025, Nov. 26, 1996]

§250.124   Production safety-system testing and records.

(a) *Inspection and testing.* The safety-system devices shall be successfully inspected and tested by the lessee at the interval specified below or more frequently if operating conditions warrant. Testing shall be in accordance with API RP 14C, Appendix D, and the following:

(1) Testing requirements for subsurface safety devices are as follows:

(i) Each surface-controlled subsurface safety device installed in a well, including such devices in shut-in and injection wells, shall be tested in place for proper operation when installed or reinstalled and thereafter at intervals not exceeding 6 months. If the device does not operate properly, or if a liquid leakage rate in excess of 200 cubic centimeters per minute or a gas leakage rate in excess of 5 cubic feet per minute is observed, the device shall be removed, repaired and reinstalled, or replaced. Testing shall be in accordance with API RP 14B to ensure proper operation.

(ii) Each subsurface-controlled SSSV installed in a well shall be removed, inspected, and repaired or adjusted, as necessary, and reinstalled or replaced at intervals not exceeding 6 months for those valves not installed in a landing nipple and 12 months for those valves installed in a landing nipple.

(iii) Each tubing plug installed in a well shall be inspected for leakage by opening the well to possible flow at intervals not exceeding 6 months. If a liquid leakage rate in excess of 200 cubic centimeters per minute or a gas leakage rate in excess of 5 cubic feet per minute is observed, the device shall be removed, repaired and reinstalled, or replaced. An additional tubing plug may be installed in lieu of removal.

(iv) Injection valves shall be tested in the manner as outlined for testing tubing plugs in paragraph (a)(1)(iii) of this section. Leakage rates outlined in paragraph (a)(1)(iii) of this section shall apply.

(2) All PSV's shall be tested for operation at least once every 12 months. These valves shall be either bench-tested or equipped to permit testing with an external pressure source. Weighted disk vent valves used as PSV's on atmospheric tanks may be disassembled and inspected in lieu of function testing.

(3) The following safety devices shall be tested at least once each calendar month, but at no time shall more than 6 weeks elapse between tests:

(i) All PSH and PSL,

(ii) All LSH and LSL controls,

(iii) All automatic inlet SDV's which are actuated by a sensor on a vessel or compressor, and

(iv) All SDV's in liquid discharge lines and actuated by vessel low-level sensors.

(4) All SSV's and USV's shall be tested for operation and for leakage at least once each calendar month, but at no time shall more than 6 weeks elapse between tests. The testing shall be in accordance with the test procedures specified in API RP 14H, Section 4, Table 2. If the SSV or USV does not operate properly or if any fluid flow is observed during the leakage test, the valve shall be repaired or replaced.

(5) All flowline Flow Safety Valves (FSV) shall be checked for leakage at least once each calendar month, but at no time shall more than 6 weeks elapse between tests. The FSV's shall be tested for leakage in accordance with the test procedure specified in API RP 14C, appendix D, section D4, table D2, subsection D. If the leakage measured exceeds a liquid flow of 200 cubic centimeters per minute or a gas flow of 5 cubic feet per minute, the FSV's shall be repaired or replaced.

(6) The TSH shutdown controls installed on compressor installations which can be nondestructively tested shall be tested every 6 months and repaired or replaced as necessary.

(7) All pumps for firewater systems shall be inspected and operated weekly.

(8) All fire- (flame, heat, or smoke) detection systems shall be tested for operation and recalibrated every 3 months provided that testing can be performed in a nondestructive manner. Open flame or devices operating at temperatures which could ignite a methane-air mixture shall not be used. All combustible gas-detection systems shall be calibrated every 3 months.

(9) All TSH devices shall be tested at least once every 12 months, excluding those addressed in paragraph (a)(6) of this section and those which would be destroyed by testing. Burner safety low and flow safety low devices shall also be tested at least once every 12 months.

(10) The ESD shall be tested for operation at least once each calendar month, but at no time shall more than 6 weeks elapse between tests. The test shall be conducted by alternating ESD stations monthly to close at least one wellhead SSV and verify a surface-controlled SSSV closure for that well as indicated by control circuitry actuation.

(11) Prior to the commencement of production, the lessee shall notify the District Supervisor when the lessee is ready to conduct a preproduction test and inspection of the integrated safety system. The lessee shall also notify the District Supervisor upon commencement of production in order that a complete inspection may be conducted.

(b) *Records.* The lessee shall maintain records for a period of 2 years for each subsurface and surface safety device installed. These records shall be maintained by the lessee at the lessee's field office nearest the OCS facility or other locations conveniently available to the District Supervisor. These records shall be available for review by a representative of MMS. The records shall show the present status and history of each device, including dates and details of installation, removal, inspection, testing, repairing, adjustments, and reinstallation.

[53 FR 10690, Apr. 1, 1988, as amended at 55 FR 47753, Nov. 15, 1990; 62 FR 5331, Feb. 5, 1997]

§250.125   Safety device training.

Personnel installing, inspecting, testing, and maintaining these safety devices and personnel operating the production platforms to be qualified in accordance with subpart O.

§250.126   Quality assurance and performance of safety and pollution prevention equipment.

(a) Safety and pollution prevention equipment installed on the OCS shall meet the quality assurance requirements identified in paragraph (c) of this section. The MMS will consider approval of other quality assurance programs for the manufacture of safety and pollution prevention equipment. Quality assurance programs proposed for consideration by MMS shall be submitted to the Deputy Associate Director for Offshore Operations; Minerals Management Service; Mail Stop 647; 381 Elden Street; Herndon, Virginia 22070-4817.

(b) (1) By August 29, 1988, each lessee shall submit to the Deputy Associate Director for Offshore Operations a list of all certified and noncertified SSV's (both actuator and valve), USV's (both actuator and valve), and SSSV's (including safety valve locks and landing nipples) in the lessee's inventory as of April 1, 1988. The list shall indicate which items of the safety and pollution-prevention equipment were manufactured under American National Standards Institute/American Society of Mechanical Engineers Standard ANSI/ASME SPPE-1, Quality Assurance and Certification of Safety and Pollution-Prevention Equipment Used in Offshore Oil and Gas Operations and shall identify each listed item by manufacturer, serial number, model, the date the item entered inventory, and whether the item is in service on the OCS or in stock for installation on the OCS.

(2) Lists received from lessees under paragraph (b)(1) of this section shall be maintained by MMS. An SSSV shall be removed from the list when it is removed from service for a failure or malfunction. A USV or an SSV shall be removed from the list when it is removed from service for remanufacture (repairs employing machining, welding, heat treating, or other manufacturing operations). Lessees shall notify the Deputy Associate Director for Offshore Operations when an SSSV is removed from service for a failure or malfunction or a USV or SSV is removed from service for remanufacture (repairs employing machining, welding, heat treating, or other manufacturing operations).

(c) Safety and pollution-prevention equipment used in the OCS shall meet one of the following:

(1) Be identified on the list submitted under paragraph (b) of this section by a lessee of the lease on which the item is to be installed.

(2) Be certified by the manufacturer as having been produced under a quality assurance program that meets the requirements of ASME/ANSI SPPE-1-1988 and addenda a, b, c, and d, or

(3) Be certified by the manufacturer as having been produced under a quality assurance program that meets the requirements of API Spec Q1 and the technical specification API Spec 14A for SSSV's. For SSV's and USV's the manufacturer must meet API Spec 6A and API Spec 6A VI, or API Spec 14D.

(d) The installation, inspection, maintenance, testing, removal, redress, field repair, and documentation of safety and pollution-prevention equipment used in the OCS shall be in accordance with API RP 14B for an SSSV and API RP 14H for an SSV or USV. A remanufactured SSV or USV shall meet the requirements of paragraph (c) of this section.

(e) Each lessee shall report the failure of safety and pollution-prevention equipment certified in accordance with paragraph (c)(2) or (3) of this section.

(1) Equipment certified under paragraph (c)(2) shall be reported in accordance with section OE-2670 of ASME/ANSI SPPE-1-1988 and addenda a, b, c, and d.

(2) Equipment certified under paragraph (c)(3) of this section, must be reported in accordance with Appendix C of API Spec 14A or Appendix L of API Spec 6A or Appendix C of API Spec 14D, as appropriate.

(3) Equipment certified under both paragraphs (c)(2) and (c)(3) of this section must be reported in accordance with both section OE-2670 of ASME/ANSI SPPE-1-1988 and Appendix C of API Spec 14A or Appendix L of API Spec 6A or Appendix C of API Spec 14D, as appropriate.

(f) Each lessee shall report the installation of equipment certified in accordance with either paragraph (c)(2) or (c)(3) of this section, to the manufacturer of the equipment on the shipping and receiving report that comes with the equipment.

[53 FR 10690, Apr. 1, 1988; 53 FR 12227, Apr. 13, 1988, as amended at 54 FR 50617, Dec. 8, 1989; 55 FR 10617, Mar. 22, 1990; 55 FR, 37710, Sept. 13, 1990; 55 FR 47753, Nov. 15, 1990; 61 FR 60025, Nov. 26, 1996]

§250.127   Hydrogen sulfide.

Production operations in zones known to contain hydrogen sulfide (H_2S) or in zones where the presence of H_2S is unknown, as defined in §250.67 of this part, shall be conducted in accordance with that section and other relevant requirements of subpart H, Production Safety Systems.

SUBPART O—WELL CONTROL AND PRODUCTION SAFETY TRAINING

SOURCE: 65 FR 49490, Aug. 14, 2000, unless otherwise noted.

Sec. 250.1500   Definitions.

Terms used in this subpart have the following meaning:

Employee means direct employees of the lessees who are assigned well control or production safety duties.

I or *you* means the lessee engaged in oil, gas, or sulphur operations in the Outer Continental Shelf (OCS).

Lessee means a person who has entered into a lease with the United States to explore for, develop, and produce the leased minerals. The term lessee also includes an owner of operating rights for that lease and the MMS-approved assignee of that lease.

Production safety means production operations as well as the installation, repair, testing, maintenance, or operation of surface or subsurface safety devices.

Well control means drilling, well completion, well workover, and well servicing operations. For purposes of this subpart, well completion/well workover means those operations following the drilling of a well that are intended to establish or restore production to a well. It includes small tubing operations but does not include well servicing. Well servicing means snubbing, coil tubing, and wireline operations.

Sec. 250.1501   What is the goal of my training program?

The goal of your training program must be safe and clean OCS operations. To accomplish this, you must ensure that your employees and contract personnel engaged in well control or production safety operations understand and can properly perform their duties.

Sec. 250.1502   Is there a transition period for complying with the regulations in this subpart?

(a) During the period October 13, 2000 until October 15, 2002 you may either:

   (1) Comply with the provisions of this subpart. If you elect to do so, you must notify the Regional Supervisor; or

   (2) Comply with the training regulations in 30 CFR 250.1501 through 250.1524 that were in effect on June 1, 2000 and are contained in the 30 CFR, parts 200 to 699, edition revised as of July 1, 1999, as amended on December 28, 1999 (64 FR 72794).

(b) After October 15, 2002, you must comply with the provisions of this subpart.

Sec. 250.1503   What are my general responsibilities for training?

(a) You must establish and implement a training program so that all of your employees are trained to competently perform their assigned well control and production safety duties. You must verify that your employees understand and can perform the assigned well control or production safety duties.

(b) You must have a training plan that specifies the type, method(s), length, frequency, and content of the training for your employees. Your training plan must specify the method(s) of verifying employee understanding and performance. This plan must include at least the following information:

   (1) Procedures for training employees in well control or production safety practices;

   (2) Procedures for evaluating the training programs of your contractors;

   (3) Procedures for verifying that all employees and contractor personnel engaged in well control or production safety operations can perform their assigned duties;

(4)   Procedures for assessing the training needs of your employees on a periodic basis;

(5)   Recordkeeping and documentation procedures; and

(6)   Internal audit procedures.

(c)   Upon request of the Regional or District Supervisor, you must provide:

(1)   Copies of training documentation for personnel involved in well control or production safety operations during the past 5 years; and

(2)   A copy of your training plan.

Sec. 250.1504   May I use alternative training methods?

You may use alternative training methods. These methods may include computer-based learning, films, or their equivalents. This training should be reinforced by appropriate demonstrations and "hands-on" training. Alternative training methods must be conducted according to, and meet the objectives of, your training plan.

Sec. 250.1505   Where may I get training for my employees?

You may get training from any source that meets the requirements of your training plan.

Sec. 250.1506   How often must I train my employees?

You determine the frequency of the training you provide your employees. You must do all of the following:

(a)   Provide periodic training to ensure that employees maintain understanding of, and competency in, well control or production safety practices;

(b)   Establish procedures to verify adequate retention of the knowledge and skills that employees need to perform their assigned well control or production safety duties; and

(c)   Ensure that your contractors' training programs provide for periodic training and verification of well control or production safety knowledge and skills.

Sec. 250.1507   How will MMS measure training results?

MMS may periodically assess your training program, using one or more of the methods in this section.

(a)   Training system audit. MMS or its authorized representative may conduct a training system audit at your office. The training system audit will compare your training program against this subpart. You must be prepared to explain your overall training program and produce evidence to support your explanation.

(b)   Employee or contract personnel interviews. MMS or its authorized representative may conduct interviews at either onshore or offshore locations to inquire about the types of training that were provided, when and where this training was conducted, and how effective the training was.

(c)   Employee or contract personnel testing. MMS or its authorized representative may conduct testing at either onshore or offshore locations for the purpose of evaluating an individual's knowledge and skills in perfecting well control and production safety duties.

(d)   Hands-on production safety, simulator, or live well testing. MMS or its authorized representative may conduct tests at either onshore or offshore locations. Tests will be designed to evaluate the competency of your employees or contract personnel in performing their assigned well control and production safety duties. You are responsible for the costs associated with this testing, excluding salary and travel costs for MMS personnel.

Sec. 250.1508   What must I do when MMS administers written or oral tests?

MMS or its authorized representative may test your employees or contract personnel at your worksite or at an onshore location. You and your contractors must:

    (a) Allow MMS or its authorized representative to administer written

    or oral tests; and

    (b) Identify personnel by current position, years of experience in present position, years of total **oil** field experience, and employer's name (e.g., operator, contractor, or sub-contractor company name).

Sec. 250.1509   What must I do when MMS administers or requires hands-on, simulator, or other types of testing?

If MMS or its authorized representative conducts, or requires you or your contractor to conduct hands-on, simulator, or other types of testing, you must:

    (a) Allow MMS or its authorized representative to administer or witness the testing;

    (b) Identify personnel by current position, years of experience in present position, years of total **oil** field experience, and employer's name (e.g., operator, contractor, or sub-contractor company name); and

    (c) Pay for all costs associated with the testing, excluding salary and travel costs for MMS personnel.

Sec. 250.1510   What will MMS do if my training program does not comply with this subpart?

If MMS determines that your training program is not in compliance, we may initiate one or more of the following enforcement actions:

    (a) Issue an Incident of Noncompliance (INC);

    (b) Require you to revise and submit to MMS your training plan to address identified deficiencies;

    (c) Assess civil/criminal penalties; or

    (d) Initiate disqualification procedures.

UNITED STATES DEPARTMENT OF THE INTERIOR
MINERALS MANAGEMENT SERVICE

NTL No. 2001-N03

Effective Date: November 05, 2001
Rescission Date: October 15, 2002

NOTICE TO LESSEES AND OPERATORS OF FEDERAL
OIL, GAS, AND SULPHUR LEASES IN THE OUTER CONTINENTAL SHELF

30 CFR 250, Subpart O - Well Control and Production Safety Training

This Notice to Lessees and Operators (NTL) supersedes NTL No. 2000-N03. It restates Minerals Management Service policy during the transition period of the revised 30 CFR 250, Subpart O, training regulations published in the Federal Register on August 14, 2000 (65 FR 49485). This NTL also addresses questions and concerns that have been raised regarding § 250.1504 on alternative training methods.

The new performance-based Subpart O, Well Control and Production Safety Training, regulations took effect on October 13, 2000. To allow sufficient time for you to develop and implement your training program, § 250.1502 provides a 2-year transition period from October 13, 2000 until October 15, 2002. During the transition period, you may either:

(1) Continue to comply with the previous regulations that were in effect prior to October 13, 2000. If you have employees whose current Subpart O certificates expire during the 2-year transition period, you must retrain those employees in accordance with the previous regulations, or

(2) Notify the MMS Regional Supervisor, in writing, that you have implemented the new Subpart O regulations. You are only required to notify the MMS Regional Supervisor during the 2-year transition period. After October 15, 2002, you are required to comply with the new rule and, therefore, no notification would be necessary.

In addition to the superseded NTL No. 2000-N03, on August 17, 2000, we sent a letter to all MMS-approved training schools and organizations to clarify the new regulations. We explained that after the new regulations became effective, MMS would no longer accredit training schools or organizations. However, we provided that all training schools or organizations with valid MMS-accreditation would continue to be accredited until October 15, 2002, according to their training curriculum and plan as currently approved by MMS. Therefore, if you are operating under option (1) above, you have the same training schools or organizations from which to choose training for your employees during the transition period.

All certificates or cards issued by MMS approved training schools or organizations expire on October 15, 2002. After this date, we will not recognize as valid any cards or certificates issued by training schools or organizations which indicate or imply compliance with the requirements of 30 CFR 250, Subpart O, training regulations. However, training cards or other evidence of employee competency may still factor into the training plans required under § 250.1503 of the new regulations.

Recently, MMS has received questions concerning hands-on training as it applies to alternative training methods. As stated in § 250.1504 in the new Subpart O regulations, alternative training methods (e.g., computer based, films, or equivalent) should be reinforced by appropriate demonstrations and "hands-on" training. Hands-on training procedures should be identified in your training plan. It is not the intent of the new regulations to allow alternative training to "stand alone" without such reinforcement.

Paperwork Reduction Act of 1995 (PRA) Statement: The collection of information referred to in this NTL provides clarification, description, or interpretation of requirements in the 30 CFR 250, Subpart O, Well Control and Production Safety Training regulations. The Office of Management and Budget (OMB) has approved the information collection requirements and assigned OMB control number 1010-0128. This NTL does not impose additional information collection requirements subject to the PRA.

Contact: If you have any questions about this transition policy, you may contact Joseph Levine at (703) 787-1032.

Date: 11/05/01

Carolita U. Kallaur
Associate Director for
 Offshore Minerals Management

Addendum to Appendix D:

30 CFR Part 250
Subpart O

Training Requirements for Oil and Gas and Sulphur Operations in the Outer Continental Shelf

Subpart O—Well Control and Production Safety Training

§ 903.11 Are certain PHAs eligible to submit a streamlined Annual Plan?

(a) Yes, the following PHAs may submit a streamlined Annual Plan, as described in paragraph (b) of this section:

(1) PHAs that are determined to be high performing PHAs as of the last annual or interim assessment of the PHA before the submission of the 5-Year or Annual Plan;

(2) PHAs with less than 250 public housing units (small PHAs) and that have not been designated as troubled in accordance with section 6(j)(2) of the 1937 Act; and

(3) PHAs that only administer tenant-based assistance and do not own or operate public housing.

(b) All streamlined plans must provide information on how the public may reasonably obtain additional information on the PHA policies contained in the standard Annual Plan, but excluded from their streamlined submissions.

(c) A streamlined plan must include the information provided in this paragraph (c) of this section. The Secretary may reduce the information requirements of streamlined Plans further, with adequate notice.

(1) For high performing PHAs, the streamlined Annual Plan must include the information required by § 903.7(a), (b), (c), (d), (g), (h), (m), (n), (o), (p) and (r). The information required by § 903.7(m) must be included only to the extent this information is required for PHA's participation in the public housing drug elimination program and the PHA anticipates participating in this program in the upcoming year.

(2) For small PHAs that are not designated as troubled or that are not at risk of being designated as troubled under section 6(j)(2) of the 1937 Act the streamlined Annual Plan must include the information required by § 903.7(a), (b), (c), (d), (g), (h), (k), (m), (n), (o), (p) and (r). The information required by § 903.7(k) must be included only to the extent that the PHA participates in homeownership programs under section 8(y). The information required by § 903.7(m) must be included only to the extent this information is required for the PHA's participation in the public housing drug elimination program and the PHA anticipates participating in this program in the upcoming year.

(3) For PHAs that administer only tenant-based assistance, the streamlined Annual Plan must include the information required by § 903.7(a), (b), (c), (d), (e), (f), (k), (l), (o), (p) and (r).

Dated: August 7, 2000.

Harold Lucas,

Assistant Secretary for Public and Indian Housing.

[FR Doc. 00–20550 Filed 8–11–00; 8:45 am]

BILLING CODE 4210–33–P

DEPARTMENT OF THE INTERIOR

Minerals Management Service

30 CFR Part 250

RIN 1010–AC41

Oil and Gas and Sulphur Operations in the Outer Continental Shelf—Subpart O—Well Control and Production Safety Training

AGENCY: Minerals Management Service (MMS), Interior.

ACTION: Final rule.

SUMMARY: This rule amends MMS regulations governing the training of lessee and contractor personnel engaged in oil and gas and sulphur operations in the OCS. MMS is making this amendment to enhance safety, allow the development of new and innovative training techniques, to impose fewer prescriptive requirements on the oil and gas industry, and provide increased training flexibility.

EFFECTIVE DATE: October 13, 2000.

FOR FURTHER INFORMATION CONTACT: Wilbon Rhome or Joseph Levine, Operations and Analysis Branch, at (703) 787–1032.

SUPPLEMENTARY INFORMATION: On April 20, 1999, we published the proposed rule in the **Federal Register** (64 FR 19318). During the 90-day comment period, which ended on July 19, 1999, MMS held a workshop.

Background

On February 5, 1997, we published a final rule in the **Federal Register** (62 FR 5320) concerning the training of lessee and contractor employees engaged in drilling, well completion, well workover, well servicing, or production safety system operations in the OCS. The final rule streamlined the previous regulations by 80 percent, provided the flexibility to use alternative training methods, and simplified the training options at 30 CFR 250, Subpart O—Training.

The February 5, 1997, final rule did not sufficiently address developing a performance-based training system, so we planned to publish a proposed rule to better address this issue. Before considering any further revisions to the rule, we decided to hold a workshop in

Houston, Texas. The purpose of the workshop was to discuss the development of a performance-based training system for OCS oil and gas activities.

On April 4, 1997, we published a **Federal Register** notice (62 FR 18070) announcing the workshop. We stated that the goal of the meeting was to develop a procedure that ensures that lessee and contractor employees are trained in well control or production safety system operations by creating a less prescriptive training program, focusing on results and not on processes.

To improve the regulations at 30 CFR 250, Subpart O—Training, the workshop notice asked attendees to be prepared to present and discuss comments on the following four performance measures and indicators that could be used as part of a performance-based program:

• MMS Written Test;

• MMS Hands-On and Simulator Testing;

• Audits, Interviews, or Cooperative Reviews; and

• Incident of Noncompliance (INC), Civil Penalty, and Event Data.

On June 10, 1997, we conducted a public workshop in Houston, Texas, which received excellent participation from industry and training schools. Approximately 190 people attended the workshop, representing a diverse cross section of the oil and gas industry.

The next step in the development of a performance-based training system was accomplished by publishing a proposed rule on April 20, 1999. The rule focused on the development of a performance-based training program. The proposed rule required lessee and contract employees to develop their own training programs tied to the job duties of their personnel. This final rule will primarily focus on training results rather than on the process by which employees are trained. By developing appropriate performance measures, MMS can evaluate the effectiveness of a lessee's training programs by:

• written testing;

• hands-on testing;

• training system audits; or

• employee interviews.

This approach requires lessees to be responsible for the quality and the level of training their employees receive.

Differences Between Proposed and Final Rules

In addition to the changes we made to the final rule in response to comments, we also reworded certain complex sections for further clarity. In many instances, the changes improve MMS's internal work processes to better serve its external customers. Following are the major changes by section.

• We replaced the tables in proposed § 250.1504. In the proposed rule, the tables listed the minimum "knowledge and job skill elements" employees must have to competently perform their assigned well control and production safety duties. The elements were far too prescriptive for a performance-based rule. The new 30 CFR 250.1503(a) is more performance-based, stating that: "You" must establish and implement a training program so that all of your employees are trained to competently perform their assigned well control and production safety duties. The knowledge and job skill elements that an employee must possess in order to perform assigned well control or production safety duties are the responsibility of the lessee.

• We added § 250.1502, establishing a 2-year transition period to ensure a smooth transition from the existing rule to the new requirement.

• We deleted proposed § 250.1502(c) that stated that both lessees and contractors are required to develop

training plans. We now specify that only lessees are required to develop a training plan.

• We modified proposed § 250.1503(b)(1) through (7) to add clarity and specificity so that lessees understand they are responsible for ensuring that all personnel working on their leases are trained and can competently perform their assigned well control or production safety duties. We also wanted contractors to understand that the lessees will review their training program for contract personnel.

• We replaced proposed § 250.1510 with § 250.1503(c). In proposed § 250.1510, we explained why it may be necessary for lessees to provide a training plan to the MMS. In § 250.1503(c), we describe what documentation the lessee must provide to MMS upon request of the Regional or District Supervisor.

• We deleted proposed § 250.1512 and moved the requirements to § 250.1509 in the final rule. Under the current system, MMS-approved training schools conduct hands-on, simulator, or other types of testing that must be

passed by the employees before they can work on the OCS. Under the final rule, § 250.1509 outlines the requirements involved if MMS conducts, or requires the lessees to conduct, these tests. We are changing the requirement in the proposed rule that the lessees pay all costs associated with testing. This final rule specifies that the lessees are responsible for paying the testing costs, excluding salary and travel costs for MMS personnel.

Response to Comments

MMS received 25 comments on the proposed rule. The comments were received from six production operators, six drilling contractors, two trade organizations, one standard setting organization, nine training schools, and one congressional office. We reviewed all the comments and, in some instances, we revised the final language based on these comments. MMS grouped the major comments and organized them by the proposed regulation section number or subject, as highlighted in the comment table.

COMMENT TABLE

Requirement/Proposed rule	Comment	MMS response
Preamble	The transition period is inadequate. Lessees will not be able to implement a satisfactory program within a 90-day timeframe.	Agree—MMS added a section establishing a 2-year transition period to ensure the smoothest transition from the existing rule to the new requirement. New 30 CFR 250.1502.
Preamble	The stated training plan development time of 2.2 hours is an understatement.	Agree—We noted and corrected. Plan development time averages 40–60 hours.
§ 250.1501	MMS should delete the requirement "experienced," as this would preclude "new hire employees." The word "experienced" does not necessarily relate to "competent," which is the primary goal of MMS' training program.	Agree—We deleted the requirement "experienced."
§ 250.1502	Several commenters stated that contractors would need to assure each individual lessee they work for that their personnel have been trained according to the specific program requirements that have been developed by that lessee. Contractors may have to modify their program to fit each lessee's definition of an acceptable program, possibly requiring the contractor to alter its training program every time a rig changes to a different customer.	Agree—Contractors may have to address the lessees' training plans. These differences may exist regardless of the system that is in place. It is the responsibility of the lessees to ensure that those differences do not impact the safety of operations.
§ 250.1502	Several commenters asked for clarification concerning which personnel are to be trained. The expanded scope of the rule from the prior regulations seems to imply that the catering staff, marine, helicopter, and other nonessential third-party "contract or" personnel must also be trained by the lessee.	Agree—MMS did not mean to imply that catering staff, marine, helicopter and other nonessential third-party "contractor" personnel be trained by the lessee. According to this rule, only personnel engaged in well control or production safety operations must be trained.
§ 250.1502	One commenter wanted MMS to remove the requirement that hot tapping practices and procedures be included in the lessee's training plan.	Agree—The focus of this rule has been limited to well control and production safety training.
§ 250.1502(a)	MMS' current prescriptive training requirements should be maintained.	Disagree—MMS believes lessees should be responsible for developing procedures that ensure their workers are properly trained prior to working on the OCS rather than having MMS prescribe them.
§ 250.1502(c)	One commenter stated that MMS should clarify if both lessees and contractors are required to develop training plans.	Agree—We now specify that lessees are required to develop a training plan. Lessees will be responsible for ensuring that all personnel working on their leases are trained and can competently perform their assigned well control or production safety duties. New 30 CFR 250.1503.

COMMENT TABLE—Continued

Requirement/Proposed rule	Comment	MMS response
§ 250.1502(c)	A 5-year record retention requirement for documentation for all employees is costly and unwarranted.	Disagree—MMS may need at least 5 years of training records to make an assessment of your training program and look at safety trends. New 30 CFR 250.1503(c)(1).
§ 250.1504	Several commenters suggested that the knowledge and job skill elements included in the tables are far too prescriptive for a rule that MMS intends to be "performance-based".	Agree—MMS believes that the tables are too prescriptive for a performance-based rule. We have elected to delete the tables.
§ 250.1509	Clarity that an employee needs to be kept current on information related to his or her particular job.	Agree—Wording has been changed to reflect periodic training of employees in relation to their specific job. New 30 CFR 250.1506.
§ 250.1510	Several commenters pointed out that the proposed rule does not contain requirements regarding course duration, class size, or periodic retraining. Some in industry may take this as a sign to extend the training frequency of their employees from 2 to 6 years, or to reduce well control certification to a one-time course and test.	Disagree—As part of the final rule, lessees will be required to develop a training plan defining their program. Minimum information to be included in the plan is included in the final rule. MMS will monitor company training programs to determine their effectiveness. New 30 CFR 250.1503.
§ 250.1510(b)(3)	Several commenters urged MMS not to use written tests as an indicator of an employee's competency or the effectiveness of an employee's training, and one commenter stated that tests should be administered orally because many offshore workers have difficulty reading regulations or company operating manuals.	Agree in part—MMS realizes that failing a written test does not mean an employee does not know his or her job. A written test is one of many tools MMS may use in assessing the performance of a company's training program. MMS may elect to conduct oral tests according to the lessee's training plan. New CFR 250.1508(a)
§ 250.1512	Several commenters stated the requirements for hands-on, simulator, or other types of testing will cause a disruption in operations if conducted offshore. This type of testing will not provide a valid indicator of the lessee's performance or the effectiveness of its training program.	Disagree—Whenever possible, MMS will try to accommodate this concern and minimize any potential disruptions. However, to assist in addressing personnel competency, hands-on, simulator, or other types of testing may be conducted in an offshore environment. Therefore, we retained the option for either onshore or offshore testing. New CFR 250.1507(d)
§ 250.1512	Several commenters stated that MMS should delete the requirement that lessees and contractors pay for all costs associated with hands-on, simulator, or other types of testing.	Disagree—MMS may use hands-on, simulator, or other types of tests as a method for evaluating the effectiveness of a training program. Whenever possible, MMS will make efforts to minimize costs associated with testing. The final rule clarifies that lessees will not be responsible for paying the salary and travel costs of MMS personnel. New 30 CFR 250.1507(d).
§ 250.1512	Several commenters stated that MMS should not use an authorized representative to administer or witness MMS hands-on, simulator, or live well testing. They believe that MMS should bear the burden of guaranteeing impartiality and controlling costs during these tests.	Disagree—MMS does not have the equipment or expertise to conduct hands-on, simulator, or live well testing. For that reason, the final rule includes a provision that either the MMS or its authorized representative would administer or witness the testing if we find it necessary. New CFR 250.1509(a).
Testing-out	One commenter urged MMS not to move in the direction of testing-out, especially in positions critical to operational safety, such as well control.	Disagree—MMS and much of industry sees value in training, even for advanced employees who can pass the test. However, under a performance-based system, certain lessees may choose to implement the testing-out options for some of their personnel. MMS will measure these results according to the requirements in § 250.1507 to ensure the competency of these employees.
General	One commenter stated that statistics on incidents in OCS waters overwhelmingly support the success of MMS' current training program. With today's environment in the oil and gas industry, this is not the time to experiment with a new type of training regulation.	Disagree—MMS believes that this final rule provides companies the opportunity to develop their own programs tailored to the needs of their employees. The changes in the final rule are expected to decrease incidents and improve company performance by holding lessees accountable for the competency of their employees.
WellCAP	Several commenters stated that MMS should consider referencing the International Association of Drilling Contractors (IADC) WellCAP training program, or its associated documents in the final rule. WellCAP is ideally positioned to act as an industry benchmark in the absence of MMS' school-based system, providing training uniformity and an acceptable level of quality to well control training worldwide.	Agree—MMS commends IADC for the WellCAP program and acknowledges the value WellCAP could bring in providing minimum well control training requirements to lessees and contractors worldwide. MMS intends to publish a proposed rule that proposes the incorporation of WellCAP or a comparable third party certification program into Subpart O.

Procedural Matters

Regulatory Planning and Review (Executive Order 12866)

This document is a significant rule and is subject to review by the Office of Management and Budget (OMB) under Executive Order 12866.

(1) This rule will not have an effect of $100 million or more on the economy. It will not adversely affect in a material way the economy, productivity, competition, jobs, the environment, public health or safety, or State, local, or tribal governments or communities. The rule does not add any new cost to the oil and gas industry, and it will not reduce the level of safety to personnel or the environment.

(2) This rule will not create a serious inconsistency or otherwise interfere with an action taken or planned by another agency. The Department of the Interior (DOI) has several Memoranda of Understanding (MOUs) with the U.S. Coast Guard that define the responsibilities of each agency with respect to activities on the OCS. The MOUs are effective in avoiding inconsistency or interfering with any action taken by another agency.

(3) This rule does not alter the budgetary effects or entitlements, grants, user fees, or loan programs or the rights or obligations of their recipients. This rule will not affect programs such as listed here. This is a training rule that applies to the lessees working on the OCS. There are no entitlements, grants, or user fees that apply.

(4) Although moving towards performance-based rules is a fairly new concept, this rulemaking will not raise any legal issues. However, there may be certain novel policy issues to consider, thus, this rule is significant and is subject to review by OMB. We held a public workshop before proposing this change.

Federalism (Executive Order 13132)

According to Executive Order 13132, this rule does not have Federalism implications. This rule does not substantially and directly affect the relationship between the Federal and State governments. This is a training rule that applies to lessees working on the OCS and amends current MMS regulations to provide increased training flexibility. Thus, this rule will not directly affect the relationship between the Federal and State Governments. This rule does not impose costs on State or localities because the rule applies only to the lessees working on the OCS.

Civil Justice Reform (Executive Order 12988)

According to Executive Order 12988, the Office of the Solicitor has determined that this rule does not unduly burden the judicial system and meets the requirements of sections 3(a) and 3(b)(2) of the Order.

Small Business Regulatory Enforcement Fairness Act (SBREFA)

This rule is not a major rule under (5 U.S.C. 804(2)) SBREFA. This rule:

(a) Does not have an annual effect on the economy of $100 million or more. The estimated yearly gross cost to the oil and gas industry to train its employees is $5,945,250. Based on a 12-year cycle, well-control students would normally take six basic courses (½ course per year), and production safety system students would take four basic courses (⅓ course per year). Therefore, the annual training cost to train 15,000 students in well control would be $3,975,000 ($530 × ½ course per year × 15,000 students). The annual training cost to train 15,000 students in a production safety system would be $1,955,250 ($395 × ⅓ course per year × 15,000 students). The total annual cost is $5,930,250. There may be additional costs to the lessees or contractors with poor performance records if MMS or its authorized representative conducts, or requires the lessee or contractor to conduct hands-on, simulator, or other types of testing. They will be required to pay for all costs associated with the testing, excluding salary and travel costs for MMS personnel.

We estimate that not more than 50 employees (industry-wide) per year, at a cost of $300 per employee, will be required to take the MMS hands-on, simulator, or other types of testing. The total cost for those employees should not exceed $15,000 per year.

We feel that the cost of complying with the final rule would be somewhat less than this amount.

(b) Will not cause a major increase in costs or prices for consumers, individual industries, Federal, State, or local government agencies, or geographic regions. Based on our experience, the training industry should not change significantly under a performance-based system. Because of lower overhead and competitive pricing in the industry, costs should remain stable; and

(c) Does not have significant adverse effects on competition, employment, investment, productivity, innovation, or ability of United States-based enterprises to compete with foreign-based enterprises.

Unfunded Mandates Reform Act (UMRA) of 1995 (Executive Order 12866)

This rule does not impose an unfunded mandate on State, local, or tribal governments or the private sector of more than $100 million per year. The rule does not have a significant or unique effect on State, local, or tribal governments or the private sector. A statement containing the information required by the UMRA (2 U.S.C. 1531 *et seq.*) is not required.

Paperwork Reduction Act (PRA) of 1995

We examined the proposed rule and these final regulations under section 3507(d) of the PRA. Because of the changes proposed to the current 30 CFR 250, Subpart O regulations, we submitted the information collection requirements to OMB for approval as part of the proposed rulemaking process. As the final rule contains minor changes in the collection of information, before publication, we again submitted the information collection to OMB for approval. In response to comments, we concluded that we significantly underestimated the burden for the primary paperwork aspect of the rule that requires lessees to develop "training plans" (§ 250.1503(b) and (c)). In our resubmission to OMB, the burden for this requirement is 60 hours per plan. The following two new requirements (associated hour burden is shown in parenthesis) are the only differences in the information collected under the final rule from that approved for the proposed rule:

• § 1502—Notify MMS if lessees implement the revised final regulations before the end of the 2-year transition period (1 hour).

• § 1503(c)—Provide copies of the training plan to MMS, if requested (5 hours).

The PRA provides that an agency may not conduct or sponsor, and a person is not required to respond to, a collection of information unless it displays a currently valid OMB control number. OMB has approved the collection of information required in the final rule under OMB control number 1010–0128.

The title of this collection of information was changed to "30 CFR 250, Subpart O Well Control and Production Safety Training" to correspond with the revised title of the subpart. Responses are mandatory. The frequency of submission varies according to the requirement but is generally "on occasion." We estimate there are approximately 130 respondents to this collection of information.

We use the collection of information required by these regulations to ensure that workers in the OCS are properly trained with the necessary skills to perform their jobs in a safe and pollution-free manner. In some instances, MMS will conduct oral interviews of offshore employees to evaluate the effectiveness of a company's training program. This information is necessary to verify training compliance with the requirements.

Reporting and Recordkeeping "Hour" Burden: The approved annual burden of this collection of information is 5,739 hours. Based on $50 per hour, we estimate the total "hour" burden cost to respondents to be $286,950.

Reporting and Recordkeeping "Non-Hour Cost" Burden: There are no "non-hour cost" burdens in the final regulations.

It should be noted that this final rule will not take full effect for 2 years from the effective date of the rule, but it allows for early implementation at the discretion of lessees. Therefore, we will continue to maintain approved information collections for the current Subpart O regulations (under OMB control number 1010–0078) as well as for these final regulations during the transition period.

Regulatory Flexibility (RF) Act

DOI certifies that this document will not have a significant economic effect on a substantial number of small entities under the RF Act (5 U.S.C. 601 *et seq.*). The Small Business Administration (SBA) defines a small business as having:

• Annual revenues of $5 million or less for exploration service and field service companies; and

• Fewer than 500 employees for drilling companies and for companies that extract oil, gas, or natural gas liquids.

Under SBA's Standard Industrial Classification (SIC) code 1381, Drilling Oil and Gas Wells, MMS estimates that there is a total of 1,380 firms that drill oil and gas wells onshore and offshore. Of these, approximately 130 companies are offshore lessees/operators, based on current estimates. According to SBA estimates, 39 companies qualify as large firms, leaving 91 companies qualified as small firms with fewer than 500 employees.

As explained in the PRA section, companies will be required to develop training plans. We estimate that the burden for developing these plans is approximately 60 hours each. If 91 lessees are small businesses, the burden would be 5,460 hours. At an average

hourly cost of $50, the impact of this requirement is $273,000 on small businesses. Once the plan has been developed, there are no new costs for implementation.

The costs for an alternative training program would simply offset the current cost of sending employees to accredited schools. Alternative training provides both added flexibility and cost savings for companies who train their employees either onshore or offshore, at a centralized location, or during their off hours on a platform or drilling rig. It is expected that they would receive the same quality of training that they have been receiving for years. We estimate that the company may spend 5–10 ($250–$500) hours annually to update the plans. Thus, the annual cost for updating plans for small businesses is approximately $22,750 to $45,500. The cost for this update will be minimal.

A positive effect for the lessees under the new rule is that they will have increased options concerning where to get their training. This will change how a company does business. This should not result in any additional training costs or economic burdens. Under the final rule, the oil and gas industry will have the flexibility to tailor its training program to the specific needs of each company. Lessees will be given the added flexibility to determine the type of training, methodology (classroom, computer, team, on-the-job), length of training, frequency and subject matter content for their training program.

In addition to lessees, MMS currently regulates the training schools. There are 52 MMS-accredited training schools. We have approved 26 schools to teach production safety courses, 22 schools to teach well control courses, and 4 schools to teach both well control and production courses. The training companies best fit under the SIC 8249, and the criterion for small businesses is $5 million in revenue. Based on this criterion, 25 training companies will fall into the small business category.

Under these final regulations, we will no longer be accrediting training schools or imposing any regulatory burden. However, lessee personnel and the employees of contractors hired by the lessee will have to be trained and found competent in the duties associated with their particular job. Training schools that teach a broad range of vocational courses, in addition to MMS accreditation courses, and who provide quality training at a competitive price, should experience no significant change in their normal business, except the schools will no longer be burdened with MMS reporting and recordkeeping requirements.

Training schools that were previously MMS-accredited will benefit because their plans are in place and approved by MMS. Additionally, schools that have established a loyal customer-base will not be affected by the implementation of this rule. Therefore, this new provision will not cause prices to increase or decrease. Based on our experience, the failure rate of the schools in the offshore training industry should not change significantly under a performance-based program. Under the current regulations, we maintain a database that tracks training schools approved by the agency. Based on information from this database, less than 2 percent of the schools approved by MMS go out of business each year. Under the new rule, we expect this to remain the same. MMS experience has shown that because of lower overhead and competitive pricing, small training schools are just as capable as the larger schools at adapting to change.

Your comments are important. The Small Business and Agriculture Regulatory Enforcement Ombudsman and 10 Regional Fairness Boards were established to receive comments from small business about Federal agency enforcement actions. The Ombudsman will annually evaluate the enforcement activities and rate each agency's responsiveness to small business. If you wish to comment on the enforcement actions of MMS, call toll-free (888) 734–3247.

Takings Implication Assessment (Executive Order 12630)

According to Executive Order 12630, the rule does not have significant takings implications. MMS determined that this rule does not represent a governmental action capable of interference with constitutionally protected property rights. Thus, a Takings Implication Assessment is not required under Executive Order 12630, Governmental Actions and Interference with Constitutionally Protected Property Rights.

National Environmental Policy Act (NEPA) of 1969

This rule does not constitute a major Federal action significantly affecting the quality of the human environment. A detailed statement under the NEPA is not required.

List of Subjects in 30 CFR Part 250

Continental shelf, Environmental impact statements, Environmental protection, Government contracts, Investigations, Mineral royalties, Oil and gas development and production, Oil and gas exploration, Oil and gas reserves, Penalties, Pipelines, Public lands—mineral resources, Public lands—rights-of-way, Reporting and recordkeeping requirements, Sulphur development and production, Sulphur exploration, Surety bonds.

Dated: July 14, 2000.

Sylvia V. Baca,

Assistant Secretary, Land and Minerals Management.

For the reasons stated in the preamble, MMS amends 30 CFR part 250 as follows:

PART 250—OIL AND GAS AND SULPHUR OPERATIONS IN THE OUTER CONTINENTAL SHELF

1. The authority citation for part 250 continues to read as follows:

Authority: 43 U.S.C. 1331 *et seq.*

2. Subpart O is revised to read as follows:

Subpart O—Well Control and Production Safety Training

Sec.
250.1500   Definitions.
250.1501   What is the goal of my training program?
250.1502   Is there a transition period for complying with the regulations in this subpart?
250.1503   What are my general responsibilities for training?
250.1504   May I use alternative training methods?
250.1505   Where may I get training for my employees?
250.1506   How often must I train my employees?
250.1507   How will MMS measure training results?
250.1508   What must I do when MMS administers written or oral tests?
250.1509   What must I do when MMS administers or requires hands-on, simulator, or other types of testing?
250.1510   What will MMS do if my training program does not comply with this subpart?

§ 250.1500   Definitions.

Terms used in this subpart have the following meaning:

Employee means direct employees of the lessees who are assigned well control or production safety duties.

I or *you* means the lessee engaged in oil, gas, or sulphur operations in the Outer Continental Shelf (OCS).

Lessee means a person who has entered into a lease with the United States to explore for, develop, and produce the leased minerals. The term lessee also includes an owner of operating rights for that lease and the MMS-approved assignee of that lease.

Production safety means production operations as well as the installation,

repair, testing, maintenance, or operation of surface or subsurface safety devices.

Well control means drilling, well completion, well workover, and well servicing operations. For purposes of this subpart, well completion/well workover means those operations following the drilling of a well that are intended to establish or restore production to a well. It includes small tubing operations but does not include well servicing. Well servicing means snubbing, coil tubing, and wireline operations.

§ 250.1501   What is the goal of my training program?

The goal of your training program must be safe and clean OCS operations. To accomplish this, you must ensure that your employees and contract personnel engaged in well control or production safety operations understand and can properly perform their duties.

§ 250.1502   Is there a transition period for complying with the regulations in this subpart?

(a) During the period October 13, 2000 until October 15, 2002 you may either:

(1) Comply with the provisions of this subpart. If you elect to do so, you must notify the Regional Supervisor; or

(2) Comply with the training regulations in 30 CFR 250.1501 through 250.1524 that were in effect on June 1, 2000 and are contained in the 30 CFR, parts 200 to 699, edition revised as of July 1, 1999, as amended on December 28, 1999 (64 FR 72794).

(b) After October 15, 2002, you must comply with the provisions of this subpart.

§ 250.1503   What are my general responsibilities for training?

(a) You must establish and implement a training program so that all of your employees are trained to competently perform their assigned well control and production safety duties. You must verify that your employees understand and can perform the assigned well control or production safety duties.

(b) You must have a training plan that specifies the type, method(s), length, frequency, and content of the training for your employees. Your training plan must specify the method(s) of verifying employee understanding and performance. This plan must include at least the following information:

(1) Procedures for training employees in well control or production safety practices;

(2) Procedures for evaluating the training programs of your contractors;

(3) Procedures for verifying that all employees and contractor personnel

engaged in well control or production safety operations can perform their assigned duties;

(4) Procedures for assessing the training needs of your employees on a periodic basis;

(5) Recordkeeping and documentation procedures; and

(6) Internal audit procedures.

(c) Upon request of the Regional or District Supervisor, you must provide:

(1) Copies of training documentation for personnel involved in well control or production safety operations during the past 5 years; and

(2) A copy of your training plan.

§ 250.1504   May I use alternative training methods?

You may use alternative training methods. These methods may include computer-based learning, films, or their equivalents. This training should be reinforced by appropriate demonstrations and "hands-on" training. Alternative training methods must be conducted according to, and meet the objectives of, your training plan.

§ 250.1505   Where may I get training for my employees?

You may get training from any source that meets the requirements of your training plan.

§ 250.1506   How often must I train my employees?

You determine the frequency of the training you provide your employees. You must do all of the following:

(a) Provide periodic training to ensure that employees maintain understanding of, and competency in, well control or production safety practices;

(b) Establish procedures to verify adequate retention of the knowledge and skills that employees need to perform their assigned well control or production safety duties; and

(c) Ensure that your contractors' training programs provide for periodic training and verification of well control or production safety knowledge and skills.

§ 250.1507   How will MMS measure training results?

MMS may periodically assess your training program, using one or more of the methods in this section.

(a) *Training system audit.* MMS or its authorized representative may conduct a training system audit at your office. The training system audit will compare your training program against this subpart. You must be prepared to explain your overall training program and produce evidence to support your explanation.

(b) *Employee or contract personnel interviews.* MMS or its authorized representative may conduct interviews at either onshore or offshore locations to inquire about the types of training that were provided, when and where this training was conducted, and how effective the training was.

(c) *Employee or contract personnel testing.* MMS or its authorized representative may conduct testing at either onshore or offshore locations for the purpose of evaluating an individual's knowledge and skills in perfecting well control and production safety duties.

(d) *Hands-on production safety, simulator, or live well testing.* MMS or its authorized representative may conduct tests at either onshore or offshore locations. Tests will be designed to evaluate the competency of your employees or contract personnel in performing their assigned well control and production safety duties. You are responsible for the costs associated with this testing, excluding salary and travel costs for MMS personnel.

§ 250.1508 What must I do when MMS administers written or oral tests?

MMS or its authorized representative may test your employees or contract personnel at your worksite or at an onshore location. You and your contractors must:

(a) Allow MMS or its authorized representative to administer written or oral tests; and

(b) Identify personnel by current position, years of experience in present position, years of total oil field experience, and employer's name (e.g., operator, contractor, or sub-contractor company name).

§ 250.1509 What must I do when MMS administers or requires hands-on, simulator, or other types of testing?

If MMS or its authorized representative conducts, or requires you or your contractor to conduct hands-on, simulator, or other types of testing, you must:

(a) Allow MMS or its authorized representative to administer or witness the testing;

(b) Identify personnel by current position, years of experience in present position, years of total oil field experience, and employer's name (e.g., operator, contractor, or sub-contractor company name); and

(c) Pay for all costs associated with the testing, excluding salary and travel costs for MMS personnel.

§ 250.1510 What will MMS do if my training program does not comply with this subpart?

If MMS determines that your training program is not in compliance, we may initiate one or more of the following enforcement actions:

(a) Issue an Incident of Noncompliance (INC);

(b) Require you to revise and submit to MMS your training plan to address identified deficiencies;

(c) Assess civil/criminal penalties; or

(d) Initiate disqualification procedures.

[FR Doc. 00–20352 Filed 8–11–00; 8:45 am]

BILLING CODE 4310–MR–P

DEPARTMENT OF DEFENSE

Office of the Secretary

32 CFR Part 199

Civilian Health and Medical Program of the Uniformed Services (CHAMPUS); Enhancement of Dental Benefits Under the TRICARE Retiree Dental Program

AGENCY: Office of the Secretary, DoD.

ACTION: Interim final rule with request for comments.

SUMMARY: This interim final rule implements section 704 of the National Defense Authorization Act for Fiscal Year 2000, to allow additional benefits under the retiree dental insurance plan for Uniformed Services retirees and their family members that may be comparable to those under the Dependents Dental Program. The Department is publishing this rule as an interim final rule in order to comply timely with the desire of Congress to meet the needs of retirees for additional dental coverage. Public comments are invited and will be considered for possible revisions to this rule at the time of publication of the final rule.

DATES: Effective August 14, 2000. Comments must be received on or before October 13, 2000.

ADDRESSES: Forward comments to: TRICARE Management Activity (TMA), Special Contracts and Operations Office, 16401 East Centretech Parkway, Aurora, CO 80011–9043.

FOR FURTHER INFORMATION CONTACT: Linda Winter, Special Contracts and Operations Office, TMA, (303) 676–3682.

SUPPLEMENTARY INFORMATION:

I. Background

The TRICARE Retiree Dental Program (TRDP), a voluntary dental insurance plan completely funded by enrollees' premiums, was implemented in 1998 to provide benefits for basic dental care and treatment based on the authority of 10 U.S.C. 1076c. Under the enabling legislation, the benefits that can be provided are limited to "basic dental care and treatment, involving diagnostic services, preventative services, basic restorative services (including endodontics), surgical services, and emergency services." Accordingly, the implementing regulation, 32 CFR 199.22, limited coverage to the most common dental procedures necessary for maintenance of good dental health and did not include coverage of major restorative services, prosthodontics, orthodontics or other procedures considered to be outside of the "basic dental care and treatment" range.

Although the program was viewed as a major advance in offering dental coverage to retired members of the Uniformed Services and their family members at a very reasonable cost, there were still concerns that the enabling legislation was too restrictive in scope and that there should be expansion of services to better meet the needs of retirees.

Congress responded to these concerns by amending 10 U.S.C. 1076c with section 704 of the National Defense Authorization Act for Fiscal Year 2000, Pub. L. 106–065, to allow the Secretary of Defense to offer additional coverage. Under provisions of the amendment, the TRDP benefits may be "comparable to the benefits authorized under section 1076a" of title 10, the Dependents Dental Plan, commonly known as the TRICARE Family Member Dental Plan. Thus, in addition to the original basic services described above, which continue to be mandated, coverage of "orthodontic services, crowns, gold fillings, bridges, complete or partial dentures, and such other services as the Secretary of Defense considers to be appropriate" [10 U.S.C. 1076a(d)(3)] may be covered by the TRDP.

The language of section 704 of the National Defense Authorization Act for Fiscal Year 2000 is permissive and does not mandate such coverage. However, because of the many requests for additional TRDP coverage regardless of the inevitable increase in premiums, the DoD is proposing to expand the current coverage through a contractual arrangement. The premium cost of the enhanced coverage will remain the responsibility of the enrollees.

II. Provisions of the Interim Final Rule for Enhancement of TRDP Benefits

This interim final rule allows expansion of the TRDP benefits to be

Cross-Reference Relating Elements in *Practical Well Control* to IADC WellCAP Well-Control Accreditation Program

Note: the labels **I**, **F**, and **S** represent introductory, fundamental, and supervisory level course material. An **X** under a letter indicates that the adjacent material applys to that level.

I F S

    I. Cause of kicks

X X X   • *Identify causes of kicks*, pp. 2-3 – 2-9

X X X   • *Define a kick*, p. 1-1

     A. Unintentional flow or "kick" from a formation

      1. Failure to keep hole full, pp. 2-5 – 2-7

      2. Swabbing effect of pulling pipe, pp. 2-7, 2-8

       a. Hole and pipe geometry, p. 2-7

       b. Well depth, p. 2-8

       c. Mud rheology, p. 2-7

       d. Hole conditions and formation properties, p. 2-7

       e. Pipe pulling and running speed, p. 2-7

       f. Bottomhole assembly (BHA) configuration, p. 2-7

  X X   • *Describe the piston effect (suction and how increased drag may be associated with swab)*, p. 2-7

  X X   • *Describe the effect of the following on surge and swab pressures:*

      3. Loss of circulation, p. 2-9

      4. Insufficient density of drilling fluid, brines, cement, etc., pp. 2-7, 2-8

      5. Abnormally pressured formation, p. 2-6

      6. Lowering pipe too rapidly in hole, i.e.: surge, pp. 2-8, 2-9

      7. Annular gas flow after cementing (i.e., cementing intermediate casing), pp. 2-9, 2-10

  X X   • *Describe how fluid density can unintentionally be reduced, i.e., barite removed by centrifuge, dilution, cement settling, temperature effects on fluids, settling of mud materials, etc.,* p. 2-4

X X X   B. Intentional flow or "kick" from a formation

      1. Drill stem test, pp. 8-10, 8-11

      2. Completion, pp. 8-2, 8-4, 8-9, 8-10, 8-20 – 8-23

    II. Kick detection

X X X   A. Kick indicators

X X X   • *Identify kick indicators*

      1. Gain in pit volume (rapid increase in fluid volume at the surface), pp. 2-13, 2-14

      2. Increase in return fluid-flow rate (with no pump strokes per minute increase), pp. 2-12, 13

I F S

I   F   S

I F S

X X   16. Bottoms up strokes, App. C-1, eq. 44

X X   17. Surface-to-bit strokes, App. C-12, eqs. 34, 35

X X   18. Bit-to-shoe strokes, p. 4-8

X X   19. Total circulating strokes, including surface equipment, p. 4-8

X X   20. Pump output (look up from chart values only), App. B-11, 12, tabs. B-6, B-7

  X   21. Equivalent circulating density based on given annular pressure drop data, App. C-3, eq. 5

  X   22. Relationship between pump pressure and pump speed, pp. 4-1, 4-2

X X   23. Relationship between pump pressure and mud density, pp. 4-8, 4-9

X X   24. Maximum allowable annulus surface pressure, p. 5-3

  X   25. Effect of water depth on formation strength calculation, p. 1-1

  X   26. Gas laws, PV = K, App. C-5, eq. 28

X X   27. Weighting material required to increase density per volume, App. C-8, eq. 55

X X   28. Volume increase due to increase in density, App. C-7, eq. 57

X   29. Volume to be bled off, corresponding pressure increase (volumetric method), App. C-3, eq. 59

X X   30. Initial circulating pressure, App. C-6, eq. 33

X X   31. Final circulating pressure, App. C-4, eq. 48

X X   32. Riser volume and fluid required to displace, App. C-11, eqs. 41, 42

X X   33. Choke and kill line volumes, App. C-2, eq. 43

X X   34. Choke and kill line strokes, App. C-2, eq. 46

X X   35. Choke and kill line circulation times, App. C-2, eq. 47

X X   36. Pressure drop per step, pp. 6-5, 6, 6-8

     C. Conversion of pressure to an equivalent mud weight, pp. 1-6, 1-7

X X   1. Required mud weight

           a. Fluid density increase required to balance formation pressure, pp. 3-8, 3-9

X X   • *Calculate fluid density increase required to balance formation pressure, pp. 3-8, 3-9*

  X   2. Equivalent circulating density (ECD), p. 1-8

           a. ECD loss during flow check while drilling, p. 1-8

           b. No ECD loss during tripping flow check, p. 1-8

  X   • *Calculate the effect of circulating friction pressure losses on surface and downhole pressures, p. 1-8*

X X   D. Volume-height relationship and effect on pressure

X X   • *Calculate height of given volume of fluid, pp. 4-5, 6, eqs. 39, 40*

X X   E. Drop in pump pressure as fluid density increases during well-control operations, pp. 4-8, 4-9

X X   • *Describe why pump pressure drops as fluid density increases during a constant bottomhole pressure method, pp. 4-8, 4-9*

X X   F. Maximum wellbore pressure limitations

X X   1. Surface (e.g., wellhead, BOP, casing), p. 10-10

X X   2. Subsurface (e.g., perforations, casing shoe, open-hole formation), pp. 5-1 – 5-7

X X   • *Describe the consequences of exceeding maximum wellbore pressure limitations, pp. 5-1 – 5-7*

     IV. Procedures

  X   A. Alarm limits

  X   • *Demonstrate the procedures for setting well-control monitoring indicators, including, where applicable*

         1. High and low pit level, p. 10-18

         2. Return flow sensor, pp. 10-18, 19

         3. Trip tank level, p. 10-20

         4. Others (i.e., H$_2$S and flammable/explosive gas sensors), App. A-1

     B. Prerecorded well-control information

I F S
X X   • *Identify appropriate prerecorded information, pp. 6-2, 3*
                   1. Standpipe pressure at slow pump rates, pp. 4-1, 2
X X   • *Record standpipe pressure at slow pump rate, pp. 4-1, 2*
X X   • *Read at choke console, pp. 10-11 – 10-13*
X X   • *Recognize an error in gauge readings based on discrepancies between readings, p. 6-1*
   X   2. Well configuration, pp. 6-3, fig. 6.1, 6-7, fig. 6.3
   X   3. Fracture gradient, pp. 6-3, fig. 6.1, 6-7, fig. 6.3
   X   4. Maximum safe casing pressures, pp. 6-3, fig. 6.1, 6-7, fig. 6.3
                    a. Wellhead rating, p. 3-11
                    b. Casing burst rating, App. B-15, Tab. B.12, p. 3-11
                    c. Pipe/tubing collapse, p. 3-11
                    d. Subsurface weak zone (optional), p. 5-1
          C. Flow checks
X X   • *Describe the procedure to perform a flow check in the following situations*
X X X   1. When drilling, p. 2-10
                    a. Normal flow back, p. 2-10
X X   • *Recognize and measure normal flow back*
                    b. Not normal flow back, p. 2-10
X X   • *Recognize a flow that is different from normal flow back, p. 2-10*
X X   • *Take action based on recognition of flow, p. 2-10*
                    c. Loss of equivalent circulating density (ECD) with pumps off, p. 1-8
X X X   2. When tripping
                    a. Well is hydrostatically balanced (no ECD loss considerations), p. 1-8
X X   • *Explain how to establish that well is static before starting trip, pp. 2-10 – 2-12*
X X   • *Explain why an absence of flow (flow check) is not an absolute indicator that there is no influx, p. 2-10*
                    b. Use and purpose of trip sheet, pp. 2-11, 2-12
X X   • *Demonstrate understanding that the primary indicator of influx is the trip sheet (hole fill up), not a flow check, pp. 2-10 – 2-12*
          D. Shut in
X X X   • *Upon observing positive flow indicators, shut in the well in a timely and efficient manner to minimize influx, according to a specific procedure that will address the following operations*
X X   • *Design a shut-in procedure for the following operations*
X X   • *List differences between the soft vs. hard methods of well control as defined by API RP 59 for both levels (schools should include relevant information from API RP 59 in course materials or have this document available for reference; NOTE: API RP 59 is copyrighted by API and cannot be duplicated without the express written consent of API) pp. 3-3, 3-4*
(Note: the following is not intended to prescribe the exact sequence of events.)
X X   1. While drilling
                    a. Individual responsibilities (at option of contractor and operator)
                    b. Pick up with pump on, p. 3-1
                    c. Space out, p. 3-1
                    d. Shut pump off, p. 3-1
                    e. Flow check, p. 3-1
                    f. Close in BOP, p. 3-1
                    g. Close choke, p. 3-1
                    h. Notify supervisors, p. 3-1
X X   2. While tripping
                    a. Individual responsibilities (at option of contractor and operator)
                    b. Close off drill string bore given variety of tubular used, p. 3-2
                    c. Close BOP, p. 3-2

I   F   S

                        d. Notify supervisors, p. 3-2

X X   3. While running casing

                        a. Individual responsibilities (at option of contractor and operator)

                        b. Install device to stop potential flow through casing, p. 3-3

                        c. Close appropriate BOP or divert as appropriate, p. 3-3

                        d. Chose choke as applicable, p. 3-3

                        e. Notify supervisor, p. 3-3

X X   4. While cementing

                        a. Individual responsibilities (at option of drilling and service contractors and operator)

                        b. Space out, including consequences of irregular tubular lengths, p. 3-3

                        c. Shut pump off, p. 3-3

                        d. Close BOP, p. 3-3

                        e. Close choke, p. 3-3

                        f. Notify supervisor, p. 3-3

   X   5. Wireline operations

                        a. Individual responsibilities (at option of drilling and service contractors and operator)

                        b. Close BOP with consideration for cutting/closure around wire, p. 3-4

X X   6. During other rig activities

                        a. Individual responsibilities (at option of contractor and operator)

                        b. Use of surface equipment to shut in well, p. 3-4

                        c. Notify supervisors, p. 3-4

                        d. Close choke, p. 3-4

X X X   7. Verification of shut in, pp. 3-4, 5

X X X   • *For any shut in, verify well closure by demonstrating that the following flow paths are closed*

                        a. Annulus

                           1) Through BOP, p. 3-5

                           2) At the flow line, p. 3-5

                        b. Drill string

                           1) Pump pressure relief valves, p. 3-5

                           2) Standpipe manifold, p. 3-5

                        c. Wellhead/BOP

                           1) Casing valve (not applicable on subsea stack), p. 3-5

                           2) Broaching to the surface (outside of wellbore), p. 3-5

                        d. Choke manifold

                           1) Choke, p. 3-5

                           2) Overboard lines, p. 3-5

         E. Well monitoring during shut in

   X   • *Explain or demonstrate recommended procedures to use for well monitoring during well shut in*

X X X   1. Recordkeeping

                        a. Time of shut in, p. 6-3, fig. 6.1

                        b. Drill pipe and casing pressures

                           1) At initial shut in, p. 6-3, fig. 6.1

                           2) At regular intervals, p. 6-11, fig. 6.8

                        c. Estimated pit gain, p. 6-3, fig. 6.1

X X X   • *Read, record, and report well shut-in recordkeeping parameters*

  X X   2. Principles of bleeding volume from a shut-in well

                        a. Trapped pressure, p. 1-8

  X X   • *Identify at least two causes of trapped pressure, pp. 1-8, 9-9*

                           1) Causes, pp. 1-8, 9-9

                           2) Relief of, pp. 1-8, 9-9

I F S

    b. Pressure increase at surface and downhole from
       1) Gas migration, pp. 3-11 – 3-13
       2) Gas expansion, pp. 3-11 – 3-13

X X   • *Describe the effects of trapped pressure on wellbore pressure, pp. 1-8, 9-9*

  X   • *List two consequences on surface pressure resulting from shutting in on a gas versus a liquid kick of equivalent volume, pp. 2-2, 3*

X X   • *Perform choke manipulation to achieve specific pressure or volume objectives*

  X   • *Demonstrate procedure for relieving trapped pressure without creating underbalance, pp. 1-8, 9-9*

X X   3. Determining shut-in drill pipe pressure (SIDPP) when using a drill pipe float, pp. 7-5, 7-6

X X   • *If a float valve is in use (ported or nonported), demonstrate the procedure to open the float to obtain shut-in drill pipe pressure (SIDPP), pp. 7-5, 7-6*

  X   4. Gas, oil, or saltwater kick differences on surface pressures
       a. Density differences, p. 3-6

  X   5. Situations in which drill pipe shut-in pressure exceeds casing shut-in pressure
       a. Cuttings loading, p. 3-5
       b. Inaccurate gauge readings, pp. 3-5, 6
       c. Density of influx fluid greater than drilling fluid, p. 3-5
       d. Flow through drill string, p. 3-5
       e. Blockage downhole, p. 3-5

  X   • *List two situations in which shut-in drill pipe pressure would exceed shut-in casing pressure, p. 3-5*
      6. Maximum safe annulus pressure, pp. 5-1 – 5-7

  X   • *List hazards if closed-in annulus pressure exceeds maximum safe pressure, p. 5-1*

  X   • *Describe at least one method for controlling bottomhole pressure (BHP) while gas is migrating, pp. 6-15 – 6-17*
      7. Pressure between casing strings, p. 7-3

  X   • *Identify two causes of pressure between casing strings, p. 7-3*

  X   • *Describe potential hazards of pressure trapped between casing strings and actions required, p. 7-3*

    F. Response to massive or total loss of circulation

X X   • *Identify at least two methods of responding to massive or total loss of circulation during a well kill operation, pp. 7-15, 7-16*

X X   • *Upon observing loss of circulation, perform the following*

X X   1. During drilling, fill annulus with fluid in use, p. 7-14

X X   2. Notify supervisor immediately, p. 7-14

  X   3. Use of bridging materials (e.g., cement, gunk plugs, lost circulation material, etc.), pp. 7-15, 7-16

  X   4. Elimination of overbalance, p. 7-14

  X   G. Tripping

X X   1. Procedures for keeping hole filled
       a. Using rig pump, p. 2-12
       b. Using trip tank, pp. 2-11, 2-12
       c. Using recirculating tank (continuous fill), p. 2-12

X X   2. Methods of measuring and recording hole fill volumes, pp. 2-11, 2-12

X X   3. Wet trip calculations
       a. Return to mud system, p. 2-5
       b. No return to mud system, p. 2-5

X X   4. Dry trip calculations, pp. 2-6, 2-7

X X   5. Slugs, p. 2-6

X X   6. Trip margin, p. 2-8

X X X   • *Perform the following with regard to hole fill-up on trips*

I F S

                    a. Measure hole fill-up, pp. 2-11, 2-12

X X   b. Recognize discrepancy from calculated fill-up, pp. 2-11, 2-12

X X   c. Take appropriate action, p. 2-12

               7. At flow, go to shut in, pp. 2-10, 2-11

               8. At no flow and short fill-up, go back to bottom, p. 2-10

       • *Demonstrate, explain, or perform the following with regard to tripping*

X X X   a. Procedure and lineup to keep hole filled, pp. 2-10, 2-11, 10-20

    X   b. Calculate correct fill-up volumes

                  1) Wet trip, p. 2-5

                  2) Dry trip, pp. 2-6, 7

  X   c. Explain trip margin, p. 2-8

  X   d. Explain the effects of slugs on hole fill-up, pp. 2-6, 2-7

  X   e. Measurement of displacement volumes while tripping into the hole

                  1) With check valve in drill string, pp. 2-6, 2-7

                  2) Without check valve in drill string, p. 2-6

      H. Well-control drills (types and frequency)

  X   • *Describe the steps involved in conducting the following types of drills*

  X   1. Pit drills, p. 11-2

  X   2. Trip drills, p. 11-2

  X   3. Personnel evacuation, p. 11-1

  X   4. Diverter drills as they relate to shallow gas hazards, pp. 9-2 – 9-4

      I. Formation competency

X X   • *Describe or perform proper hookup and procedures for conducting a formation leak-off test or competency test for a given configuration*

X X   1. Pressure integrity test (testing to a specific limit), pp. 5-6, 7

X X   2. Leak-off test (testing to formation injectivity), pp. 5-3 – 5-6

X X   • *Identify from a plot the point at which leak off begins*, pp. 5-4 – 5-6

    X   3. Interpret data from formation tests, p. 5-5, fig.5.3

    X   4. Effect of fluid density change as applicable, pp. 5-6, 7

       • *Describe or perform*

X X   a. Leak-off test (at least one method), p. 5-3

    X   1) Calculate equivalent mud weight for leak-off test pressure, p. 5-6, eq. 51

X X   b. Formation pressure integrity test, p. 5-6

    X   2) Calculate surface pressure required to test to an equivalent mud weight, p. 5-6

    X   • *Describe how formation competency test results may be affected by fluid density change*, p. 5-6

              5. Preparing the well for leak-off testing, pp. 5-3, 4

      J. Stripping operations

X X X   • *Define the basic purpose of suitability, and method for stripping*, p. 7-7

       • *Demonstrate stripping procedures*

X X   1. Line up for bleeding volume to stripping tank, pp. 7-7 – 7-10

X X   2. Stripping through BOP, pp. 7-7 – 7-10

X X   3. Measurement of volume bled from the well, p. 7-9, eq. 63

    X   4. Calculations relating volumes and pressures to be bled for a given number of drill string stands run into the hole, p. 7-9, eq. 64

    X   5. Stripping with and without volumetric control, pp. 7-9, 10

      K. Shallow gas hazards

             1. Mechanisms and timing of events, pp. 9-1, 9-2

X X X   • *Explain why, at shallow depths, it is relatively easy to become underbalanced (e.g., hole sweeps, gas cutting, swabbing, lost circulation)*, pp. 9-1, 9-2

X X X   • *Explain the limited reaction time for kick detection*, p. 9-1

I F S

                2. Kill procedures
                   a. Shut in, p. 9-1
                   b. Use of diverters
                      1) With drill pipe, pp. 9-2, 9-3
                      2) Running casing, pp. 9-2, 9-3
                   c. Riserless drilling, p. 9-4

X X   • *Explain the well-control procedural options available (i.e., shut in vs. divert), pp. 9-1, 9-2*
X   3. Pilot holes, p. 9-1
X   • *Explain the use of pilot holes*
X   4. During and after cementing conductor and surface casing, p. 9-1
             5. Setting barite or cement plugs, p. 9-1
X   • *Describe the technique or procedure for preparing and setting barite or cement plugs, pp. 7-15, 7-16*

       V. Gas characteristics and behavior
X X X A. Gas types
           1. Hydrocarbon, p. 2-1
           2. Toxic
              a. H_2S, p. 2-1
              b. CO_2, p. 2-1
       B. Density
X X X   1. Gas, p. 2-1
X X X • *Recognize the relatively low density of gas and its effect on the hydrostatic column, p. 2-2*
X X   • *Describe how the presence of gas affects wellbore pressure*
X X   2. Gas and mud mixtures, pp. 2-2, 2-3
X X   • *Explain the effect of gas cutting on bottomhole pressure and the use of pit level to recognize hydrostatic loss, p. 2-1*
X   • *Describe the conditions where gas cutting may have little effect on hydrostatic head and bottomhole pressure, pp. 2-16, 2-17*
       C. Migration
X X X • *Explain the consequences of gas migration*
X X X   1. If the well is left shut in while gas is migrating, pp. 2-3, 3-11
X X X   2. If the well is allowed to remain open with no control, p. 2-3
X X   3. If bottomhole pressure is controlled, p. 6-1
       D. Expansion
X X   1. While in well, p. 1-7
X X   • *Explain the relationship between pressure and volume of gas in the wellbore, pp. 3-6, 7*
X X   • *Explain why a gas kick must expand as it is circulated out to keep bottomhole pressure constant, p. 1-7*
X X   2. Through surface equipment, p. 1-7
X X   • *Explain the consequence of gas moving through the choke from a high-pressure to a low-pressure area, p. 1-7*
       E. Compressibility and phase behavior
X X   • *Describe how hydrocarbon gas may not migrate and the consequences for well control*
           1. Hydrocarbon gas can be either in a liquid or gaseous form when it enters the wellbore, depending on its pressure and temperature, p. 3-7
           2. Hydrocarbon gas entering as a liquid may not migrate or expand until it is circulated up the wellbore, p. 3-7
           3. Liquids can move down the annulus and come up the drill string, p. 2-3
       F. Solubility in mud
X X   1. Combinations of gas and liquid in which solubility issues may apply
              a. H_2S and water, p. 2-1
              b. CO_2 and water, p. 2-1

I   F   S

                 c. H_2S and oil-base mud, p. 2-1

                 d. Methane and oil-base mud, p. 2-1

                 e. CO_2 and oil-base mud, p. 2-1

X   X   •   *Identify combinations of gas and liquid in which solubility issues may apply (H_2S and water, CO_2 and water, H_2S and oil-base mud, methane and oil-base mud, and CO_2 and oil-base mud*

X   X   2.   Gases dissolved in mud behave like liquids, p. 2-1

X   X   •   *Describe the difficulty of detecting kicks with soluble gases while drilling or tripping*, p. 2-1

    X   3.   Dissolved gases evolve out of mud at some point in the wellbore, p. 2-1

    X   •   *Describe how dissolved gas affects wellbore pressures when it comes out of solution*, p. 2-1

    X   •   *Describe the sequential consequences of gas evolving from the mud system*, p. 2-1

         VI.   Fluids

X   X   X   A.   Types of drilling fluids

            1.   Water based mud, pp. 8-5, 8-6

            2.   Oil-base mud (OBM), synthetic oil based mud (SOBM), p. 8-4

            3.   Cement, p. 8-6

            4.   Completion fluids, pp. 8-6, 7

    X   B.   Fluid property effects on pressure losses

            1.   Density, p. 1-8

            2.   Viscosity, p. 1-8

            3.   Changes in mud properties caused by contamination with formation fluids, p. 2-19

X   X   C.   Fluid density measuring techniques

            1.   Mud balance, pp. 1-4, 1-5

            2.   Pressurized mud balance, pp. 1-4 – 1-6

X   X   •   *Measure fluid density*

    X   D.   Mud properties following weight-up and dilution

            1.   Gel strengths, pp. 6-14, 6-15

            2.   Plastic viscosity and yield point, pp. 6-14, 6-15

        VII.   Constant bottomhole pressure well-control methods

X   X   X   A.   Objectives of well-control methods

            1.   Circulate kick safely out of the well, p. 6-1

            2.   Re-establish primary well control by restoring hydrostatic balance, p. 6-1

            3.   Avoid additional kicks, p. 6-1

            4.   Avoid excessive surface and downhole pressures so as not to induce an underground blowout, p. 6-1

         B.   Principles of constant bottomhole pressure methods

X   X   X   1.   Well shut-in will stop influx when bottomhole pressure (BHP) equals formation pressure, p. 6-1

X   X   2.   Circulating out a kick with choke back-pressure to keep BHP equal to or slightly greater than formation pressure, p. 6-1

X   X   •   *Explain how pump and choke manipulation relates to maintaining constant bottomhole pressure*, p. 6-1

            3.   Bottom of drill string must be at the kicking formation (or bottom of the well) to effectively kill the kick and be able to resume normal operations, p. 3-5

         C.   Example steps of a constant bottomhole pressure well-control method: driller's or wait and weight, pp. 6-2 – 6-12

    X   1.   Well kill and control calculations and procedures

    X   a.   Proficiency in both of these constant bottomhole pressure well-control methods, pp. 6-2 – 6-12

                 1)   Driller's and/or wait and weight method

I F S

a) Bring pump up to slow kill rate while opening choke, pp. 6-2, 6-5, 6-6
b) Maintain surface pressure while circulating according to method, pp. 6-2, 6-5, 6-6
c) Increase mud weight in pits to kill weight, pp. 6-2, 6-5, 6-6
d) Line up pump to kill mud, pp. 6-2, 6-5, 6-6
e) Line up choke manifold and auxiliary well-control equipment, pp. 6-2, 6-5, 6-6
f) Pump kill mud weight until kill mud completely fills the wellbore, pp. 6-2, 6-5, 6-6
g) Circulate until all kicks are removed from the well, pp. 6-2, 6-5, 6-6
h) Shut off pumps, pp. 6-2, 6-5, 6-6

X X X i) Close choke and observe pressure gauges (SIDPP and SICP = 0 psi), pp. 6-2, 6-5, 6-6

X X X • *Read, record, and report drill pipe and annulus pressures*
j) If hydrostatic balance is restored, open BOPs, check for flow, pp. 6-2, 6-5, 6-6
k) Resume operations, pp. 6-2, 6-5, 6-6

X X b. Preparing the kill sheet, pp. 6-3, 6-7

X 2. Organize the specific responsibilities of rig crew during a well kill or control procedure, pp. 11-1, 11-2

X X X • *List the phases of at least one constant bottomhole pressure well-control method, pp. 6-2, 6-5, 6-6*

X X • *Explain how these steps relate to maintaining bottomhole pressure equal to or greater than formation pressure, pp. 6-2, 6-5, 6-6*

X X • *Demonstrate proficiency in at least one constant bottomhole pressure well-control method (Driller to act under direction of supervisor)*

X • *Demonstrate or describe the process of organizing the specific responsibilities of the rig crew during the execution of a well kill operation, pp. 11-1, 11-2*

D. Well-control kill sheets

X X • *Correctly fill out a kill sheet for one well-control method, pp. 6-1, fig. 6.1, 6-5, 6-6, figs. 6.4, 6.5, 6.6, 6.7*
1. Well-control calculations
a. Drill string and annular volumes, strokes, and times, pp. 4-3 – 4-7
b. Fluid density increase required to balance increased formation pressure, pp. 3-8, 9

X X • *Determine weight up material required and corresponding volume increase, pp. 6-12 – 6-14*
c. Initial and final circulating pressure as appropriate for methods taught, pp. 4-2, 4-3, 4-8, 4-9

X X 2. Maximum wellbore pressure limitations
a. Surface, p. 6-4
b. Subsurface, p. 6-4

X X • *Describe the consequences of exceeding maximum wellbore pressure*
a. Surface, p. 6-4
b. Subsurface, p. 6-4
3. Selection of kill rate for pump

X a. Allowing for friction losses, p. 4-1

X X b. Barite delivery rate, p. 4-1

X c. Choke operator reaction time, p. 4-1

X X d. Pump limitations, p. 4-1
e. Surface fluid handling capacity, p. 4-1

X • *Identify factors affecting selection of kill rate for mud pump, p. 4-1*

E. Well-control procedures

X X 1. Procedure to bring pump on and off line and change pump speed while holding bottomhole pressure constant using the choke
a. Use of casing pressure gauge, p. 6-2

I F S

                b. Lag time response on drill pipe pressure gauge, p. 6-2

X X   • *Demonstrate bringing the pump on and off line and chaning pump speed while holding bottomhole pressure constant using choke*

  X   2. Initial circulation pressure

              a. Using recorded shut-in drill pipe pressure and reduced circulating pressure, p. 4-2

              b. Without a prerecorded value for reduced circulating pressure, p. 4-2

              c. Adjustment for difference in observed vs. calculated circulating pressures, pp. 7-1 – 7-3

  X   • *Determine correct initial circulating pressure*

              a. Using recorded shut-in drill pipe pressure and reduced circulating pressure, p. 4-2

              b. Without a prerecorded value for reduced circulating pressure, p. 4-2

              c. Adjustment for difference in observed vs. calculated circulating pressures, pp. 4-2, 6-1

  X   3. Choke adjustment during well kill procedure, p. 6-1

  X   • *Operate choke to achieve specific pressure objectives relative to selected constant bottomhole pressure methods*

              a. Changes in surface pressure as a result in changes in hydrostatic head or circulating rates, p. 6-1

                1) Drop in pump pressure as fluid density increases in drill string during well-control operations, p. 6-5

                2) Increase in pump pressure with increased pump rate and vice versa, p. 4-1

  X   • *Describe why pump pressure drops as fluid density increases during a constant bottomhole pressure method*, p. 6-4

              b. Pressure response time, p. 6-2

                1) Casing pressure gauge (immediate), p. 6-2

                2) Drill pipe pressure gauge (lag time), p. 6-2

  X   4. Handling of problems during well-control operations

  X   • *Given a scenario detailing a well-control problem, identify the problem and demonstrate or describe an appropriate response*

X X   a. Pump failure, p. 7-3

  X   b. Changing pumps, pp. 7-3, 4

X X   c. Plugged or washed nozzles, pp. 7-1, 2

  X   d. Washout or parting of the drill string, p. 7-1

  X   e. BOP failure

                1) Flange failure, p. 7-3

                2) Weephole leakage, p. 7-3

                3) Failure to close, p. 7-4

                4) Failure to seal, p. 7-4

X X   f. Plugged or washed choke, p. 7-2

X X   g. Fluid losses, p. 7-2

  X   h. Flow problems downstream of choke, p. 7-4

  X   i. Hydrates, p. 9-6

  X   j. Malfunction of remote choke system, p. 7-4

  X   k. Mud-gas separator, p. 7-4

  X   l. Problems with surface pressure gauges, p. 7-4

X X   m. Annulus pack off, p. 7-4

         5. Considerations using a diverter, pp. 9-1 – 9-4

X X X • *Describe the difference between diverting vs. conventional well kills*, p. 9-1

  X   • *List at least two conditions under which the use of a diverter may be applicable*, pp. 9-1, 9-2

  X   • *List at least two potential hazards when using a diverter*, pp. 9-1, 9-2

I F S

I F S

           1) Blind, p. 10-6

           2) Blind/shear, p. 10-6

           3) Pipe, p. 10-6

           4) Variable bore pipe, p. 10-6

           5) Ram elements, p. 10-6

        c. Drilling spool or integral body, p. 10-1

        d. Valves, p. 10-1

      2. Functions, pp. 10-2 – 10-7

X X X   • *Identify flow path for normal drilling operations, p. 10-2*

X X X   • *Identify flow path for well-control operations, p. 10-2*

   X X   • *Identify areas exposed to high and low pressure during shut in and pumping operations, p. 10-2*

X X X   • *Identify and confirm line-up for equipment pressure testing, shut in, and pumping operations, p. 10-2*

   X X   • *Demonstrate ability to shut in the well in the event of primary equipment failure*

   X X   • *Given a BOP stack configuration, identify potential flow paths for kill operations, p. 10-2*

   X X   • *Given a BOP stack configuration, identify shut in, monitoring, and circulation operations which are possible and those which are not, p. 10-2*

   X X   • *Given a BOP stack configuration, select an appropriate BOP to effect closure on a given tubular, pp. 10-2 – 10-5*

X X X   C. Manifolds and piping

       1. Standpipe, p. 10-3

       2. Choke, pp. 10-11 – 10-13

   X X   D. Valving

   X X   1. BOP stack, pp. 10-2 – 10-7

   X X   2. Drill string

        a. Full opening valves (drill pipe safety valves, kelly cocks, and kelly valves), pp. 10-7 – 10-10

        b. Check valves, p. 10-9

        c. Float valves: advantages and disadvantages, pp. 10-8, 10-9

   X X   • *Describe opening and closing a full opening safety valve, p. 10-8*

   X X   • *Describe the difference in use between a full opening safety valve and a check valve (e.g., inside BOP), pp. 10-7 – 10-10*

   X X   • *Be able to identify compatibility of thread types*

       3. Choke manifold, pp. 10-11, 12

X X X   • *Distinguish the funtion of the choke from that of other valve types, pp. 10-11, 10-12*

   X X   • *Define the function of a choke, pp. 10-11, 10-2*

        a. Adjustable choke

         1) Hydraulic (remote operated), pp. 10-11, 10-12

         2) Manual, p. 10-11

         b. Fixed choke, p. 10-11

         c. Valves to direct flow, p. 10-11

   X X   • *Describe the function of adjustable chokes, both manual and hydraulic, pp. 10-11, 10-12*

X X X   4. Mud pump pressure relief, p. 10-11

   X X   • *Identify changes in valve positions resulting from opening or closing the diverter*

      E. Auxiliary well-control equipment

X X X   • *Define function, operating principles, flowpaths, and components of mud-gas separators, pp. 10-20, 10-21*

     X   • *List two possible consequences and corrective actions of overloading the mud/gas separator, p. 10-21*

X X X   1. Mud-gas separator

        a. Gas blow through, p. 10-21

        b. Vessel rupture, p. 10-21

I F S

    X   • *Explain use of testing and completion well-control equipment, pp. 8-19 – 8-23*

        H. Pressure and function tests

  X X   1. Maximum safe working pressure

            a. Pressure ratings of all equipment, p. 10-10

            b. Reasons for derating, p. 10-10

            c. Areas exposed to both high and low pressures during shut-in and pumping operations, p. 10-10

  X X   • *Identify the maximum safe working pressure for a given set of well-control equipment upstream and downstream of the choke, p. 10-10*

    X   • *List at least two reasons for possible derating of the working pressure of well-control equipment, p. 10-10*

  X X   2. General emphasis on quality maintenance practices

            a. Correct installation, pp. 10-10, 10-11

            b. Maintenance, pp. 10-10, 10-11

            c. Wear and replacement requirements, pp. 10-10, 10-11

            d. Rings, flanges, and connectors, pp. 10-10, 10-11

  X X   • *Describe correct installation, maintenance, wear, and replacement requirements for rings, flanges, and connectors, pp. 10-10, 10-11*

  X X   3. Emphasis on quality testing practices, p. 10-11

  X X   • *Describe the need for function and pressure testing of BOP equipment, p. 10-11*

            4. Procedures for function and pressure testing all well-control equipment

X X X   • *Perform, explain, or demontrate the following*

            a. Function and testing of high-pressure well-control equipment

               1) BOP stack, pp. 10-11; API RP53, Chaps. 17, 18

               2) Manifolds, pp. 10-11; API RP53, Chaps. 17, 18

               3) Auxiliary well-control equipment, pp. 10-11; API RP53, Chaps. 17, 18

            b. Function and testing of low-pressure well-control equipment

               1) Mud-gas separator, pp. 10-11; API RP53, Chaps. 17, 18

               2) Fluid and gas pathways, pp. 10-11; API RP53, Chaps. 17, 18

            c. Pressure or function testing of diverter systems, pp. 9-4; API RP53, Chaps. 17, 18

        I. Well-control equipment arrangements

X X X   1. General arrangements for BOP, valving, manifolds, and auxiliary equipment (applicable to both written and practical testing), pp. 10-1, 10-2

            a. BOP, manifold piping, and valve line up for

               1) Drilling operations, p. 10-2

               2) Shut in, p. 10-2

               3) Well-control operations, p. 10-2

               4) Testing, p. 10-2

X X X   • *Identify flow path for well-control operations, p. 10-2*

  X X   • *Identify areas exposed to high- and low-pressure during shut-in and pumping operations, p. 10-2*

X X X   • *Identify and confirm line up for equipment pressure testing, shut-in, and pumping operations, p. 10-2*

X X X   • *Demonstrate ability to shut in the well in the event of primary equipment failure*

X X X   • *Demonstrate the correct alignment of standpipe and choke manifold valves, including downstream valves for*

               a. Drilling operations, p. 10-2

               b. Shut in, p. 10-2

               c. Well-control operations, p. 10-2

               d. Testing. p. 10-2

      IX. Government, industry, and company rules, orders, and policies

I F S

           A. Incorporate by reference

  X   1. API and ISO recommended practices, standards, and bulletins pertaining to well control, (obtain from API, ISO, and company)

X X X   2. Regional and/or local regulations where required

      X. Subsea well control (required for subsea endorsement)

X X X   A. Subsea equipment

X X X   • *Identify and describe function of each, pp. 9-2, 9-6 – 9-8*

  X X   • *State how to activate ram locks, pp. 10-6, 7*

  X X   1. Marine riser systems

          a. Drilling with riser, pp. 9-2 – 9-4

          b. Drilling without riser, p. 9-4

  X X   • *Describe reasons for drilling with and without riser, pp. 9-2 – 9-4*

       2. BOP control systems

          a. Block position, pp. 9-6, 7

          b. Pilot system, pp. 9-6, 7

          c. Subsea control pods, pp. 9-6, 9-7

  X X   • *Describe operating principles of subsea BOP stack control system, pp. 9-6 – 9-8*

       3. BOP stack

          a. Lower marine riser package, p. 9-7

          b. Configuration, p. 9-7

          c. Ram locks, pp. 9-6, 7

       4. Ball joint, p. 9-7

       5. Flex joint, p. 9-7

       6. Slip joint, p. 9-7

       7. Riser dump valve, p. 9-7

      B. Diverter systems

X X X   • *Describe principle of operation of the diverter system on a floating unit, p. 9-2*

X X X   1. Configuration and components, pp. 9-2, 3

X X X   2. Diverter line size and location, pp. 9-2, 9-3

X X X   3. Line up for diversion

          a. Valve arrangement and function, pp. 9-2 – 9-4

          b. Valve operational sequence, pp. 9-2 – 9-4

          c. Limitations of the diverter system, pp. 9-2 – 9-4

  X X   C. Kick detection issues

       1. Vessel motion, pp. 9-9, 10-18, 10-19

       2. With and without riser, pp. 9-3, 9-4

       3. Riser collapse, p. 9-9

       4. Water depth (BOP placement), pp. 9-4, 9-5

      D. Procedures

  X X   1. Choke and kill line friction

  X X   a. Measurement of choke and kill line friction, pp. 9-4, 9-5

    X   b. Compensating for choke and kill line friction

             1) Static kill line, pp. 9-4, 9-5

             2) Casing pressure adjustment, pp. 9-4, 9-5

  X X   • *Define or describe the effect of fluids of different densities in the choke and kill lines for both levels*

    X   2. Removing trapped gas in BOPs, pp. 9-5, 9-7

  X X   • *Explain the consequences of trapped gas in the subsea BOP system, pp. 9-5, 9-7*

    X   • *Describe a specific procedure for removing trapped gas from the BOP stack following a kill operation, pp. 9-5, 9-7*

          a. Use of bleed lines, p. 9-7

I F S

                        b. U-tubing of trapped gas, p. 9-5

X X   3. Clearing riser

                        a. Gas in riser, pp. 9-5 & 9-7

                        b. Displacing riser with kill weight mud, pp. 9-5, 9-7

X X   • *Describe killing a subsea riser with kill mud and the consequences of failure to fill riser with kill mud after circulating out a kick, p. 9-5*

X X   4. Hydrostatic effect of riser disconnects and reconnects, p. 9-8

X X   • *Describe possible consequence of well behavior upon removal of riser without a riser margin, p. 9-8*

X X   5. Spacing and hang off, pp. 3-2, 3-3

X X   • *Describe steps necessary to space out drill pipe and hang it off using a motion compensator, ram locks, etc. , pp. 3-2, 3-3*

  X   6. Effect of water depth on formation competency, p. 9-4

        E. Compensating for hydrostatic head changes in choke lines

  X   • *Demonstrate ability to adjust for circulating pressures to compensate for choke line friction*

  X   • *Demonstrate ability to adjust choke appropriately to compensate for rapid change in hydrostatic pressure caused by gas in long choke lines.*

        F. Hydrates

  X   • *Identify possible complications caused by hydrates, p. 9-6*

    XI. Special situations (optional)

        Note: the following items are not required but may be added at a school's discretion based on local conditions and needs. Course level marks suggest the level of student that might benefit from instruction on a particular topic.

        A. Hydrogen sulfide (H_2S)

X X X   1. Risks encountered in well-control operations involving H_2S

                        a. Toxicity, App. A-1

                        b. Explosiveness, App. A-1

                        c. Corrosivity, App. A-5, A-6

                        d. Solubility, App. A-1

X X   2. Well-control handling options

                        a. Bullheading, p. 7-17

                        b. Circulation with flaring, pp. 10-20, 10-21

X X X   • *Identify risks, App. A-1*

  X   • *Specify crew responsibilities*

X X   • *Identify well-control options, including bullheading and circulation with flaring, pp. 7-17, 10-20, 21*

  X   B. Horizontal well-control considerations

X X   1. Influx detection, pp. 7-19, 7-20

X X   2. Off-bottom kill, pp. 7-21, 7-22

  X   3. Special kill sheet, p. 7-21

      • *Explain*

                        a. Any kill sheet modifications, p. 7-21

                        b. Influx detection, pp. 7-19 – 7-21

                        c. Procedure for off-bottom kill, pp. 7-21, 22

                        d. Gas behavior in horizontal section, pp. 7-19, 20

                        e. Pressurized drilling equipment, pp. 7-20, 7-21

  X   C. Off-bottom kills

  X   • *Explain off-bottom kills, pp. 7-21, 7-22*

  X   D. Underground blowouts

        1. Indications of underground flow

                        a. At shut in, pp. 7-19, 20

                        b. During kill, pp. 7-19, 20

I F S

X X   • *Demonstrate how to recognize loss of formation integrity during shut in or circulation, pp. 7-19, 7-0*

  X   E. Combination thief and kick zones
        1. Thief zone on top, kick zone on bottom, p. 7-22
        2. Kick zone on top, thief zone on bottom, p. 7-22
  X   • *Explain problems and procedural responses for combination thief and kick zone, p. 7-22*

  X   F. False kick indicators
        1. Kelly cut, p. 2-19
        2. Background gas, p. 2-20
        3. Bottom up with oil base mud, p.2-20
        4. Fluid transfer, p. 2-20
       • *Describe or explain false kick indicators, pp. 2-19, 2-20*

  X   G. Pipe reciprocation during well kill (biaxial loading), p. 7-11
       • *Explain procedures for pipe movement during well kill, p. 7-11*

  X   H. Underbalanced drilling
        1. Producing while drilling, pp. 7-18, 19
        2. Pressurized drilling equipment
           a. Rotating annular preventer, p. 7-18
           b. Rotating head, p. 7-18
       • *Explain well-control procedures for underbalanced drilling, pp. 7-18, 7-19*

  X   I. Slim hole well-control considerations
       • *Explain well-control concerns during slim hole drilling, p. 7-19*

  X   J. Coil tubing
       • *Explain well control during coil tubing operations (tree in place), p. 10-25*

  X   K. Snubbing
  X   • *Explain well control during snubbing operations (tree in place), pp. 7-10, 7-11*

  X   L. New well-control technology
  X   M. High pressure-high temperature considerations, p. 3-14
  X   N. Tapered string, tapered hole, p. 7-19
  X   O. Wellhead component failure points
        1. Casing hangers, p. 10-10
        2. Casing isolation seals, p. 10-10
        3. Connections and fittings. p. 10-10

  X   P. Shut-in and circulating kick tolerance, p. 5-1
  X   Q. Deep water well control, pp. 9-4 – 9-6

A

abnormal pressure *n*: pressure exceeding or falling below the pressure to be expected at a given depth. Normal pressure increases approximately 0.465 pounds per square inch per foot of depth or 10.5 kilopascals per metre of depth. Thus, normal pressure at 1,000 feet is 465 pounds per square inch; at 1,000 metres it is 10,500 kilopascals. See *pressure gradient*.

accumulator *n*: the storage device for nitrogen-pressurized hydraulic fluid, which is used in operating the blowout preventers. See *blowout preventer control unit*.

accumulator bottle *n*: a bottle-shaped steel cylinder located in a blowout preventer control unit to store nitrogen and hydraulic fluid under pressure (usually at 3,000 pounds per square inch). The fluid is used to actuate the blowout preventer stack.

acid fracture *v*: to part or open fractures in productive hard limestone formations by using a combination of oil and acid or water and acid under high pressure. See *formation fracturing*.

adjustable choke *n*: a choke in which the position of a conical needle, sleeve, or plate may be changed with respect to their seat to vary the rate of flow; may be manual or automatic. See *choke*.

annular blowout preventer *n*: a large valve, usually installed above the ram preventers, that forms a seal in the annular space between the pipe and the wellbore or, if no pipe is present, in the wellbore itself. Compare *ram blowout preventer*.

annular pressure *n*: fluid pressure in an annular space, as around tubing within casing.

annular space *n*: the space between two concentric circles. In the petroleum industry, it is usually the space surrounding a pipe in the wellbore; sometimes termed the annulus.

annular velocity *n*: the rate at which mud is traveling in the annular space of a drilling well.

annulus *n*: see *annular space*.

anticline *n*: rock layers folded in the shape of an arch. Anticlines sometimes trap oil and gas.

atmospheric pressure *n*: the pressure exerted by the weight of the atmosphere. At sea level, the pressure is approximately 14.7 pounds per square inch (101.325 kilopascals), often referred to as 1 atmosphere. Also called barometric pressure.

B

background gas *n*: in drilling operations, gas that returns to the surface with the drilling mud in measurable quantities but does not cause a kick. Increases in background gas may indicate that the well is about to kick or has kicked.

back off *v*: to unscrew one threaded piece (such as a section of pipe) from another.

back-off *n*: the procedure whereby one threaded piece (such as a pipe) is unscrewed from another.

back-pressure *n*: 1. the pressure maintained on equipment or systems through which a fluid flows. 2. in reference to engines, a term used to describe the resistance to the flow of exhaust gas through the exhaust pipe. 3. the operating pressure level measured downstream from a measuring device.

barite *n*: barium sulfate, $BaSO_4$; a mineral frequently used to increase the weight or density of drilling mud. Its relative density is 4.2 (i.e., it is 4.2 times denser than water). See *barium sulfate, mud*.

barium sulfate *n*: a chemical compound of barium, sulfur, and oxygen ($BaSO_4$), which may form a tenacious scale that is very difficult to remove. Also called barite.

barrel (bbl) *n*: measure of volume for petroleum products in the United States. One barrel is the equivalent of 42 U.S. gallons or 0.15899 cubic metres (9,702 cubic inches). One cubic metre equals 6.2897 barrels.

bell nipple *n*: a short length of pipe (a nipple) installed on top of the blowout preventer. The top end of the nipple is flared, or belled, to guide drill tools into the hole and usually has side connections for the fill line and mud return line.

BHP *abbr*: bottomhole pressure.

bleed *v*: to drain off liquid or gas, generally slowly, through a valve called a bleeder. To bleed down, or bleed off, means to release pressure slowly from a well or from pressurized equipment.

bleed line *n*: a pipe through which pressure is bled, as from a pressurized tank, vessel, or other pipe.

blind ram *n*: an integral part of a blowout preventer, which serves as the closing element on an open hole. Its ends do not fit around the drill pipe but seal against each other and shut off the space below completely. See *ram*.

blind ram preventer *n*: a blowout preventer in which blind rams are the closing elements. See *blind ram*.

blowout *n*: an uncontrolled flow of gas, oil, or other well fluids into the atmosphere or into an underground formation. A blowout, or gusher, can occur when formation pressure exceeds the pressure applied to it by the column of drilling fluid.

blowout preventer *n*: one of several valves installed at the wellhead to prevent the escape of pressure either in the annular space between the casing and the drill pipe or in open hole (i.e., hole with no drill pipe) during drilling or completion operations. Blowout preventers on land rigs are located beneath the rig at the land's surface; on jackup or platform rigs, at the water's surface; and on floating offshore rigs, on the seafloor. See *annular blowout preventer, inside blowout preventer, ram blowout preventer*.

blowout preventer control panel *n*: controls, usually located near the driller's position on the rig floor, that are manipulated to open and close the blowout preventers. See *blowout preventer*.

blowout preventer control unit *n*: a device that stores hydraulic fluid under pressure in special containers and provides a method to open and close the blowout preventers quickly and reliably. Usually, compressed air and hydraulic pressure provide the opening and closing force in the unit. See *blowout preventer*.

blowout sticking *n*: jamming or wedging of the drill string in the borehole by sand or shale that is driven uphole by formation fluids during a blowout.

BOP *abbr*: blowout preventer.

BOP stack *n*: the assembly of blowout preventers installed on a well.

bottomhole pressure *n*: 1. the pressure at the bottom of a borehole. It is caused by the hydrostatic pressure of the wellbore fluid and, sometimes, by any back-pressure held at the surface, as when the well is shut in with blowout preventers. When mud is being circulated, bottomhole pressure is the hydrostatic pressure plus the remaining circulating pressure required to move the mud up the annulus. 2. the pressure in a well at a point opposite the producing formation, as recorded by a bottomhole pressure bomb.

bottomhole pressure test *n*: a test that measures the reservoir pressure of the well, obtained at a specific depth or at the midpoint of the producing zone. A flowing bottomhole pressure test measures pressure while the well continues to flow; a shut-in bottomhole pressure test measures pressure after the well has been shut in for a specified period of time. See *bottomhole pressure*.

bottoms up *n*: a complete trip from the bottom of the wellbore to the top.

bottoms-up time *n*: the time required for mud to travel up the borehole from the bit to the surface.

Boyle's law *n*: the principle that states that at a fixed temperature, the volume of an ideal gas or gases varies inversely with its absolute pressure. As gas pressure increases, gas volume decreases proportionately.

bridging materials *n pl*: the fibrous, flaky, or granular material added to a cement slurry or drilling fluid to aid in sealing formations in which lost circulation has occurred. See *lost circulation material*.

bullheading *n*: 1. forcing gas back into a formation by pumping into the annulus from the surface. 2. any pumping procedure in which fluid is pumped into the well against pressure.

C

cased hole *n*: a wellbore in which casing has been run.

casing *n*: steel pipe placed in an oil or gas well to prevent the wall of the hole from caving in, to prevent movement of fluids from one formation to another, and to improve the efficiency of extracting petroleum if the well is productive. A joint of casing is available in three length ranges: a joint of range 1 casing is 16 to 25 feet (4.8 to 7.6 metres) long; a joint of range 2 casing is 25 to 34 feet (7.6 to 10.3 metres) long; and a joint of range 3 casing is 34 to 48 feet (10.3 to 14.6 metres) long. Diameters of casing manufactured to API specifications range from 4.5 to 20 inches (11.4 to 50.8 centimetres). Casing is also made of many types of steel alloy, which vary in strength, corrosion resistance, and so on.

casing burst pressure *n*: the amount of pressure that, when applied inside a string of casing, causes the wall of the casing to fail. This pressure is critically important when a gas kick is being circulated out, because gas on the way to the surface expands and exerts more pressure than it exerted at the bottom of the well.

casing point *n*: the depth in a well at which casing is set, generally the depth at which the casing shoe rests.

casing pressure *n*: the pressure in a well that exists between the casing and the tubing or the casing and the drill pipe.

casing seat *n*: the location of the bottom of a string of casing that is cemented in a well. Typically, a casing shoe is made up on the end of the casing at this point.

casing string *n*: the entire length of all the joints of casing run in a well. Most casing joints are manufactured to specifications established by API, although non-API specification casing is available for special situations. Casing manufactured to API specifications is available in three length ranges. A joint of range 1 casing is 16 to 25 feet (4.8 to 7.6 metres) long; a joint of range 2 casing is 25 to 34 feet (7.6 to 10.3 metres) long;

and a joint of range 3 casing is 34 to 48 feet long (10.3 to 14.6 metres). The outside diameter of a joint of API casing ranges from 4.5 to 20 inches (11.43 to 50.8 centimetres).

cement *n*: a powder consisting of alumina, silica, lime, and other substances that hardens when mixed with water. Extensively used in the oil industry to bond casing to the walls of the wellbore.

cement plug *n*: a portion of cement placed at some point in the wellbore to seal it.

change rams *v*: to take rams out of a blowout preventer and replace them with rams of a different size or type. When the size of a drill pipe is changed, the size of the pipe rams must be changed to ensure that they seal around the pipe when closed (unless variable-bore pipe rams are in use).

Charles's law *n*: an ideal gas law that states that at constant pressure the volume of a fixed mass or quantity of gas varies directly with the absolute temperature; that is, at a constant pressure, as temperature rises, gas volume increases proportionately.

check valve *n*: a valve that permits flow in one direction only. If the gas or liquid starts to reverse, the valve automatically closes, preventing reverse movement. Commonly referred to as a one-way valve.

choke *n*: a device with an orifice installed in a line to restrict the flow of fluids. Chokes are used to control the rate of flow of the drilling mud out of the hole when the well is closed in with the blowout preventer and a kick is being circulated out of the hole.

choke line *n*: a pipe attached to the blowout preventer stack out of which kick fluids and mud can be pumped to the choke manifold when a blowout preventer is closed in on a kick.

choke manifold *n*: an arrangement of piping and special valves, called chokes. In drilling, mud is circulated through a choke manifold when the blowout preventers are closed. In well testing, a choke manifold attached to the wellhead allows flow and pressure control for test components downstream.

circulate-and-weight method *n*: see *concurrent method*.

circulating components *n pl*: the equipment included in the drilling fluid circulating system of a rotary rig. Basically, the components consist of the mud pump, rotary hose, swivel, drill stem, bit, and mud return line.

circulating density *n*: see *equivalent circulating density*.

circulating fluid *n*: also called drilling mud. See *drilling fluid, mud*.

circulating head *n*: an accessory attached to the top of the drill pipe or tubing to form a connection with the

mud system to permit circulation of the drilling mud. In some cases, it is also a rotating head.

circulating pressure *n*: the pressure generated by the mud pumps and exerted on the drill stem.

circulation *n*: the movement of drilling fluid out of the mud pits, down the drill stem, up the annulus, and back to the mud pits.

clay hydration *n*: the swelling that occurs when clays in the formation take on water.

closed-in pressure *n*: see *shut-in pressure*.

close in *v*: to close the blowout preventers on a well to control a kick. The blowout preventers close off the annulus so that pressure from below cannot flow to the surface.

concurrent method *n*: a method for killing well pressure in which circulation is commenced immediately and mud weight is brought up in steps, or increments, usually a point at a time. Also called circulate-and-weight method.

condition *v*: to treat drilling mud with additives to give it certain properties. To condition and circulate mud is to ensure that additives are distributed evenly throughout a system by circulating the mud while it is being conditioned.

connate water *n*: water retained in the pore spaces, or interstices, of a formation from the time the formation was created. Compare *interstitial water*.

connection gas *n*: the relatively small amount of gas that enters a well when the mud pump is stopped for a connection to be made. Since bottomhole pressure decreases when the pump is stopped, gas may enter the well.

constant choke-pressure method *n*: a method of killing a well that has kicked, in which the choke size is adjusted to maintain a constant casing pressure. This method does not work unless the kick is all or nearly all salt water, a situation that seldom exists. If the kick is gas, this method does not maintain a constant bottomhole pressure, because gas expands as it rises in the annulus. In any case, it is *not* a recommended well-control procedure.

constant pit-level method *n*: a method of killing a well in which the mud level in the pits is held constant while the choke size is reduced and the pump speed slowed. It is not effective, and therefore is not recommended, because casing pressure increases to the point at which the formation fractures or the casing ruptures, and control of the well is lost.

crystallization *n*: the formation of crystals from solutions or melts.

D

degasser *n*: the device used to remove unwanted gas from a liquid, especially from drilling fluid.

diverter *n*: a device used to direct fluids flowing from a well away from the drilling rig. When a kick is encountered at shallow depths, the well often cannot be shut in safely; therefore, a diverter is used to allow the well to flow through a side outlet (a diverter line).

diverter line *n*: a side outlet on a rig that directs flow away from the rig.

drag *n*: friction between the rotating bit and the nonmoving formation.

driller's BOP control panel *n*: a series of controls on the rig floor that the driller manipulates to open and close the blowout preventers.

driller's method *n*: a well-killing method involving two complete and separate circulations. The first circulates the kick out of the well; the second circulates heavier mud through the wellbore.

drilling break *n*: a sudden increase in the drill bit's rate of penetration. It sometimes indicates that the bit has penetrated a high-pressure zone and thus warns of the possibility of a kick.

drilling fluid *n*: circulating fluid, one function of which is to lift cuttings out of the wellbore and to the surface. Other functions are to cool the bit and to counteract downhole formation pressure. Although a mixture of barite, clay, water, and chemical additives is the most common drilling fluid, wells can also be drilled by using air, gas, water, or oil-base mud as the drilling mud. See *mud*.

drilling mud *n*: a specially compounded liquid circulated through the wellbore during rotary drilling operations. See *drilling fluid, mud*.

drilling rate *n*: the speed with which the bit drills the formation; usually called the rate of penetration.

drilling under pressure *n*: the continuation of drilling operations while maintaining a seal at the top of the wellbore to prevent the well fluids from blowing out.

drill pipe float *n*: a check valve installed in the drill stem that allows mud to be pumped down the drill stem but prevents flow back up the drill stem. Also called a float.

drill pipe pressure *n*: the amount of pressure exerted inside the drill pipe as a result of circulating pressure, entry of formation pressure into the well, or both.

drill pipe pressure gauge *n*: an indicator, mounted in the mud circulating system, that measures and indicates the amount of pressure in the drill stem. See *drill stem*.

drill pipe safety valve *n*: a special valve used to close off the drill pipe to prevent backflow during a kick. It has threads to match the drill pipe in use.

drill stem *n*: all members in the assembly used for rotary drilling from the swivel to the bit, including the kelly, drill pipe and tool joints, drill collars, stabilizers, and various specialty items. Compare *drill string*.

drill stem safety valve *n*: a special valve normally installed below the kelly. Usually, the valve is open so that drilling fluid can flow out of the kelly and down the drill stem. It can, however, be manually closed with a special wrench when necessary. In one case, the valve is closed and broken out, still attached to the kelly to prevent drilling mud in the kelly from draining onto the rig floor. In another case, when kick pressure inside the drill stem exists, the drill stem safety valve is closed to prevent the pressure from escaping up the drill stem. Also called lower kelly cock, mud saver valve.

drill string *n*: the column, or string, of drill pipe with attached tool joints that transmits fluid and rotational power from the kelly to the drill collars and bit. Often, especially in the oil patch, the term is loosely applied to both drill pipe and drill collars. Compare *drill stem*.

drill under pressure *v*: to carry on drilling operations with a mud whose density is such that it exerts less pressure on bottom than the pressure in the formation while maintaining a seal (usually with a rotating head) to prevent the well fluids from blowing out under the rig. Drilling under pressure is advantageous in that the rate of penetration is relatively fast; however, the technique requires extreme caution.

E-F

ECD *abbr*: equivalent circulating density.

entrained gas *n*: formation gas that enters the drilling fluid in the annulus. See *gas-cut mud*.

equivalent circulating density (ECD) *n*: the increase in bottomhole pressure expressed as an increase in pressure that occurs only when mud is being circulated. Because of friction in the annulus as the mud is pumped, bottomhole pressure is slightly, but significantly, higher than when the mud is not being pumped. ECD is calculated by dividing the annular pressure loss by 0.052, dividing that by true vertical depth, and adding the result to the mud weight. Also called circulating density, mud-weight equivalent.

explosive fracturing *n*: when explosives are used to fracture a formation. At the moment of detonation, the explosion furnishes a source of high-pressure gas to force fluid into the formation. The rubble prevents fracture healing, making the use of proppants unnecessary. Compare *hydraulic fracturing*.

fault *n*: a break in the earth's crust along which rocks on one side have been displaced (upward, downward, or laterally) relative to those on the other side.

fill line *n*: see *fill-up line*.

fill the hole *v*: to pump drilling fluid into the wellbore while the pipe is being withdrawn to ensure that the wellbore remains full of fluid even though the pipe is withdrawn. Filling the hole lessens the danger of a kick or of caving of the wall of the wellbore.

fill-up line *n*: the smaller of the side fittings on a bell nipple, used to fill the hole when drill pipe is being removed from the well.

fill-up rate *n*: the frequency with which the hole is filled with drilling fluid to replace the pipe removed from the hole.

filter cake *n*: 1. compacted solid or semisolid material remaining on a filter after pressure filtration of mud with a standard filter press. Thickness of the cake is reported in thirty-seconds of an inch or in millimetres. 2. the layer of concentrated solids from the drilling mud or cement slurry that forms on the walls of the borehole opposite permeable formations; also called wall cake or mud cake.

filter loss *n*: the amount of fluid that can be delivered through a permeable filter medium after being subjected to a set differential pressure for a set length of time.

filter press *n*: a device used in the testing of filtration properties of drilling mud. See *mud*.

filtrate *n*: 1. a fluid that has been passed through a filter. 2. the liquid portion of drilling mud that is forced into porous and permeable formations next to the borehole.

final circulating pressure *n*: the pressure at which a well is circulated during well-killing procedures after kill-weight mud has filled the drill stem. This pressure is maintained until the well is completely filled with kill-weight mud.

flow check *n*: a method of determining whether a kick has occurred. The mud pumps are stopped for a short period to see whether mud continues to flow out of the hole; if it does, a kick may be occurring.

formation breakdown pressure *n*: the pressure at which a formation will fracture.

formation competency test *n*: a test used to determine the amount of pressure required to cause a formation to fracture.

formation fluid *n*: fluid (such as gas, oil, or water) that exists in a subsurface rock formation.

formation fracture gradient *n*: a plot of pressure versus depth that reveals the pressure at which a formation will fracture at a given depth.

formation fracture pressure *n*: the point at which a formation will crack from pressure in the wellbore.

formation fracturing *n*: a method of stimulating production by opening new flow channels in the rock surrounding a production well. Often called a frac job. Under extremely high hydraulic pressure, a fluid (such as distillate, diesel fuel, crude oil, dilute hydrochloric acid, water, or kerosene) is pumped downward through production tubing or drill pipe and forced out below a packer or between two packers. The pressure causes cracks to open in the formation, and the fluid penetrates the formation through the cracks. Sand grains, aluminum pellets, walnut shells, or similar materials (propping agents) are carried in suspension by the fluid into the cracks. When the pressure is released at the surface, the fracturing fluid returns to the well. The cracks partially close on the pellets, leaving channels for oil to flow around them to the well. See *explosive fracturing*, *hydraulic fracturing*.

formation pressure *n*: the force exerted by fluids in a formation, recorded in the hole at the level of the formation with the well shut in. Also called reservoir pressure or shut-in bottomhole pressure.

formation strength *n*: the ability of a formation to resist fracture from pressures created by fluids in a borehole.

frac job *n*: see *formation fracturing*.

fracture *n*: a crack or crevice in a formation, either natural or induced.

fracture pressure *n*: the pressure at which a formation will break down, or fracture.

friction loss *n*: a reduction in the pressure of a fluid caused by its motion against an enclosed surface (such as a pipe). As the fluid moves through the pipe (or the walls of an open hole), friction between the fluid and the wall and within the fluid itself creates a pressure loss. The faster the fluid moves, the greater are the losses.

G

gas *n*: a compressible fluid that completely fills any container in which it is confined. Technically, a gas will not condense when it is compressed and cooled, because a gas can exist only above the critical temperature for its particular composition. Below the critical temperature, this form of matter is known as a vapor, because liquid can exist and condensation can occur. Sometimes the terms "gas" and "vapor" are used interchangeably. The latter, however, should be used for those streams in which condensation can occur and that originate from, or are in equilibrium with, a liquid phase.

gas buster *n*: see *mud-gas separator*.

gas-cut mud *n*: a drilling mud that contains entrained formation gas, giving the mud a characteristically fluffy texture. When entrained gas is not released before the fluid returns to the well, the weight or density of the fluid column is reduced. Because a large amount of gas in mud lowers its density, gas-cut mud must be treated to reduce the chance of a kick.

gas cutting *n*: a process in which gas becomes entrained in a liquid.

gas detection analyzer *n*: a device used to detect and measure any gas in the drilling mud as it is circulated to the surface.

geopressured shales *n pl*: impermeable shales, highly compressed by overburden pressure, that are characterized by large amounts of formation fluids and abnormally high pore pressure.

geostatic pressure *n*: the pressure to which a formation is subjected by its overburden. Also called ground pressure, lithostatic pressure, rock pressure.

geothermal gradient *n*: the increase in the temperature of the earth with increasing depth. It averages about 1°F per 60 feet, but may be considerably higher or lower.

guide shoe *n*: a short, heavy, cylindrical section of steel filled with concrete and rounded at the bottom, which is placed at the end of the casing string. It prevents the casing from snagging on irregularities in the borehole as it is lowered. A passage through the center of the shoe allows drilling fluid to pass up into the casing while it is being lowered and allows cement to pass out during cementing operations. Also called casing shoe.

gunk plug *n*: a slurry in crude or diesel oil containing any of the following materials or combinations: bentonite, cement, attapulgite, and guar gum (never with cement). Used primarily in combating lost circulation.

H

hang off *v*: to close a ram blowout preventer around the drill pipe when the annular preventer has previously been closed to offset the effect of heave on floating offshore rigs during well-control procedures.

hard shut-in *n*: in a well-control operation, closing the BOP without first opening an alternate flow path up the choke line. When the BOP is closed, pressure in the annulus cannot be read on the casing pressure gauge.

head *n*: the height of a column of liquid required to produce a specific pressure. See *hydraulic head*.

hole geometry *n*: the shape and size of the wellbore.

hydraulic control pod *n*: a device used on floating offshore drilling rigs to provide a way to actuate and control subsea blowout preventers from the rig. Hydraulic lines from the rig enter the pods, through which fluid is sent toward the preventer. Usually two pods, one yellow and one blue, are used, each to safeguard and back up the other. Also called blue pod, yellow pod.

hydraulic fracturing *n*: an operation in which a specially blended liquid is pumped down a well and into a formation under pressure high enough to cause the formation to crack open, forming passages through which oil can flow into the wellbore. Sand grains, aluminum pellets, glass beads, or similar materials are carried in suspension into the fractures. When the pressure is released at the surface, the fractures partially close on the proppants, leaving channels for oil to flow through to the well. Compare *explosive fracturing*.

hydraulic head *n*: the force exerted by a column of liquid expressed by the height of the liquid above the point at which the pressure is measured. Although "head" refers to distance or height, it is used to express pressure, since the force of the liquid column is directly proportional to its height. Also called head or hydrostatic head. Compare *hydrostatic pressure*.

Hydril *n*: the registered trademark of a prominent manufacturer of oilfield equipment, especially annular blowout preventers.

hydrogen sulfide *n*: a flammable, colorless gaseous compound of hydrogen and sulfur (H_2S), which in small amounts has the odor of rotten eggs. Sometimes found in petroleum, it causes the foul smell of petroleum fractions. In dangerous concentrations, it is extremely corrosive and poisonous, causing damage to skin, eyes, breathing passages, and lungs and attacking and paralyzing the nervous system, particularly that part controlling the lungs and heart. In large amounts, it deadens the sense of smell. Also called hepatic gas or sulfureted hydrogen.

hydrostatic pressure *n*: the force exerted by a body of fluid at rest. It increases directly with the density and the depth of the fluid and is expressed in pounds per square inch or kilopascals. The hydrostatic pressure of fresh water is 0.433 pounds per square inch per foot of depth (9.792 kilopascals per metre). In drilling, the term refers to the pressure exerted by the drilling fluid in the wellbore.

I

IADC *abbr*: International Association of Drilling Contractors.

ICP *abbr*: initial circulating pressure, used in drilling reports.

initial circulating pressure (ICP) *n*: the pressure at which a well that has been closed in on a kick is circulated when well-killing procedures are begun.

inside blowout preventer *n*: any one of several types of valve installed in the drill stem to prevent a blowout through the stem. Flow is possible only downward, allowing mud to be pumped in but preventing any flow back up the stem. Also called an internal blowout preventer.

inside BOP *abbr*: an inside blowout preventer.

internal blowout preventer *n*: see *inside blowout preventer*.

internal preventer *n*: see *inside blowout preventer*.

International Association of Drilling Contractors (IADC) *n*: an organization of drilling contractors that sponsors or conducts research on education, accident prevention, drilling technology, and other matters of interest to drilling contractors and their employees. Its official publication is *The Drilling Contractor*. Address: Box 4287; Houston, TX 77210; (281) 578-7171.

interstitial water *n*: water contained in the interstices, or pores, of reservoir rock. In reservoir engineering, it is synonymous with connate water. Compare *connate water*.

invert-emulsion mud *n*: an oil mud in which fresh or salt water is the dispersed phase and diesel, crude, or some other oil is the continuous phase. See *oil mud*.

K

kelly cock *n*: a valve installed at one or both ends of the kelly that is closed when a high-pressure backflow begins inside the drill stem. The valve is closed to keep pressure off the swivel and rotary hose.

kick *n*: an entry of water, gas, oil, or other formation fluid into the wellbore during drilling. It occurs because the pressure exerted by the column of drilling fluid is not great enough to overcome the pressure exerted by the fluids in the formation drilled. If prompt action is not taken to control the kick, or kill the well, a blowout may occur.

kick fluids *n pl*: oil, gas, water, or any combination that enters the borehole from a permeable formation.

kick tolerance *n*: a calculated estimate of the size of potential kick that could fracture an exposed formation and lead to serious well-control problems .

kill *v*: to control a kick by taking suitable preventive measures (e.g., to shut in the well with the blowout preventers, circulate the kick out, and increase the weight of the drilling mud).

kill fluid *n*: drilling mud of a weight great enough to equal or exceed the pressure exerted by formation fluids.

kill line *n*: a pipe attached to the blowout preventer stack, into which mud or cement can be pumped to overcome the pressure of a kick. Sometimes used when normal kill procedures (circulating kill fluids down the drill stem) are not sufficient.

kill rate *n*: the speed, or velocity, of the mud pump used when killing a well. Usually measured in strokes per minute, it is considerably slower than the rate used for normal operations.

kill-rate pressure *n*: the pressure exerted by the mud pump (and read on the standpipe or drill pipe pressure gauge) when the pump's speed is reduced to a speed lower than that used during normal drilling. A kill rate pressure or several kill rate pressures are established for use when a kick is being circulated out of the wellbore.

kill sheet *n*: a printed form that contains blank spaces for recording information about killing a well. It is provided to remind personnel of the necessary steps to take to kill a well.

kill string *n*: the drill string through which kill fluids are circulated when handling a kick.

L

leak-off test *n*: a gradual pressurizing of the casing after the blowout preventers have been installed to permit estimation of the formation fracture pressure at the casing seat.

log *n*: a systematic recording of data, such as a driller's log, mud log, electrical well log, or radioactivity log. Many different logs are run in wells to discern various characteristics of downhole formation. *v*: to record data.

log a well *v*: to run any of the various logs used to ascertain downhole information about a well.

logging devices *n pl*: any of several electrical, acoustical, mechanical, or radioactivity devices that are used to measure and record certain characteristics or events that occur in a well that has been or is being drilled.

loss of circulation *n*: see *lost circulation*.

lost circulation *n*: the quantities of whole mud lost to a formation, usually in cavernous, fissured, or coarsely permeable beds. Evidenced by the complete or partial failure of the mud to return to the surface as it is being circulated in the hole. Lost circulation can lead to a blowout and, in general, can reduce the efficiency of the drilling operation. Also called lost returns.

lost circulation additives *n pl*: materials added to the mud in varying amounts to control or prevent lost circulation. Classified as fiber, flake, or granular.

lost circulation material (LCM) *n*: a substance added to cement slurries or drilling mud to prevent the loss of cement or mud to the formation. See *bridging materials*.

lost circulation plug *n*: cement set across a formation that is taking excessively large amounts of drilling fluid during drilling operations.

lost returns *n pl*: see *lost circulation*.

lubricate *v*: 1. to apply grease or oil to moving parts. 2. to lower or raise tools in or out of a well with pressure inside the well. The term comes from the fact that a lubricant (grease) is often used to provide a seal against well pressure while allowing wireline to move in or out of the well.

M

macaroni string *n*: a string of tubing or pipe, usually 0.75 or 1 inch (1.9 or 2.54 centimetres) in diameter.

MASP *abbr*: maximum allowable surface pressure.

matrix acidizing *n*: an acidizing treatment using low or no pressure to improve the permeability of a formation without fracturing it.

maximum allowable surface pressure (MASP) *n*: the maximum amount of pressure that is allowed to appear on the casing pressure gauge during a well-killing operation. This pressure is often determined by a leak-off test and if exceeded can lead to formation fracture at the casing shoe and a subsequent underground blowout or broaching.

MD *abbr*: measured depth.

measured depth (MD) *n*: the total length of the wellbore, measured in feet along its actual course through the earth. Measured depth can differ from true vertical depth, especially in directionally drilled wellbores. Compare *true vertical depth*.

Minerals Management Service (MMS) *n*: an agency of the U.S. Department of the Interior that establishes requirements through the Code of Federal Regulations (CFR) for drilling while operating on the Outer Continental Shelf of the United States. The agency regulates rig design and construction, drilling procedures, equipment, qualification of personnel, and pollution prevention. Address: 1849 C St., NW; Washington, DC 20240; (202) 208-3500.

mud *n*: the liquid circulated through the wellbore during rotary drilling and workover operations. In addition to its function of bringing cuttings to the surface, drilling mud cools and lubricates the bit and drill stem, protects against blowouts by holding back subsurface pressures, and deposits a mud cake on the wall of the borehole to prevent loss of fluids to the formation. Although it was originally a suspension of earth solids (especially clays) in water, the mud used in modern drilling operations is a more complex, three-phase mixture of liquids, reactive solids, and inert solids. The liquid phase may be fresh water, diesel oil, or crude oil and may contain one or more conditioners. See *drilling fluid*.

mud additive *n*: any material added to drilling fluid to change some of its characteristics or properties.

mud analysis logging *n*: a continuous examination of the drilling fluid circulating in the wellbore for the purpose of discovering evidence of oil or gas regardless of the quantities entrained in the fluid. When this service is utilized, a portable mud logging laboratory is set up at the well. Also called mud logging.

mud column *n*: the borehole when it is filled or partially filled with drilling mud.

mud conditioning *n*: the treatment and control of drilling mud to ensure that it has the correct properties. Conditioning may include the use of additives, the removal of sand or other solids, the removal of gas, the addition of water, and other measures to prepare the mud for conditions encountered in a specific well.

mud density recorder *n*: a device that automatically records the weight or density of drilling fluid as it is being circulated in a well.

mud-flow indicator *n*: a device that continually measures and may record the flow rate of mud returning from the annulus and flowing out of the mud return line. If the mud does not flow at a fairly constant rate, a kick or lost circulation may have occurred.

mud-flow sensor *n*: see *mud-flow indicator*.

mud-gas separator *n*: a device that removes gas from the mud coming out of a well when a kick is being circulated out.

mud gradient *n*: pressure exerted with depth by drilling fluid. Often expressed in pounds per square inch per foot. Also called pressure gradient.

mud-level recorder *n*: a device that measures and records the height (level) of the drilling fluid in the mud pits. The level should remain fairly constant during the drilling of a well. If it rises, the possibility of a kick exists. Conversely, if it falls, loss of circulation may have occurred.

mud log *n*: a record of information derived from examination of drilling fluid and drill bit cuttings. See *mud logging*.

mud logger *n*: an employee of a mud logging company who performs mud logging.

mud logging *n*: the recording of information derived from examination and analysis of formation cuttings made by the bit and of mud circulated out of the hole. A portion of the mud is diverted through a gas-detecting device. Cuttings brought up by the mud are examined under ultraviolet light to detect the presence of oil or gas. Mud logging is often carried out in a portable laboratory set up at the well.

mud pit *n*: originally, an open pit dug in the ground to hold drilling fluid or waste materials discarded after the treatment of drilling mud. For some drilling operations, mud pits are used for suction to the mud pumps, settling of mud sediments, and storage of reserve mud. Steel tanks are much more commonly used for these purposes now, but they are still usually referred to as pits, except offshore, where "mud tanks" is preferred.

mud program *n*: a plan or procedure, with respect to depth, for the type and properties of drilling fluid to be used in drilling a well. Some factors that influence the mud program are the casing program and such formation characteristics as type, competence, solubility, temperature, and pressure.

mud pump *n*: a large, high-pressure reciprocating pump used to circulate the mud on a drilling rig. A typical mud pump is a two-cylinder double-acting or a three-cylinder single-acting piston pump whose pistons travel in replaceable liners and are driven by a crankshaft actuated by an engine or a motor. Also called a slush pump.

mud return line *n*: a trough or pipe that is placed between the surface connections at the wellbore and the shale shaker and through which drilling mud flows on its return to the surface from the hole. Also called flow line.

mud system *n*: the composition and characteristics of the drilling mud used on a particular well.

mud tank *n*: one of a series of open tanks, usually made of steel plate, through which the drilling mud is cycled to remove sand and fine sediments. Additives are mixed with the mud in the tanks, and the fluid is temporarily stored there before being pumped back into the well. Modern rotary drilling rigs are generally provided with three or more tanks, fitted with built-in piping, valves, and mud agitators. Also called mud pits.

mud weight *n*: a measure of the density of a drilling fluid expressed as pounds per gallon, pounds per cubic foot, or kilograms per cubic metre. Mud weight is directly related to the amount of pressure the column of drilling mud exerts at the bottom of the hole.

mud-weight equivalent *n*: see *equivalent circulating density*.

mud weight recorder *n*: an instrument installed in the mud pits that has a recorder mounted on the rig floor to provide a continuous reading of the mud weight.

N-O-P

nipple up *v*: in drilling, to assemble the blowout preventer stack on the wellhead at the surface.

normal circulation *n*: the smooth, uninterrupted circulation of drilling fluid down the drill stem, out the bit, up the annular space between the pipe and the hole, and back to the surface. Compare *reverse circulation*.

normal formation pressure *n*: formation fluid pressure equivalent to about 0.465 pounds per square inch per foot (10.5 kilopascals per metre) of depth from the surface. If the formation pressure is 4,650 pounds per square inch (32,062 kilopascals) at 10,000 feet (3,048 metres), it is considered normal.

OCS *abbr*: Outer Continental Shelf.

OCS orders *n pl*: rules and regulations, set by the Minerals Management Service (MMS) of the U.S. Department of the Interior, that govern oil operations in U.S. waters on the Outer Continental Shelf. Now supplanted by rules published in the Code of Federal Regulations (CFR), Part 250.

oil-base mud *n*: a drilling or workover fluid in which oil is the continuous phase and which contains from less than 2 percent and up to 5 percent water. This water is spread out, or dispersed, in the oil as small droplets.

oil-emulsion mud *n*: a water-base mud in which water is the continuous phase and oil is the dispersed phase. The oil is spread out, or dispersed, in the water in small droplets, which are tightly emulsified so that they do not settle out. Because of its lubricating abilities, an oil-emulsion mud increases the drilling rate and ensures better hole conditions than other muds. Compare *oil mud*.

oil mud *n*: a drilling mud, e.g., oil-base mud and invert-emulsion mud, in which oil is the continuous phase. It is useful in drilling certain formations that may be difficult or costly to drill with water-base mud. Compare *oil-emulsion mud*.

open *adj*: 1. of a wellbore, having no casing. 2. of a hole, having no drill pipe or tubing suspended in it.

open hole *n*: 1. any wellbore in which casing has not been set. 2. open or cased hole in which no drill pipe or tubing is suspended. 3. the portion of the wellbore that has no casing.

overburden pressure *n*: the pressure exerted by the rock strata on a formation of interest. It is usually considered to be about 1 pound per square inch per foot (22.621 kilopascals per metre).

permeability *n*: 1. a measure of the ease with which a fluid flows through the connecting pore spaces of rock or cement. The unit of measurement is the millidarcy. 2. fluid conductivity of a porous medium. 3. ability of a fluid to flow within the interconnected pore network of a porous medium.

pipe ram *n*: a sealing component for a blowout preventer that closes the annular space between the pipe and the blowout preventer or wellhead.

pipe ram preventer *n*: a blowout preventer that uses pipe rams as the closing elements. See *pipe ram.*

pit gain *n*: an increase in the average level of mud maintained in each of the mud pits, or tanks. If no mud or other substances have been added to the mud circulating in the well, then a pit gain is an indication that formation fluids have entered the well and that a kick has occurred.

pit level *n*: height of drilling mud in the pits.

pit-level indicator *n*: one of a series of devices that continuously monitor the level of the drilling mud in the mud tanks. The indicator usually consists of float devices in the mud tanks that sense the mud level and transmit data to a recording and alarm device (a pit-volume recorder) mounted near the driller's position on the rig floor. If the mud level drops too low or rises too high, the alarm sounds to warn the driller that he or she may either be losing circulation or taking a kick.

pit-level recorder *n*: see *pit-level indicator.*

pit-volume recorder *n*: the gauge at the driller's position that records data from the pit-level indicator.

Pit Volume Totalizer (PVT) *n*: trade name for a type of pit-level indicator. See *pit-level indicator.*

plug *n*: any object or device that blocks a hole or passageway (such as a cement plug in a borehole).

plug back *v*: to place cement in or near the bottom of a well to exclude bottom water, to sidetrack, or to produce from a formation higher in the well. Plugging back can also be accomplished with a mechanical plug set by wireline, tubing, or drill pipe.

plugging material *n*: a substance used to block off zones temporarily or permanently while treating or working on other portions of the well.

positive choke *n*: a choke in which the orifice size must be changed to change the rate of flow through the choke.

pounds per cubic foot *n*: a measure of the density of a substance (such as drilling fluid).

pounds per gallon (ppg) *n*: a measure of the density of a fluid (such as drilling mud).

pounds per square inch gauge (psig) *n*: the pressure in a vessel or container as registered on a gauge attached to the container. This reading does not include the pressure of the atmosphere outside the container.

pounds per square inch per foot *n*: a measure of the amount of pressure in pounds per square inch that a column of fluid (such as drilling mud) exerts on the bottom of the column for every foot of its length. For example, 10 pounds per gallon mud exerts 0.52 pounds per square inch per foot, so a column of 10 pounds per gallon mud that is 1,000 feet long exerts 520 pounds per square inch at the bottom of the column. See *pressure gradient.*

pressure *n*: the force that a fluid (liquid or gas) exerts uniformly in all directions within a vessel, pipe, hole in the ground, and so forth, such as that exerted against the inner wall of a tank or that exerted on the bottom of the wellbore by a fluid. Pressure is expressed in terms of force exerted per unit of area, as pounds per square inch, or in kilopascals.

pressure drop *n*: a loss of pressure that results from friction sustained by a fluid passing through a line, valve, fitting, or other device.

pressure gauge *n*: an instrument that measures fluid pressure and usually registers the difference between atmospheric pressure and the pressure of the fluid by indicating the effect of such pressures on a measuring element (e.g., a column of liquid, pressure in a Bourdon tube, a weighted piston, or a diaphragm).

pressure gradient *n*: 1. a scale of pressure differences in which there is a uniform variation of pressure from point to point. For example, the pressure gradient of a column of water is about 0.433 pounds per square inch per foot (9.794 kilopascals per metre) of vertical elevation. The normal pressure gradient in a formation is equivalent to the pressure exerted at any given depth by a column of 10 percent salt water extending from that depth to the surface (0.465 pounds per square inch per foot or 10.518 kilopascals per metre). 2. the change (along a horizontal distance) in atmospheric pressure. Isobars drawn on weather maps display the pressure gradient.

pressure-integrity test *n*: a method of determining the amount of pressure that is allowed to appear on the casing pressure gauge as a kick is circulated out of a well. In general, it is determined by slowly pumping mud into the well while it is shut in and observing the pressure at which the formation begins to take mud.

pressure loss *n*: 1. a reduction in the amount of force a fluid exerts against a surface, such as the walls of a pipe. It usually occurs because the fluid is moving against the surface and is caused by the friction between the fluid and the surface. 2. the amount of pressure indicated by a drill pipe pressure gauge when drilling fluid is being circulated by the mud pump. Pressure losses occur as the fluid is circulated.

preventer *n*: shortened form of blowout preventer. See *blowout preventer*.

preventer packer *n*: in annular and ram blowout preventers, the rubber or rubberlike material that contacts itself or drill pipe to form a seal against well pressure.

psi *abbr*: pounds per square inch.

psi/ft *abbr*: pounds per square inch per foot.

pump pressure *n*: fluid pressure arising from the action of a pump.

pump rate *n*: the speed, or velocity, at which a pump is run. In drilling, the pump rate is usually measured in strokes per minute.

R

ram *n*: the closing and sealing component on a blowout preventer. One of three types—blind, pipe, or shear—may be installed in several preventers mounted in a stack on top of the wellbore. Blind rams, when closed, form a seal on a hole that has no drill pipe in it; pipe rams, when closed, seal around the pipe; shear rams cut through drill pipe and then form a seal.

ram blowout preventer *n*: a blowout preventer that uses rams to seal off pressure on a hole that is with or without pipe. Also called a ram preventer.

rate of penetration (ROP) *n*: a measure of the speed at which the bit drills into formations, usually expressed in feet (metres) per hour or minutes per foot (metre).

reduced circulating pressure (RCP) *n*: the amount of pressure generated on the drill stem when the mud pumps are run at a speed (or speeds) slower than the speed used when drilling ahead. An RCP or several RCPs are established for use when a kick is being circulated out of the hole.

remote BOP control panel *n*: a device placed on the rig floor that can be operated by the driller to direct air pressure to actuating cylinders that turn the control valves on the main BOP control unit, located a safe distance from the rig.

remote choke panel *n*: a set of controls, usually placed on the rig floor, that is manipulated to control the amount of drilling fluid being circulated through the choke manifold. This procedure is necessary when a kick is being circulated out of a well. See *choke manifold*.

returns *n pl*: the mud, cuttings, and so forth, that circulate up the hole to the surface.

reverse circulation *n*: the course of drilling fluid downward through the annulus and upward through the drill stem, in contrast to normal circulation in which the course is downward through the drill stem and upward through the annulus. Seldom used in open hole, but frequently used in workover operations. Also referred to as "circulating the short way," since returns from bottom can be obtained more quickly than in normal circulation.

reverse drilling break *n*: a sudden decrease in the rate of penetration. When drilling with an oil mud and diamond bit, and an abnormally high-pressure formation is penetrated, the penetration rate may decrease rather than increase. An increase is a drilling break.

rotating blowout preventer *n*: see *rotating head*.

rotating head *n*: a sealing device used to close off the annular space around the kelly in drilling with pressure at the surface, usually installed above the main blowout preventers. A rotating head makes it possible to drill ahead even when there is pressure in the annulus that the weight of the drilling fluid is not overcoming; the head prevents the well from blowing out. It is used mainly in the drilling of formations that have low permeability. The rate of penetration through such formations is usually rapid.

S

safety valve *n*: a valve installed at the top of the drill stem to prevent flow out of the drill pipe if a kick occurs during tripping operations.

saturation point *n*: the point at which, at a certain temperature and pressure, no more solid material will dissolve in a liquid.

set point *n*: the depth of the bottom of the casing when it is set in the well.

setting depth *n*: the depth at which the bottom of the casing extends in the wellbore when it is ready to be cemented.

shale *n*: a fine-grained sedimentary rock composed mostly of consolidated clay or mud. Shale is the most frequently occurring sedimentary rock.

shallow gas *n*: natural gas deposit located near enough to the surface that a conductor or surface hole will penetrate the gas-bearing formations. Shallow gas is potentially dangerous because, if encountered while drilling, the well usually cannot be shut in to control it. Instead, the flow of gas must be diverted. See *diverter*.

shear ram *n*: the component in a blowout preventer that cuts, or shears, through drill pipe and forms a seal against well pressure. Shear rams are used in floating offshore drilling operations to provide a quick method of moving the rig away from the hole when there is no time to trip the drill stem out of the hole.

shear ram preventer *n*: a blowout preventer that uses shear rams as closing elements.

shut in *v*: to close in a well in which a kick has occurred.

shut-in *adj*: shut off to prevent flow. Said of a well when valves are closed at both inlet and outlet.

shut-in bottomhole pressure (SIBHP) *n*: the pressure at the bottom of a well when the surface valves on the well are completely closed. It is caused by formation fluids at the bottom of the well.

shut-in bottomhole pressure test *n*: a bottomhole pressure test that measures pressure after the well has been shut in for a specified period of time. See *bottomhole pressure test*.

shut-in casing pressure (SICP) *n*: pressure of the annular fluid on the casing at the surface when a well is shut in.

shut-in drill pipe pressure (SIDPP) *n*: pressure of the drilling fluid on the inside of the drill stem. It is used to measure the difference between hydrostatic pressure and formation pressure when a well is shut in after a kick and the mud pump is off and to calculate the required mud-weight increase to kill the well.

shut-in pressure (SIP) *n*: the pressure when the well is completely shut in, as noted on a gauge installed on the surface control valves. When drilling is in progress, shut-in pressure should be zero, because the pressure exerted by the drilling fluid should be equal to or greater than the pressure exerted by the formations through which the wellbore passes. On a flowing, producing well, however, shut-in pressure should be above zero.

SIBHP *abbr*: shut-in bottomhole pressure; used in drilling reports.

SICP *abbr*: shut-in casing pressure.

SIDPP *abbr*: shut-in drill pipe pressure

snub *v*: 1. to force pipe or tools into a high-pressure well that has not been killed (i.e., to run pipe or tools into the well against pressure when the weight of pipe is not great enough to force the pipe through the BOPs). Snubbing usually requires an array of wireline blocks and wire rope that forces the pipe or tools into the well through a stripper head or blowout preventer until the weight of the string is sufficient to overcome the lifting effect of the well pressure on the pipe in the preventer. In workover operations, snubbing is usually accomplished by using hydraulic power to force the pipe through the stripping head or blowout preventer. 2. to tie up short with a line.

snubber *n*: a device that mechanically or hydraulically forces pipe or tools into the well against pressure.

snubbing line *n*: 1. a line used to check or restrain an object. 2. a wire rope used to put pipe or tools into a well while the well is closed in. See *snub*.

snubbing unit *n*: a device used to apply additional force to the drill stem or tubing when it is necessary to put them into the hole against high pressure.

soft shut-in *n*: in well-control operations, closing the BOPs with the choke and HCR, or fail-safe, valves open. Compare *hard shut-in*.

space-out *n*: the act of ensuring that a pipe ram preventer will not close on a drill pipe tool joint when the drill stem is stationary. A pup joint is made up in the drill string to lengthen it sufficiently.

stack *n*: a vertical arrangement of blowout prevention equipment. Also called preventer stack. See *blowout preventer*.

strip a well *v*: to move the drill stem, tubing, and other tools into or out of the hole with the well closed in. If the weight of the pipe is sufficient to overcome the upward force of well pressure, then the pipe can be stripped in. Compare *snub*.

stripper head *n*: a blowout prevention device consisting of a gland and packing arrangement bolted to the wellhead. It is often used to seal the annular space between tubing and casing.

stripper rubber *n*: 1. a rubber disk surrounding drill pipe or tubing that removes mud as the pipe is brought out of the hole. 2. the pressure-sealing element of a stripper blowout preventer. See *stripper head*.

stripping *n*: to run the drill stem into or remove it from the well under pressure.

stripping in *n*: the process of lowering the drill stem into the wellbore when the well is shut in on a kick and when the weight of the drill stem is sufficient to overcome the force of well pressure.

stripping out *n*: the process of raising the drill stem out of the wellbore when the well is shut in on a kick.

strip pipe *v*: to remove the drill stem from the hole while the blowout preventers are closed.

stump pressure test *n*: a pressure test of a subsea blowout preventer stack performed on the rig floor on a test stump—a device that allows pressure to be exerted on the stack—to ensure that all the pressure-sealing elements of the stack are working properly.

subsea blowout preventer *n*: a blowout preventer placed on the seafloor for use by a floating offshore drilling rig.

subsea choke-line valve *n*: a valve mounted in the choke line of a subsea blowout preventer stack. It serves to regulate the flow of well fluids being circulated through the choke line when the well is closed in.

surface stack *n*: a blowout preventer stack mounted on top of the casing string at or near the surface of the ground or the water. Surface stacks are employed on land rigs and on bottom-supported offshore rigs.

surging *n*: a rapid increase in pressure downhole that occurs when the drill stem is lowered too fast or when the mud pump is brought up to speed after starting.

swab *v*: to pull formation fluids into a wellbore by raising the drill stem at a rate that reduces the hydrostatic pressure of the drilling mud below the bit.

swabbed show *n*: formation fluid that is pulled into the wellbore because of an underbalance of formation pressure caused by pulling the drill string too fast.

swabbing *n*: see *swabbing effect*.

swabbing effect *n*: a phenomenon characterized by formation fluids being pulled or swabbed into the wellbore when the drill stem and bit are pulled up the wellbore fast enough to reduce the hydrostatic pressure of the mud below the bit. If enough formation fluid is swabbed into the hole, a kick can result.

T

temperature gradient *n*: the increase in temperature of a well as its depth increases.

30 CFR 250 *abbr*: 30 Code of Federal Regulations, Part 250.

30 Code of Federal Regulations Part 250 *n*: the U.S. Department of the Interior, Minerals Management Service rules and regulations that must be followed by those who drill and produce oil and gas wells located in the Outer Continental Shelf.

total depth (TD) *n*: the maximum depth reached in a well.

transition zone *n*: 1. the area in which underground pressures begin to change from normal to abnormally high as a well is being deepened. 2. the areas in the drill stem near the point where drill pipe is made up on drill collars.

trip gas *n*: gas that enters the wellbore when the mud pump is shut down and pipe is being pulled from the wellbore. The gas may enter because of the reduction in bottomhole pressure when the pump is shut down, because of swabbing, or because of both.

trip margin *n*: the small amount of additional mud weight carried over that needed to balance formation pressure to overcome the pressure-reduction effects caused by swabbing when a trip out of the hole is made.

trip tank *n*: a small mud tank with a capacity of 10 to 15 barrels (1,590 to 2,385 litres), usually with 1-barrel or 0.5-barrel (159-litre or 79.5-litre) divisions, used to ascertain the amount of mud necessary to keep the wellbore full with the exact amount of mud that is displaced by drill pipe. When the bit comes out of the hole, a volume of mud equal to that which the drill pipe occupied while in the hole must be pumped into the hole to replace the pipe. When the bit goes back in the hole, the drill pipe displaces a certain amount of mud, and a trip tank can be used again to keep track of this volume.

true vertical depth (TVD) *n*: the depth of a well measured from the surface straight down to the bottom of the well. The true vertical depth of a well may be quite different from its actual measured depth, because wells are very seldom drilled exactly vertical.

tubingless completion *n*: a method of producing a well in which only production casing is set through the pay zone, with no tubing or inner production string used to bring formation fluids to the surface. This type of completion has its best application in low-pressure, dry-gas reservoirs.

TVD *abbr*: true vertical depth.

U-V-W

underground blowout *n*: an uncontrolled flow of gas, salt water, or other fluid out of the wellbore and into another formation that the wellbore has penetrated.

United States Geological Survey (USGS) *n*: a federal agency within the Department of the Interior established in 1879 to conduct investigations of the geological structure, mineral resources, and products of the United States. Its activities include assessing onshore and offshore mineral resources; providing information that allows society to mitigate the impact of floods, earthquakes, landslides, volcanoes, and droughts; monitoring the nation's groundwater and surface water supplies and people's impact thereon; and providing mapped information on the nation's landscape and land use. Address: United States Geological Survey, 119 National Center, 12201 Sunrise Valley Drive, Reston, VA 22092.

upper kelly cock *n*: a valve installed above the kelly that can be manually closed to protect the rotary hose from high pressure that may exist in the drill stem.

USGS *abbr*: United States Geological Survey.

wait-and-weight method *n*: a well-killing method in which the well is shut in and the mud weight is raised the amount required to kill the well. The heavy mud is then circulated into the well while the kick fluids are circulated out. So called because one shuts the well in and waits for the mud to be weighted before circulation begins.

water-base mud *n*: a drilling mud in which the continuous phase is water. In water-base muds, any additives are dispersed in the water. Compare *oil-base mud*.

water hammer *n*: a pressure concussion caused by suddenly stopping the flow of liquids in a closed container.

wellbore *n*: a borehole; the hole drilled by the bit. A wellbore may have casing in it or it may be open (uncased); or part of it may be cased, and part of it may be open. Also called a borehole or hole.

wellbore pressure *n*: 1. bottomhole pressure. 2. casing pressure.

well control *n*: the methods used to control a kick and prevent a well from blowing out. Such techniques include, but are not limited to, keeping the borehole completely filled with drilling mud of the proper weight or density during all operations, exercising reasonable care when tripping pipe out of the hole to prevent swabbing, and keeping careful track of the amount of mud put into the hole to replace the volume of pipe removed from the hole during a trip.

well kick *n*: see *kick*.

wettability *n*: the relative affinity between individual grains of rock and each fluid that is present in the spaces between the grains. If oil and water are both present, the water is usually in contact with the surface of each grain, and the rock is called water-wet. If, however, the oil contacts the surface, the rock is called oil-wet.

wild well *n*: a well that has blown out of control and from which oil, water, or gas is escaping with great force to the surface. Also called a gusher.

Italicized page numbers indicate illustrations or photographs.